76 Structure and Bonding

Complex Chemistry

With Contributions by
G. Blasse E. König S.B. Padhye,
P.B. Sonawane D.X. West

With 67 Figures and 27 Tables

Springer-Verlag
Berlin Heidelberg New York
London Paris Tokyo
Hong Kong Barcelona Budapest

ISBN 3-540-53499-7 Springer-Verlag Berlin Heidelberg New York
ISBN 0-387-53499-7 Springer-Verlag New York Berlin Heidelberg

Typesetting: Macmillan India Ltd, Bangalore-25; Printing: Colordruck, Berlin; Bookbinding: Lüderitz & Bauer, Berlin
51/3020-5 4 3 2 1 0 – Printed on acid-free paper

Table of Contents

Structural and Physical Correlations in the Biological Properties of Transition Metal Heterocyclic Thiosemicarbazone and *S*-alkyldithiocarbazate Complexes
D.X. West, S.B. Padhye, P.B. Sonawane 1

Nature and Dynamics of the Spin-State Interconversion in Metal Complexes
E. König 51

Optical Electron Transfer Between Metal Ions and its Consequences
G. Blasse 153

Author Index Volumes 1–76 189

Structural and Physical Correlations in the Biological Properties of Transition Metal Heterocyclic Thiosemicarbazone and *S*-alkyldithiocarbazate Complexes

Douglas X. West,[1] Subhash B. Padhye[2] and Pramila B. Sonawane[2]

[1] Department of Chemistry, Illinois State University, Normal, Illinois 61761, USA
[2] Department of Chemistry, University of Poona, Poona 411 007, India

The nature of metal complexes of heterocyclic thiosemicarbazones reported in the literature through 1989 have been reviewed with an emphasis on variations in stoichiometry and stereochemistry. A brief introduction is followed by a survey of the biological activity of the uncomplexed thiosemicarbazones. Next is a section on the chemical nature of thiosemicarbazones including a discussion of preparative methods, modes of coordination and isomerism. This is followed by a discussion of structural information and biological activity of the iron(III), iron(II), copper(II), cobalt(III), cobalt(II) and nickel(II) complexes of heterocyclic thiosemicarbazones. Short sections on the biological activity of other chelating thiosemicarbazones and metal complexes are also included.

1 Introduction 4

2 Biological Activity of Thiosemicarbazones and S-alkyldithiocarbazates 5
2.1 Antitubercular Activity 5
2.2 Antileprotic Activity 6
2.3 Antifungal Activity 7
2.4 Antiviral Activity 7
2.5 Antimalarial Activity 9
2.6 Antitumor Activity 9

3 Chemical Nature of Thiosemicarbazones and *S*-Alkyldithiocarbazates 10
3.1 Preparative Methods 10
3.2 Bonding and Stereochemistry 11

4 Transition Metal Complexes 14
4.1 Iron(III) 14
4.1.1 2-Formylpyridine Thiosemicarbazones and S-Alkyldithiocarbazates 14
4.1.2 2-Acetylpyridine Thiosemicarbazones and S-Alkyldithiocarbazates 15
4.1.3 2-Picolylketone Thiosemicarbazones and S-Alkyldithiocarbazates 17
4.1.4 2-Formylquinoline and 1-Formylisoquinoline Thiosemicarbazones and S-Alkyldithiocarbazates 18
4.1.5 2,6-Diacetylpyridine Bis(Thiosemicarbazones) and Bis(S-Alkyldithiocarbazates) 19
4.1.6 2-Acetylpyridine *N*-Oxide Thiosemicarbazones and S-Alkyldithiocarbazates 20
4.1.7 2-Formyl- and 2-Acetylthiophene Thiosemicarbazones 20

4.2 Iron(II) . . . 20
4.2.1 2-Formylpyridine Thiosemicarbazones and S-Alkyldithiocarbazates . . . 20
4.2.2 2-Acetylpyridine Thiosemicarbazones and S-Alkyldithiocarbazates . . . 21
4.2.3 2-Formylquinoline and 1-Formylisoquinoline Thiosemicarbazones and S-Alkyldithiocarbazates . . . 21
4.2.4 2,6-Diacetylpyridine Bis(Thiosemicarbazones) and Bis(S-Alkyldithiocarbazates) . . . 21
4.2.5 2-Formyl- and 2-Acetylthiophene Thiosemicarbazones . . . 21
4.3 Copper(II) . . . 22
4.3.1 2-Formylpyridine Thiosemicarbazones and S-Alkyldithiocarbazates . . . 22
4.3.2 2-Acetylpyridine Thiosemicarbazones and S-Alkyldithiocarbazates . . . 24
4.3.3 2-Picolylketone Thiosemicarbazones and S-Alkyldithiocarbazates . . . 27
4.3.4 2-Formylquinoline and 1-Formylisoquinoline Thiosemicarbazones and S-Alkyldithiocarbazates . . . 27
4.3.5 2,6-Diacetylpyridine Bis(Thiosemicarbazones) and Bis(S-Alkyldithiocarbazates) . . . 28
4.3.6 2-Acetylpyridine *N*-Oxide Thiosemicarbazones and S-Alkyldithiocarbazates 28
4.3.7 2-Formylpyrazole Thiosemicarbazones . . . 29
4.3.8 2-Formyl- and 2-Acetylfuran Thiosemicarbazones and S-Alkyldithiocarbazates . . . 29
4.3.9 Other Polydentate Thiosemicarbazones . . . 30
4.4 Cobalt(III) . . . 31
4.4.1 2-Formylpyridine Thiosemicarbazones and S-Alkyldithiocarbazates . . . 31
4.4.2 2-Acetylpyridine Thiosemicarbazones and S-Alkyldithiocarbazates . . . 31
4.4.3 2-Formylquinoline and 1-Formylisoquinoline Thiosemicarbazones and S-Alkyldithiocarbazates . . . 32
4.4.4 2,6-Diacetylpyridine Bis(Thiosemicarbazones) and Bis(S-Alkyldithiocarbazates) . . . 33
4.4.5 2-Acetylpyridine *N*-Oxide Thiosemicarbazones and S-Alkyldithiocarbazates . . . 33
4.4.6 2-Formylpyrazole Thiosemicarbazones . . . 33
4.4.7 2-Formyl- and 2-Acetylfuran Thiosemicarbazones and S-Alkyldithiocarbazates . . . 33
4.4.8 2-Formyl- and 2-Acetylthiophene Thiosemicarbazones . . . 34
4.5 Cobalt(II) . . . 34
4.5.1 2-Formylpyridine Thiosemicarbazones and S-Alkyldithiocarbazates . . . 34
4.5.2 2-Acetylpyridine Thiosemicarbazones and S-Alkyldithiocarbazates . . . 34
4.5.3 2-Formylquinoline and 1-Formylisoquinoline Thiosemicarbazones and S-Alkyldithiocarbazates . . . 35
4.5.4 2,6-Diacetylpyridine Bis(Thiosemicarbazones) and Bis(S-Alkyldithiocarbazates) . . . 35
4.5.5 2-Acetylpyridine N-Oxide Thiosemicarbazones and S-Alkyldithiocarbazates . . . 36
4.5.6 2-Formyl- and 2-Acetylfuran Thiosemicarbazones and S-Alkyldithiocarbazates . . . 36
4.5.7 Other Polydentate Thiosemicarbazones . . . 37
4.6 Nickel(II) . . . 37
4.6.1 2-Formylpyridine Thiosemicarbazones and S-Alkyldithiocarbazates . . . 37
4.6.2 2-Acetylpyridine Thiosemicarbazones and S-Alkyldithiocarbazates . . . 38
4.6.3 2-Picolylketone Thiosemicarbazones and S-Alkyldithiocarbazates . . . 39
4.6.4 2-Formylquinoline and 1-Formylisoquinoline Thiosemicarbazones and S-Alkyldithiocarbazates . . . 40
4.6.5 2,6-Diacetylpyridine Bis(Thiosemicarbazones) and Bis(S-Alkyldithiocarbazates) . . . 41
4.6.6 2-Acetylpyridine *N*-Oxide Thiosemicarbazones and S-Alkyldithiocarbazates . . . 41
4.6.7 2-Formylpyrazole Thiosemicarbazones . . . 42
4.6.8 2-Formyl- and 2-Acetylfuran Thiosemicarbazones and S-Alkyldithiocarbazates . . . 42

4.6.9 2-Formyl- and 2-Acetylthiophene Thiosemicarbazones 43
4.6.10 Other Polydentat Thiosemicarbazones 43
4.7 Other Metal ion Complexes 44

5 Conclusions and Suggestions for Further Studies 44

6 References 44

7 Appendix-Symbols for Organic Molecules 49

1 Introduction

Thiosemicarbazones possess a range of biological applications; antitumor [1], antiviral [2], antibacterial [3], antimalarial [4], and antifungal [5] activities have been studied. Heterocyclic thiosemicarbazones exercise their beneficial therapeutic properties in mammalian cells by inhibiting ribonucleotide reductase, a key enzyme in the synthesis of DNA precursors [6]. The non-heme iron subunit has been shown to be inhibited/inactivated by thiosemicarbazones [7]. Their ability to provide this inhibitory action is thought to be due to coordination of iron via their N–N–S tridentate ligating system, either by a preformed iron complex binding to the enzyme, or by the free ligand complexing with the iron-charged enzyme [8]. Studies of iron and copper complexes have shown that they can be more active in cell destruction, as well as in the inhibition of DNA synthesis, than the uncomplexed thiosemicarbazones [9]. Further, 5-hydroxy-2-formylthiosemicarbazone has been shown to cause lesions in DNA [10]. Therefore, there may be a second site of action in addition to inhibition of ribonucleotide reductase. These observations have provided an impetus to the synthesis of a large number of transition metal complexes of 2-heterocyclic thiosemicarbazones, as well as the related *S*-alkyldithiocarbazates. Recent studies have included the complexes' biological activity and an examination of structure-activity relationships is needed.

Metal complexes of ligands containing a sulfur donor in conjunction with nitrogen, oxygen or a second sulfur have been reviewed in the past [11–13]. For example, reviews of the coordination compounds of dithiophosphates [14], dithiocarbamates [15, 16], dithiolates [17], dithiodiketonates [18], and xanthates [16] have appeared. The analytical aspects [19] and the spectral and structural information of transition metal complexes of thiosemicarbazones [20, 21] have been reviewed previously. Recent developments in the structural nature of metal complexes of 2-heterocyclic thiosemicarbazones and *S*-alkyldithiocarbazates, depicted below, are correlated to their biological activities.

$$= {}^{1}\mathrm{N}\!-\!{}^{2}\mathrm{NH}\!-\!\underset{\underset{\displaystyle \mathrm{S}}{\|}}{{}^{3}\mathrm{C}}\!-\!{}^{4}\mathrm{NR}_2 \quad \text{(or SR)}$$

The superscripts refer to positions on the thiosemicarbazone (dithiocarbazate) moiety in accord with IUPAC.

2 Biological Activity of Thiosemicarbazones and S-Alkyldithiocarbazates

The extreme insolubility of most thiosemicarbazones in water causes difficulty in their oral administration in clinical practice. The efforts to overcome this limitation by introducing hydrophilic groups such as NH_2 or OH in heterocyclic ring systems, with the ultimate goal of deriving a soluble acid or sodium salt, have met with limited success [22]. Therefore, since medicinal activity of thiosemicarbazones may, in part, be related to their chelating ability, metal complexes may prove to be useful forms of thiosemicarbazones and beneficially modify their biological activity. First, a large number of these complexes involve biologically essential metal ions such as copper, iron and zinc. Second, the long-term side effects of the therapeutic agents may be avoided since metal complexes can break down and the metal ions may interact with the organism [23]. Third, the metal complex may be a vehicle for activation of the ligand as the principle cytotoxic agent [24]. Finally, complexation with metal ions may lead to reduction of the drug-resistance by several orders of magnitude [25]. In this section we briefly summarize the available data on the biological activity of the various thiosemicarbazones, but have integrated the metal complexes' biological activity with their chemical nature in Sect. 4.

Efforts to evaluate structural features essential for biological activity have included (1) replacement of the sulfur atom of the thiocarbonyl group by oxygen, selenium, an imine, an oxime, etc.; (2) modification of the sulfur center by alkylation; (3) changing the thiosemicarbazone moiety's point of attachment on the heterocyclic ring; (4) substitution on the terminal ^4N position; and (5) variation of the nature of the aldehyde or ketone. From these studies it is known that structural alterations hindering the thiosemicarbazone's ability to function as a chelating agent with transition metal ions also tend to diminish or destroy biological activity. It is clear that thiosemicarbazones having a third potential bonding site {e.g. a pyridyl nitrogen as in 2-(*N*)-heterocycles} are found to possess considerable biological activity [3, 26, 27].

2.1 Antitubercular Activity

The antimicrobial activity of thiosemicarbazones against *Mycobacterium tuberculosis* in vitro was first reported by Domagk et al. [28] and later confirmed in vivo [29]. Screening revealed that only certain substituted benzaldehyde and heterocyclic thiosemicarbazones possess antitubercular activity [30–32]. The most widely used is *p*-acetamidobenzaldehyde thiosemicarbazone (trivial name = thiacetazone), **1**.

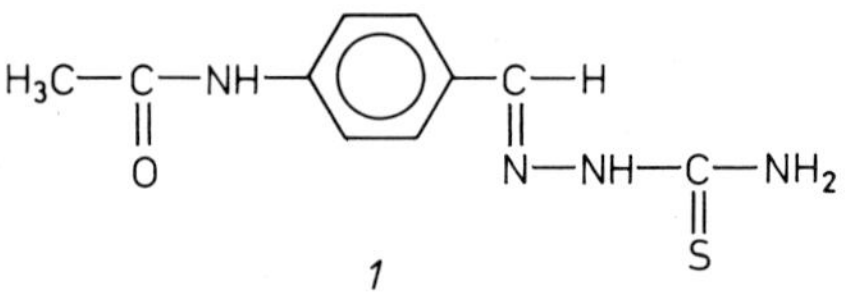

1

Therapy with thiacetazone suffers from two drawbacks. First, there is a wide range of toxic effects including hemolytic anemia, cerebral edema, excessive skin eruptions and hepatic dysfunction [33]. Second, the organism rapidly develops a resistance to the drugs, and extensive cross-resistances are observed with other second-line drugs. These difficulties are partially overcome by combining thiacetazone with other antitubercular drugs, especially isoniazide [34]. Recent reports on field trials in Asia and Africa [35, 36] of such combination therapies have revealed the usefulness of the approach although drugs other than thiacetazone have been found to be more effective for such combinations [37].

2.2 Antileprotic Activity

Since the causative organism of leprosy, one of the world's six major diseases, *Mycobacterium leprae*, is closely related to *Mycobacterium tuberculosis*, thiosemicarbazones have also been used as second-line drugs in the chemotherapy of leprosy [38]. The most widely used in leprosy treatment has been thiacetazone, and structure-activity relationships for it are similar to those observed for antitubercular thiosemicarbazones [39, 40].

Recently, thiosemicarbazone derivatives of 2-acetylpyridine were found to possess antileprotic activity [41]. The ribonucleotide diphosphate reductase (RDR) activity for this series of compounds correlates well with their observed antileprotic properties in mycobacterial systems suitable for in vitro testing. The strong metal chelating ability of tridentate thiosemicarbazones is thought to be responsible for their biological activity and any alteration that hinders this chelation leads to loss of activity. The toxicity commonly associated with these compounds can be lowered by introducing a substituent at appropriate positions (e.g. at ^{6}C in the heterocyclic ring) or by replacing the thiocarbonyl sulfur by a selenium atom [42, 43].

Recent studies have been directed towards the synthesis of heterocyclic hydrazones which have lower toxicities than thiosemicarbazones [44]. It has been proposed that the hydrazinic N–H group is essential for activity since it might be involved in a crucial radical formation step important in the mechanism of RDR. This is supported by the loss of antileprotic activity for this series of compounds when the hydrazinic hydrogen is replaced by a methyl group [44]. The heterocyclic hydrazones, like thiosemicarbazones, behave as tridentate ligands.

2.3 Antifungal Activity

In addition to their antibacterial activities, thiosemicarbazones inhibit growth of both fungi and protozoa. Wiles and Supunchuk [45] reported that heterocyclic derivatives of thiosemicarbazide are active against the growth of *Aspergillus niger* and *Chaetominum globsum* in concentrations as low as 10 ug/ml. Since then, several workers [5, 46–52] have reported the antimicrobial activity of thiosemicarbazones against selected plant pathogenic and saprophytic fungi. However, most of these studies have been limited in scope without systematic variation in the thiosemicarbazone structure and have not involved metal complexes.

2.4 Antiviral Activity

Virus replication comprises numerous biochemical transformations that might provide suitable targets for antiviral therapy. The antiviral effect of thiosemicarbazones was first demonstrated by Hamre et al. [53, 54], who showed that *p*-aminobenzaldehyde-3-thiosemicarbazone and several of its derivatives were active against vaccinia virus in mice. These studies were extended to include thiosemicarbazones of isatin, benzene, thiophene, pyridine, and quinoline derivatives, which also showed activity against vaccinia-induced encephalitis. The nature of the aldehyde/ketone moiety was not as significant as the presence of the thiosemicarbazide side chain; the latter was deemed essential for antiviral activity.

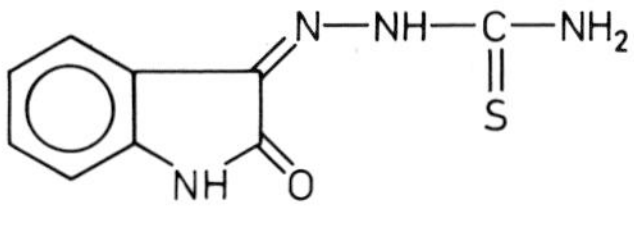

2

Following the lead that isatin-3-thiosemicarbazone, **2**, had the greatest activity against vaccinia virus, a study of the structure-activity relationship was carried out on this series of compounds by Bauer and co-workers [55–59]. The importance of the thiosemicarbazide side chain for biological activity was demonstrated; chemical modifications of this function or the aromatic ring lowered the activity. The most active compounds against vaccinia virus were found to be 2-acetylpyridine ^{4}N-methyl-, and ^{4}N-ethylthiosemicarbazones, **3**. The former, commonly known as 'methisazone', has undergone successful clinical trials as a prophylactic agent against smallpox [63–67], although these results have recently been questioned [68]. However, the beneficial effects of methisazone in the treatment of the complications of smallpox vaccination, such as progressive vaccinia and vaccinia gangrenosum, are well established. In spite of the proven requirement of the sulfur center for the biological activity of these

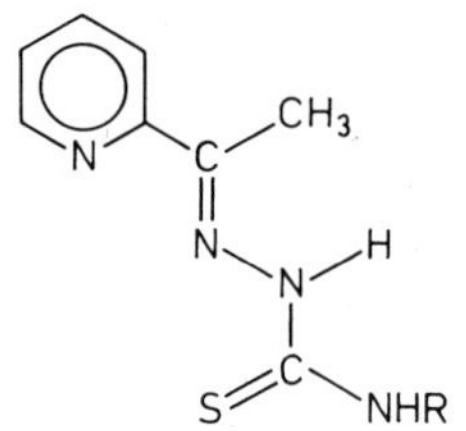

3

a: R=CH_3
b: R=C_2H_5

compounds, no studies seem to have been undertaken to elucidate the role of chelation in the antiviral activity of these thiosemicarbazones.

Besides poxvirus, thiosemicarbazones have also been tested against a variety of other viral infections including herpesvirus, adnovirus, poliovirus, rhinovirus and RNA tumor virus with mixed results [69]. For example, the isatin thiosemicarbazones, **2**, were found to be less active against herpes simplex virus (HSV) [70–73]; the 2-heterocyclic thiosemicarbazone derivatives were active against HSV-1 and HSV-2 types [74]. The activity of numerous heterocyclic thiosemicarbazones have been examined, and it was found that some inhibit the in vivo replication of the virus to a greater extent than they inhibit cellular DNA or protein synthesis [74]. These compounds are active both in vitro and in a cutaneous herpes guinea pig model [75]. Following the suggestion that activity is related to inhibition of ribonucleotide reductase [26], the inhibition of both mammalian RDR and purified HSV-1 enzymes were compared [76]. It was observed that ^{4}N-substituted 2-acetylpyridine thiosemicarbazones (e.g. **3**) inhibit the viral enzyme significantly and preferentially. This led to the development of specific inhibitor of HSV ribonucleotide reductase as an important future direction in antiviral therapy [74]. A recent study tends to support these expectations for 2-acetylpyridine thiosemicarbazone derivatives which inhibit the HSV-specified enzymes without appreciably inhibiting the cellular enzyme [77]. How chelation of a metal cofactor can induce such potent and selective inhibition needs to be determined.

The structure/activity relationships for the methisazone, **3a**, derivatives against adenoviruses and poxviruses have been shown to be similar [78]. Pearson and Zimmerman [79] demonstrated that all three types of polioviruses are inhibited by 2-acetylpyridine ^{4}N-dibutylthiosemicarbazone, which is similar to **3a**, by blocking viral RNA synthesis. A 3-substituted triazinoindole derivative of isatin was effective against several strains of rhinovirus in tissue culture; the mechanism of action is unknown [80].

In contrast to the effects obtained with viruses mentioned earlier, rous sarcoma virus (RSV) is inactivated by direct contact with **2** [81]. Evidence for the drug action by a chelate compound was obtained by using concentrations of **3a** and copper(II) sulfate, neither of which individually affected enzyme activity or transforming abilities [82]. In a later study these workers showed that several metal complexes inhibit the RNA dependent DNA polymerases and the transforming ability of RSV, the most active compound being a 1:1 copper(II) complex with 2-formylpyridine thiosemicarbazone [83]. The nickel(II) complexes have been found to be inactive against RSV [84].

2.5 Antimalarial Activity

An extensive series of thiosemicarbazones obtained from 2-acetylpyridine was tested by Klayman et al. [4, 85] for antimalarial activity against *Plasmodium berghei* in mice. The molecular features essential for activity were found to be a 2-pyridylethylidene moiety, the presence of the thiocarbonyl sulfur, and certain, bulky or cyclic substituents at the terminal ^{4}N-atom. The most active 2-acetylpyridine thiosemicarbazones were ^{4}N-phenyl- and those with azacyclic substituents. For example, ^{4}N-substituents of 4-methylpiperidine, piperazine, and azabicyclo[3.2.2.]nonyl-, **4**, were curative at a dose level as low as 20 mg/kg.

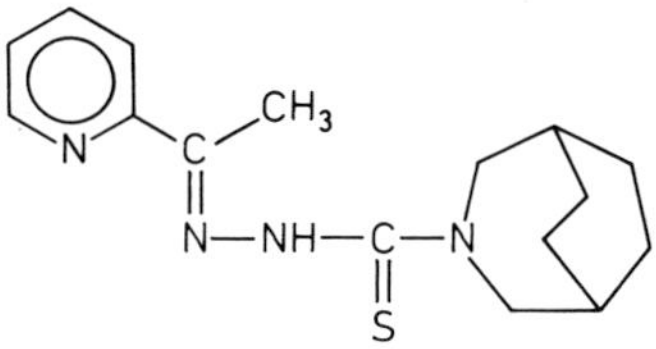

4

Also, against *Neisseria gonorrhoeae* the 2-acetylpyridine ^{4}N-dialkylthiosemicarbazones were the most active [86]. Recently, it has been shown that 2-formylpyridine thiosemicarbazone inhibited adenosine uptake in rodent erythrocytes and reticulocytes parasitized with *Plasmodium berghei* [87]. This suggests that potential interaction with the adenosine receptor may be an additional mode of action of these drugs besides chelation. Further studies are needed to substantiate these conclusions. Thiosemicarbazones of 3-formyl- and 3-acetyl-*B*-carboline effectively inhibit the promastigote form of *Leishmania donavani*; 2-formylpyridine thiosemicarbazone was considerably less active [88].

2.6 Antitumor Activity

Bis(thiosemicarbazones) [89–97] and *N*-heterocyclic thiosemicarbazones comprise two interesting series of experimental chemotherapeutic agents. 2-formylpyridine thiosemicarbazone, the first of the latter series to be examined for biological activity, showed mild antileukemic activity against 1–1210 tumor in mice [98]. However, it was found to be toxic at the therapeutic dose levels which led to synthesis of other aromatic and heterocyclic thiosemicarbazones as potential agents [80, 99, 100]. However, the only active anticancer compounds besides glyoxal bis(thiosemicarbazones) were the *N*-heterocyclic thiosemicarbazones [101], 2-formyl-3-hydroxypyridine thiosemicarbazone [102] and

2-formyl-5-hydroxypyridine thiosemicarbazone [103]. A systematic study of formyl thiosemicarbazones of sixteen different heterocyclic ring systems carried out by French and Blanz [1] revealed that the thiosemicarbazone side-chain must be adjacent to the heterocyclic nitrogen and a conjugated NNS tridentate ligand system is essential for anticancer activity [1, 94–103]. As a result, pyridine and isoquinoline ring systems have been most extensively investigated for structure-activity relationships among the antitumor compounds of this series; an account can be found in reviews by Sartorelli [8, 104] and Petering [105]. The sites chemically modified were the aldo/ketonyl carbon, the thione group and ^4N function along with the positions of attachment to the pyridine/isoquinoline moiety.

All of the active 2-heterocyclic thiosemicarbazones are potent inhibitors of the biosynthesis of DNA in mammalian cells [106]. It has further been determined that the enzymatic site for blocking of DNA synthesis is the inhibition of RDR, an obligatory enzyme for the generation of DNA precursors [107]. The kinetic studies with a partially purified enzyme and dithiothreitol as a substitute for the natural substrate thioredoxin suggests that a pre-formed iron chelate of isoquinoline thiosemicarbazone interacts with the enzyme at or near the site occupied by the dithiol substrate [107]. Structure-activity studies carried out to elucidate requirements for optimum interaction between the enzyme and inhibitor revealed that position *C*-6 of 2-acetylpyridine 3-azabicyclononylthiosemicarbazone, **4**, and the *C*-3 position of 2-formylpyridine thiosemicarbazone are equivalent with respect to orientation of the inhibitor at the enzymatic binding site [108]. Any modification at these positions as well as ^4N of the thiosemicarbazone moiety results in a decrease in activity. Introduction of a *m*-aminophenyl group at the *C*-4 position of the pyridine ring enhances inhibition of the enzyme suggesting that a hydrophobic bonding zone is adjacent to the enzyme's inhibitor binding site [109, 110].

3 Chemical Nature of Thiosemicarbazones and S-Alkyldithiocarbazates

3.1 Preparative Methods

Synthesis of 2-heterocyclic thiosemicarbazones can be summarized in three reaction sequences following the lead of Klayman et al. [5]. Condensation of equimolar quantities of a thiosemicarbazide and a 2-heterocyclic aldehyde or ketone in an alcoholic solvent is represented by Eq. (1). The product's superscripts refer to positions of substitution in the thiosemicarbazone moiety in accord with IUPAC.

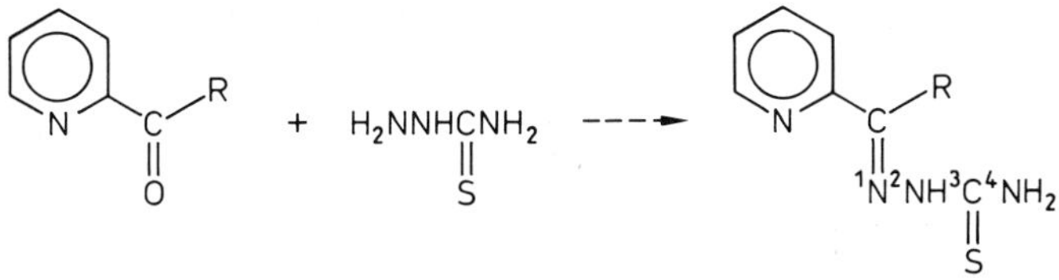

(1)

The most common method of preparing thiosemicarbazones with substituents other than hydrogen attached to ^4N involves amination of the appropriate

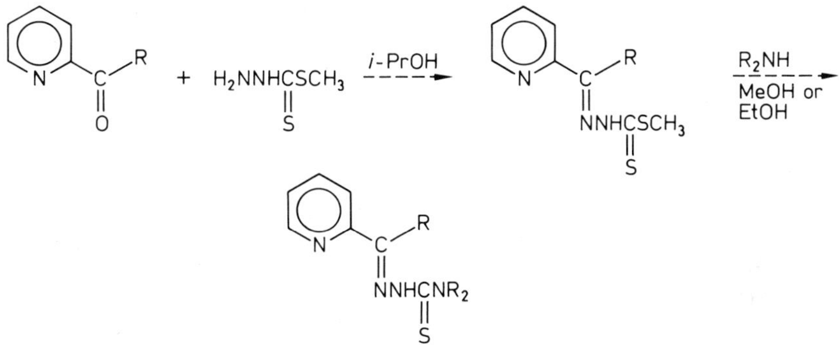

(2)

S-methyldithiocarbazate. The reaction sequence (2) starts with the ketone and hydrazine carbodithioate. A third synthetic method, Eq. (3), which has proven most useful for attaching a heterocyclic ring at ^4N on the thiosemicarbazone moiety [111], is shown below:

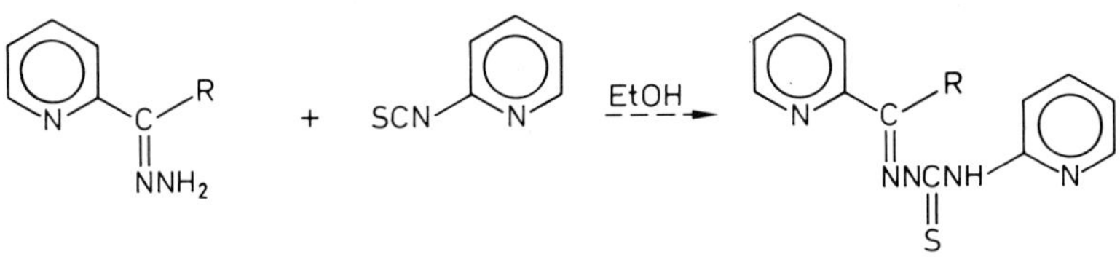

(3)

Synthesis of 2-heterocyclic thiosemicarbazones can also be accomplished by transamination with acetonitrile the most commonly used solvent [112]. ^4N-dialkyl- or ^4N-arylalkyl-functions are the best leaving groups. For example, 2-acetylpyridine ^4N-dimethylthiosemicarbazone prepared by Eq. (2) can be used as a starting material to prepare other 2-actylpyridine thiosemicarbazones.

3.2 Bonding and Stereochemistry

It has been shown [113] that the arrangement of the non-hydrogen atoms in the thiosemicarbazide molecule is nearly planar and the sulfur atom and hydrazinic NH_2 group are *trans* with respect to the C–N bond. When forming complexes while in this configuration, bonding occurs via the sulfur atom as a monodentate ligand. Gerbeleu et al. [114] have shown that coordination occurs through the hydrazido and imide nitrogens if the sulfur center is substituted. However, in most complexes [115], thiosemicarbazones coordinate as bidentate ligands via the azomethine nitrogen and thione/thiol sulfur. When an additional coordinating functionality is present in the proximity of the donating centers (e.g. 2-heterocyclic thiosemicarbazones) the ligands bond in a tridentate manner. This can be accomplished by either the neutral molecule [116] or by the monobasic anion upon loss of a hydrogen from ^{2}N [117]. There are instances reported, however, where the heterocyclic atom and the azomethine nitrogen are involved in bidentate coordination [118] and the sulfur atom is considered not to be coordinated, weakly coordinated to the same metal center, or coordinated to an adjacent metal center [119].

Besides the dentacity variation, consideration of the charge distribution is complicated in thiosemicarbazones due to the existence of thione and thiol tautomers. Although the thione form predominates in the solid state, solutions of thiosemicarbazone molecules show a mixture of both tautomers. As a result, depending upon preparative conditions (particularly solvent and pH), the metal complexes can be cationic, neutral, or anionic. Most of the earlier investigations of metal thiosemicarbazone complexes have involved ligands in the uncharged thione form, but a number of recent reports have featured complexes in which the ^{2}N-hydrogen is lost, and bonding within the thiosemicarbazone moiety is closer to that of the thiol form. Furthermore, it is possible to isolate complexes containing both the neutral and anionic forms of the ligand bonded to the same metal ion [120, 121]. Ablov and Gerbeleu [122] suggested that formation of these mixed "tautomer" complexes is promoted by trivalent central metal ions like Cr(III), Fe(III) and Co(III).

The *E* and *Z* isomers of 2-formylpyridine thiosemicarbazone, **5a**, and **5b**, respectively [123], as well as those of other heterocyclic thiosemicarbazones [124], have been separated and characterized. The differentiation of stereochem-

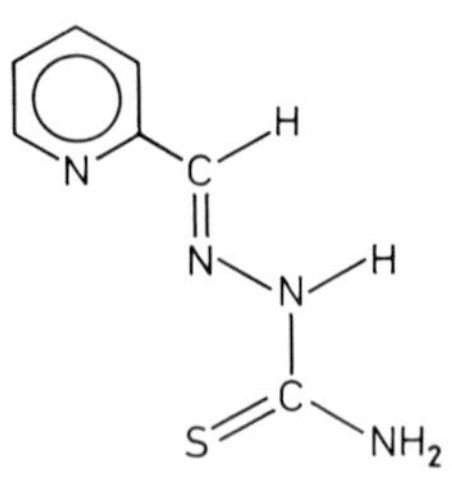

5a

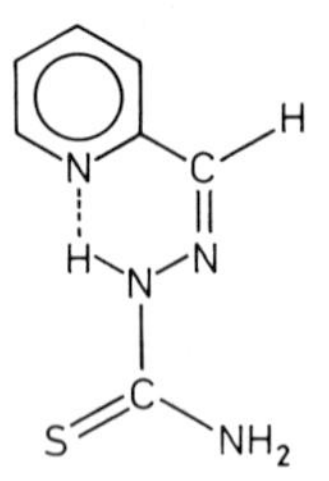

5b

istry between isomers was based upon the degree of deshielding observed for the ^{2}N proton of the *Z*-isomer ($\delta = 14.15$ ppm). More recently, it has been reported that the *E*-isomer is the stable form of 2-acetylpyridine ^{4}N-methylthiosemicarbazone, **3a**, and was not isomerized upon reflux over silica gel [125]. In contrast,

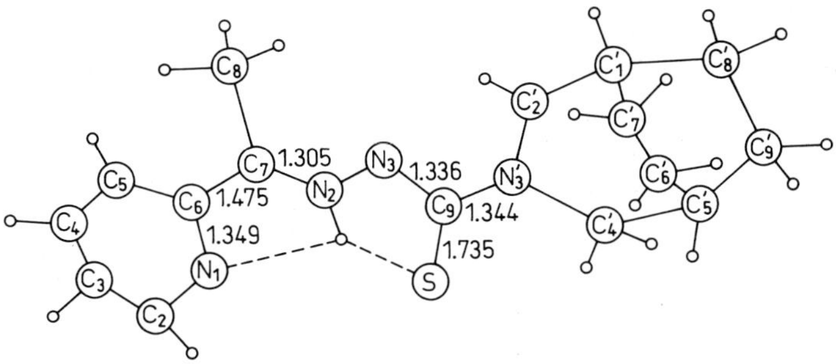

Fig. 1. 2-Acetylpyridine azabicyclo[3.2.2.]nonylthiosemicarbazone, **4**

it has been found that when the ^4N atom of the thiosemicarbazone is incorporated in a ring, a mixture of three isomers occurs. The third is the bifurcated *E*-hydrogen-bonded ring isomer of **4** depicted next.

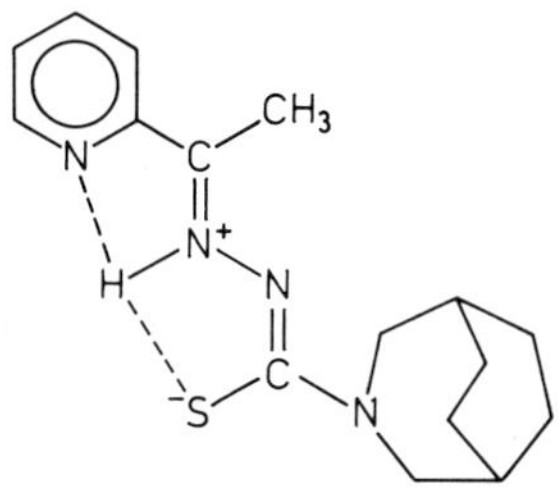

(4)

Recent reports on transition metal complexes of 2-heterocyclic thiosemicarbazones suggest that stereochemistries adopted by these complexes often depend upon the anion of the metal salt used and the nature of the ^4N-substituents. Further, as indicated previously, the charge on the ligand is dictated by the thione-thiol equilibrium which in turn is influenced by the solvent and pH of the preparative medium. Many of the reported complexes have been prepared in mixed aqueous solvents, often with bases added. However, there are few reports in which workers have varied the nature of their preparations to fully explore the potential diversity of these ligands.

The most common stoichiometries encountered with 2-heterocyclic thiosemicarbazones are six-coordinate having the general formula ML_2^{n+}, where M = Cr(III), Fe(III), Co(III) and Ni(II); L = tridentate, anionic ligand and n = 0, 1 and planar with the stoichiometry of MLX, where M = Ni(II) or Cu(II), L = tridentate, anionic ligand and A is generally a halo or pseudohalo ligand. More recently, additional mixed ligand complexes, complexes with neutral ligands, tetrahedral Co(II) and Ni(II) complexes, 5-coordinate complexes, and bimetallic compounds with MA_4^{n-} (n = 1 or 2 and A = Cl, Br, etc.) anions have all been reported.

4 Transition Metal Complexes

4.1 *Iron(III)*

4.1.1 2-Formylpyridine Thiosemicarbazones and S-Alkyldithiocarbazates

Some early studies [126] on the S-methyldithiocarbazate of 2-formylpyridine, **6**, showed it to form low spin iron(III) species, [Fe(**6**-H)$_2$]ClO$_4$ and [Fe(**6**-H)$_2$][FeCl$_4$]. The Mossbauer spectrum of the latter showed the presence of two different iron centers, and the compound was a 1:1 electrolyte in nitromethane.

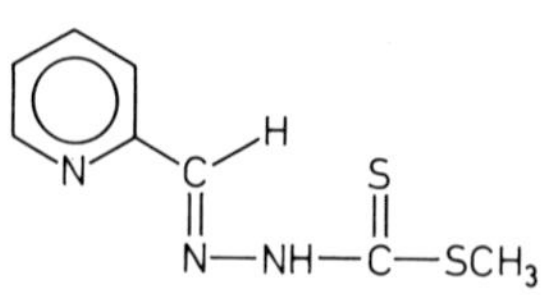

6

The Fe(III) complexes of 2-formylpyridine thiosemicarbazones, **5**, have received considerable attention because of their importance in chelation therapy [127]. Also, several heterocyclic thiosemicarbazones have been evaluated for removing excess iron from iron-loaded mice. Iron(III) complexes, as well as the copper(II) complexes of these ligands, are effective inhibitors of DNA synthesis at much lower concentrations than the uncomplexed thiosemicarbazones without apparent cytotoxicity [128]. Both iron(III) and copper(II) complexes of **5**, as well as a number of ring substituted (5-position) 2-formylpyridine thiosemicarbazones, prevent successful tumor cell transplantation [129]. A methyl substituent in the 5-position makes both metal complexes more potent agents. [Fe(**5**-H)$_2^+$] is low spin with a rhombic ESR spectrum (i.e., g = 2.185, 2.140 and 2.000) in frozen solution [130]. Earlier Raina and Srivastava [131] reported the preparation of the iron(III) complexes of **1**, [Fe(**5**-H)$_2$]A, where A = NO$_3$, Cl, N$_3$, NCS, and CN by stirring an acidified aqueous solution of the metal salt with an alcoholic solution of **5**. In DMF these complexes behaved as electrolytes indicating a [Fe(**5**-H)$_2^+$] center and coordination via the pyridyl nitrogen, azomethine nitrogen and thiol sulfur. All the complexes had magnetic moments consistent with low spin Fe(III), but the authors indicated that the chloride and thiocyanate salts did not possess ESR spectra at 77 K, while the remaining salts gave isotropic spectra. More recently other workers [118] isolated [Fe(**5**-H)$_2$]Cl from hot alcoholic solution and again showed the cation to be low spin. This complex and a zinc complex were more active against *Aspergillus fumigatus* than the uncomplexed ligand [52].

An iron(III) complex of 4-(*m*-aminophenyl)-2-formylpyridine thiosemicarbazone, **7**, having the stoichiometry [Fe(**7**)Cl_3] has been reported [132]. Its magnetic susceptibility is 2.1 B.M., consistent with low spin iron and bonding occurs via the pyridyl nitrogen, azomethine nitrogen and thione sulfur. The higher value for g_{av} is consistent with the greater number of chloro ligands as well as weaker bonding of the neutral ligand compared to the anionic form. The steric effect of the large ring substituent must prevent the formation of a [Fe(**7**-H)$_2^+$] center.

4.1.2 2-Acetylpyridine Thiosemicarbazones and S-Alkyldithiocarbazates

Treatment via chelation has been observed for 2-acetylpyridine thiosemicarbazone derivatives, which have been found to possess inhibitory activity for the RNA-polymerases of the influenza virus [133]. The iron(III) complexes were shown to be 3 to 6 times more active as inhibitors of partially purified ribonucleotide reductase (no added iron) compared to uncomplexed thiosemicarbazone [128]. Raina and Srivastava [134] prepared and characterized low spin iron(III) complexes of 2-acetylpyridine thiosemicarbazone, [Fe(**8**-H)$_2$A] (A = NO_3, OH, Cl, N_3, NCS or NO_2), which were proposed as being seven-coordinate. However, all but the azide complex are 1:1 electrolytes in DMF and their solid ESR spectra are rhombic with the g-values being about 2.20, 2.15 and 2.00. Of the six complexes, the azide ion seems to interact most strongly with the iron(III) center.

There have been no reports of complexes of ^4N-substituted thiosemicarbazones derived from 2-formylpyridine, but 2-acetylpyridine ^4N-methylthiosemicarbazone, **3a**, formed [Fe(**3a**-H)$_2$]ClO_4 and [Fe(**3a**-H)$_2$]$FeCl_4$ [117]. The nature of these two species was established by partial elemental analyses, molar conductivities, magnetic moments, electronic, infrared, mass and electron spin resonance spectra. A crystal structure of a related selenosemicarbazone complex confirmed the presence of a distorted octahedral iron(III) cation coordinated by two deprotonated anions so that each ligand is essentially planar and the azomethine nitrogens are *trans* to each other; the pyridyl nitrogen and selenium donors are both *cis*.

An iron(III) complex of 2-acetylpyridine 3-azabicyclononylthiosemicarbazone, **4**, the thiosemicarbazone that was found to have the most potent antimalarial activity of a large number of 2-acetylpyridine thiosemicarbazones tested [88], was originally formulated as 5-coordinate [Fe(**4**-H)Cl_2], [135], but more recent studies have shown it to be [Fe(**4**-H)$_2$] [$FeCl_4$] [117]. This complex has similar antimalarial activity to that of the uncomplexed ligand, but possesses enhanced antitumor activity [136].

More recently additional ^4N-substituted thiosemicarbazones {i.e., ^4N-dimethyl- [137, 138], **9**; 3-piperidinyl- [139], **10**; 3-hexamethyleneimine- [138, 140], **11**; and 3-(4-methylpiperidinyl)- [138] **12**} have all been found to yield complexes with [Fe(L-H)$_2$]$^+$ cations; there is some variation in their solid state

Table 1. ESR parameters for iron(III) thiosemicarbazone complexes at 77 K as powders

Compound	g_1 or $g_{\|\|}$	g_2	g_3 or $g_\perp$	g_{av}	Ref.
From 2-formylpyridine					
[Fe(**5**-H)$_2$]NO_3				2.107	[133]
[Fe(**5**-H)$_2$]Cl		no signal			[133]
From 2-acetylpyridine					
[Fe(**8**-H)$_2$]NO_3	2.211	2.149	1.995	2.118	[137]
[Fe(**8**-H)$_2$]Cl	2.206	2.151	1.997	2.118	[137]
[Fe(**3a**-H)$_2$]ClO_4	2.194	2.129	2.002	2.108	[120]
[Fe(**13**)(**13**-H)H_2O]$(ClO_4)_2$				2.113	[145]
[Fe(**9**-H)$_2$]ClO_4	2.166	2.125	2.004	2.098	[141]
	1.995		2.136	2.094	[142]
[Fe(**10**-H)$_2$]ClO_4	2.187	2.131	2.014	2.111	[143]
[Fe(**11**-H)$_2$]ClO_4	2.014		2.132	2.093	[144]
	2.162	2.119	2.019	2.100	[142]
[Fe(**12**-H)$_2$]ClO_4	2.180	2.111	2.013	2.101	[142]
[Fe(**4**-H)$_2$]ClO_4	2.026		2.150	2.109	[120]
[Fe(**14**-H)$(NO_3)_2$]				2.114	[146]
[Fe(**14**-H)]ClO_4	2.036		2.123	2.094	[146]
[Fe(**14b**-H)$_2$]ClO_4	2.190	2.135	2.031	2.119	[120]
[Fe(**14a**-H)$_2$]ClO_4	2.198	2.125	2.004	2.109	[120]
From 2-formylquinoline					
[Fe(**20**)Cl_3]	2.274	2.227	1.937	2.146	[135]
From 1-formylisoquinoline					
[Fe(**21**)Cl_3]	2.229	2.228	1.922	2.126	[135]
From 2-acetylpyridine N-oxide					
[Fe(**24**-H)$_2$]ClO_4	2.238	2.158	1.977	2.124	[120]
[Fe(**23**-H)$_2$]ClO_4	2.279	2.172	1.972	2.141	[120]

ESR spectra (Table 1). The ^4N-cyclohexylthiosemicarbazone, **13**, complex formed the expected $[Fe(\mathbf{13}\text{-H})_2]^+$ with $FeCl_4^-$ as the counterion [141]. However, [Fe(**13**) (**13**-H)H_2O]ClO_4 was isolated from ethanol. Bulkiness of the cyclohexyl group, and the perchlorate ion's greater ability to hydrogen bond are probably both important to the stability of this cation. The iron(III) center is considered six-coordinate with a tridentate **13**-H, bidentate **13**, and a coordinated water molecule.

Iron(III) complexes have also been prepared with 2-acetylpyridine ^4N-phenylthiosemicarbazone [142], **14**. Three have been formulated as square pyramidal [Fe(**14**-H)A_2] (A = Cl, NCS and NO_3) based, in part, on molar conductivities of 37–62 $ohm^{-1}\,cm^2\,mol^{-1}$. Their ESR spectra in frozen DMF are essentially the same as for other ^4N-substituted species, which likely indicates the presence of $[FeL_2]^+$ ions. More recently [138], [Fe(**14**-H)$_2$]ClO_4 has been isolated and its solid ESR spectrum is reported to be axial with $g_\perp > g_{\|}$. Table 1 summarizes the g-values of iron(III) heterocyclic thiosemicarbazone complexes.

Many of the complexes have rhombic spectra with g_{av} = 2.100–2.120. The iron(III) complexes, which have ^4N involved in a seven-membered ring, (**4** and

11), have been reported [117, 140] to have reversed axial spectra with $g_{||} > g_{\perp}$, although other workers report the iron(III) complex of **11** to have a rhombic spectrum [138]. Complexes having an anion other than perchlorate ion often have isotropic spectra suggesting strong interaction between iron(III) centers. It should be mentioned that 5-coordinate $[FeLA_2]$ species (A = Cl, Br, NCS etc.) would yield the same elemental analyses as $[FeL_2][FeA_4]$ and care should be taken to determine which species has been isolated [117]. Alcoholic preparations tend to give the latter complexes and their solids have a broad isotropic ESR signal, but in solution (i.e. methanol or DMF) have the same spectra as the complexes isolated with $Fe(ClO_4)_3$.

While iron(III) complexes of thiosemicarbazones with different functional groups involving ^4N have been prepared from 2-acetylpyridine, substitution on the ring has been more popular with thiosemicarbazones derived from 2-formylpyridine. The only thiosemicarbazones in which the 2-acetylpyridine ring has been substituted are **15a** and **15b**, prepared from 6-methyl-2-acetyl-pyridine [120]. Both of these iron(III) complexes have rhombic spectra and values of g_{av} are similar to those found for the 2-acetylpyridine thiosemicarbazones. Solution studies have been carried out on the iron(III) complex of 2,6-diacetylpyridine mono-thiosemicarbazone, but the solid complex was not isolated [143].

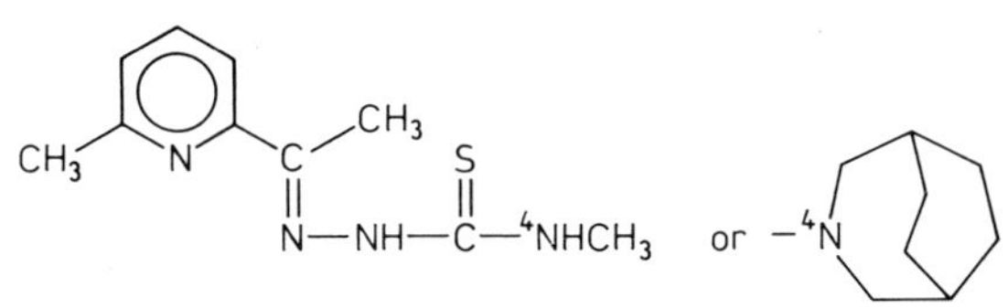

15a 15b

4.1.3 2-Picolylketone Thiosemicarbazones and S-Alkyldithiocarbazates

The S-methyldithiocarbazates of both 2-picolylphenylketone, **16**, and 2,6-lutidylphenylketone, **17**, yielded high spin iron(III) complexes, [Fe(**15**-H)$Cl_2(H_2O)$] and [Fe(**16**-H)$Cl_2(H_2O)$] [144]. Both complexes involve a tridentate ligand bonding via the pyridyl nitrogen, azomethine nitrogen and thiol sulfur.

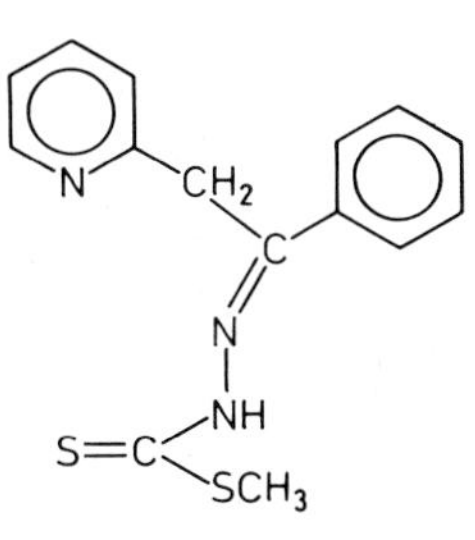

16

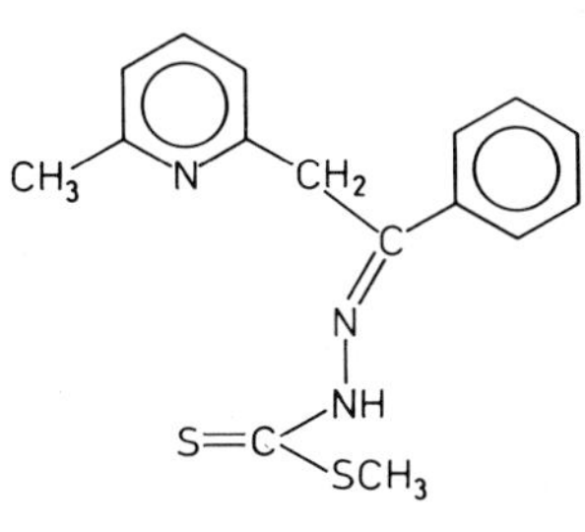

17

4.1.4 2-Formylquinoline and 1-Formylisoquinoline Thiosemicarbazones and S-Alkyldithiocarbazates

S-methyldithiocarbazates of 2-formylquinoline, **18**, and 1-formylisoquinoline, **19**, have been found to act as anionic tridentate ligands with iron(III) forming $[FeL_2][FeCl_4]$ on isolation from a dimethoxypropane-ethanol mixture [145]. No new spectral data was provided when $[Fe(\mathbf{19}\text{-}H)_2][FeCl_4]$ was included in a more recent paper [146].

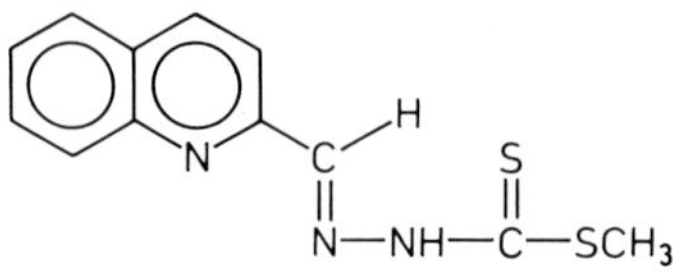

18

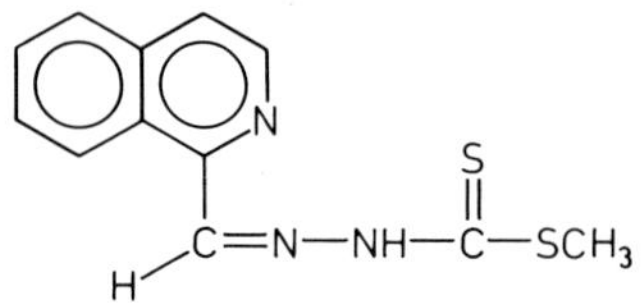

19

Complexes of 2-formylquinoline and 1-formylisoquinoline thiosemicarbazones, **20** and **21**, respectively, have been prepared, but iron(III) complexes were not included in the initial study [147]. More recently [132], an iron(III) complex of 1-formylisoquinoline thiosemicarbazone, $[Fe(\mathbf{21})Cl_3]$ has been found to have a magnetic moment of about 2.1, g_{av} of 2.146, and involve the neutral ligand bound in a tridentate manner. A complex of the same stoichiometry was reported for 4-methyl-5-amino-1-formylisoquinoline thiosemicarbazone, **22**. Its magnetic moment is about 2.1 and $g_{av} = 2.126$.

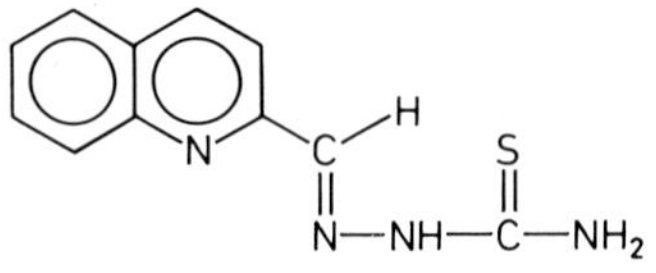

20

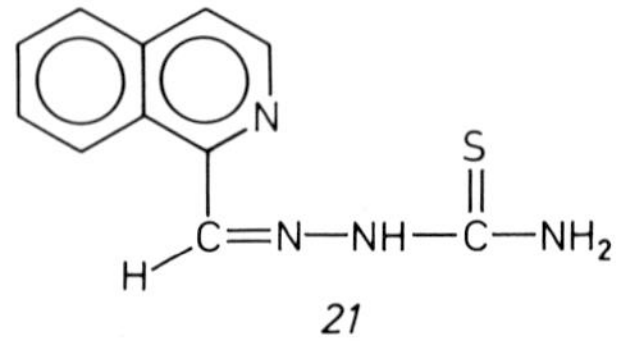

21

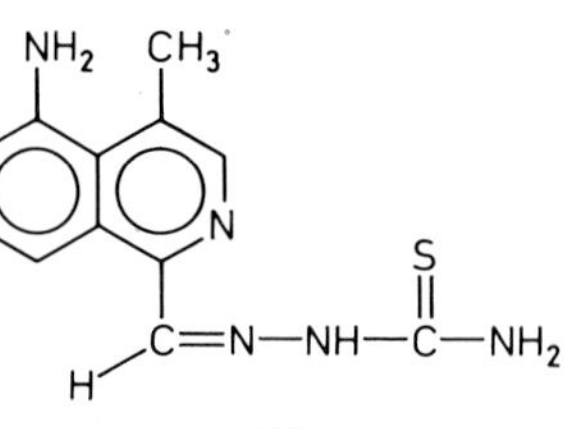

22

The iron(III) complexes of **21** and **22** were shown to be 3 to 6 times more active as inhibitors of partially purified ribonucleotide reductase than uncomplexed thiosemicarbazones [128]. The mechanism of antitumor action by these complexes still remains largely speculative, although some excellent preliminary studies have appeared. It has been postulated [148] that tridentate thiosemicarbazone ligands remove iron from ferritin to form iron(III) complexes. These complexes are rapidly reduced by hemoglobin and are only slowly reoxidized by oxygen in aqueous solution or plasma. The iron(II) complex is stable enough to exist as such in vivo, and can bind to the enzyme leading to the inhibition of RDR. Further studies are needed to elucidate the actual inhibition form.

There are several important implications of these investigations. First, there is a possibility that metal complexes may serve as effective anticancer agents against tumors. The complexes of heterocyclic thiosemicarbazones could constitute the second largest group of metal complexes, besides cisplatin derivatives, that can serve as potent anticancer agents. Secondly, it reinforces the need to conduct in vivo antitumor studies on pre-formed iron and copper complexes. By doing so, the possibility of substantial chelation of these metal ions in vivo by the free ligands can be avoided [148–150]. Without such studies, the nutritional status of iron and copper may assume a prime importance in the success of chemotherapy of the host with these ligands as has been shown for bis(thiosemicarbazones) [151].

4.1.5 2,6-Diacetylpyridine Bis (Thiosemicarbazones) and Bis (S-Alkyldithiocarbazates)

The potentially pentadentate 2,6-diacetylpyridine bis(thiosemicarbazones) have recently received attention [152] although iron(III) complexes have been isolated in only one instance [153]. A large number of 2,6-diacetylpyridine bis(^4N-azacyclic thiosemicarbazones), H_2L, were prepared and complexes involving the dibasic, anionic form of the ligand were isolated from methanolic $FeCl_3$ as the distorted six coordinate [FeLCl] complexes. Infrared spectral assignments indicate coordination via the pyridyl nitrogen, both azomethine nitrogens and both thiol sulfurs. The iron(III) complexes are low spin, having rhombic ESR spectra with g_1 about 2.244, g_2 about 2.115 and g_3 about 2.020. Many of the complexes proved to be toxic when tested for antitumor activity using mice. The iron(III) complex of the 2,6-diacetylpyridine bis(S-methyldithiocarbazate), [Fe(**23**-2H)Cl], has a magnetic moment of 5.93 B.M., consistent with high-spin octahedral iron(III) complex [154].

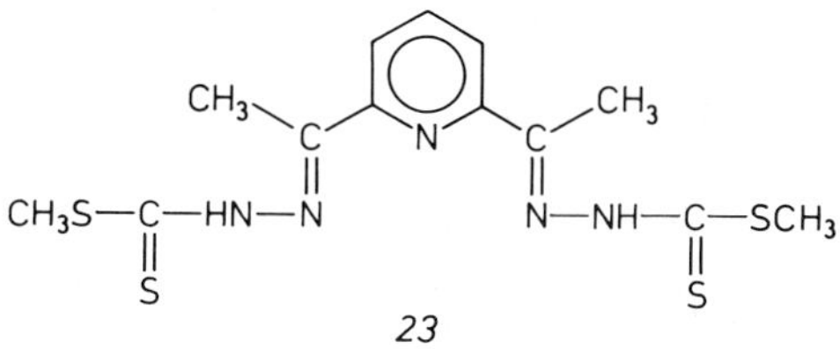

23

4.1.6 2-Acetylpyridine N-Oxide Thiosemicarbazones and S-Alkyldithiocarbazates

Iron(III) complexes of 2-acetylpyridine *N*-oxide ^4N-methyl- and 3-azabicyclo[3.2.2.]nonylthiosemicarbazone, **24** and **25**, respectively, have been isolated from both iron(III) perchlorate and chloride [117]. The perchlorate salt yields low spin, octahedral, monovalent, cationic complexes involving two deprotonated, tridentate thiosemicarbazone ligands coordinated via the *N*-oxide oxygen, azomethine nitrogen and thiol sulfur based on infrared spectral studies. Their powder ESR g-values are included in Table 1 and indicate that bonding is less covalent than for the analogous thiosemicarbazones prepared from 2-acetylpyridine, **3a** and **4**. Starting with iron(III) chloride, compounds with the same cations, but with tetrachloroferrate(III) anions, were isolated.

4.1.7 2-Formyl- and 2-Acetylthiophene Thiosemicarbazones

Besides complexes of thiosemicarbazones prepared from nitrogen heterocycles, iron(III) complexes of both 2-formylthiophene thiosemicarbazone, **26**, and 2-acetylthiophene thiosemicarbazone, **27**, have been isolated [155]. Low spin, distorted octahedral complexes of stoichiometry [Fe(**26**)$_2$A$_2$]A (A = Cl, Br, SCN) were found to be 1:1 electrolytes in nitromethane. Low spin Fe(**27**)$_3$A$_3$ (A = Cl, Br, SCN) complexes were also isolated, but their insolubility in organic solvents did not allow molar conductivity measurements. Infrared spectra indicate coordination of both via the azomethine nitrogen and thione sulfur, but not the thiophene sulfur. The thiocyanate complexes have spectral bands at 2065, 770 and 470 cm^{-1} consistent with *N*-bonded thiocyanato ligands, but ν(FeCl) and ν(FeBr) were not assigned due to the large number of bands found in the spectra of the two ligands.

4.2 Iron(II)

4.2.1 2-Formylpyridine Thiosemicarbazones and S-Alkyldithiocarbazates

Iron(II) complexes are often included in studies when complexes are prepared from a large number of different metal ions. 2-formylpyridine thiosemicarbazone, **5**, forms brown [Fe(**5**)$_2$A$_2$] (A = Cl, Br) when prepared in ethanol and [Fe(**5**-H)$_2$] from aqueous alcohol solution [156]. All of these complexes are diamagnetic. The resonance Raman and infrared spectra of [Fe(**5**-H)$_2$] were examined in detail [130] and coordination occurs via the pyridyl nitrogen, azomethine nitrogen and thiol sulfur. There is appreciable d–d sulfur-to-iron(II) π-bonding. Solution studies of iron(II) complexes of some 5-substituted-2-formylpyridine thiosemicarbazones have been reported [157], but no solids were isolated. Other ring substituted 2-formylthiosemicarbazones on complexation to iron(II) show biological activity, particularly those having a nitro group on the ring [158].

4.2.2 2-Acetylpyridine S-Alkyldithiocarbazates

No iron(II) complexes of thiosemicarbazones or *S*-alkyldithiocarbazates derived from 2-acetylpyridine have been reported to date.

4.2.3 2-Formylquinoline and 1-Formylisoquinoline Thiosemicarbazones and S-Alkyldithiocarbazates

The S-methyldithiocarbazates of both 2-formylquinoline, **18**, and 1-formylisoquinoline, **19**, yield diamagnetic green $[FeL_2]$ complexes from iron(II) sulfate [131]. Coordination for both complexes is via the ring nitrogen, azomethine nitrogen and thiol sulfur based on infrared studies.

Iron(II) complexes of general formula $[FeLA_2]$ A = Cl, Br, I, NO_3, NCS or NCSe have been formed [145] with both 2-formylquinoline thiosemicarbazone, **20**, and 1-formylisoquinoline thiosemicarbazone, **21**. All are characterized as five-coordinate, high spin iron(II) complexes and the electronic spectra are assigned consistent with distorted trigonal bipyramidal structures.

4.2.4 2,6-Diacetylpyridine Bis(Thiosemicarbazones) and Bis(S-Alkyldithiocarbazates)

The pale green [Fe(**28**)OSO_3] was isolated with pentadentate 2,6-diacetylpyridine thiosemicarbazone, **28**, and is high spin with a monodentate sulfato ligand completing the coordination sphere [148]. The diffuse reflectance spectrum shows d–d bands at 9090 and 14 290 cm^{-1} and therefore, Dq = 1151 cm^{-1}. Iron(II) complexes, [FeL], prepared with a large number of 2,6-diacetylpyridine bis(^4N-azacyclic thiosemicarbazones), have recently been reported [149]. All the FeL complexes are 5-coordinate, high-spin with coordination via the pyridyl nitrogen, the two azomethine nitrogens and two thiol sulfurs.

4.2.5 2-Formyl- and 2-Acetylthiophene Thiosemicarbazones

Both 2-formylthiophene thiosemicarbazone, **26**, and 2-acetylthiophene thiosemicarbazone, **27**, form six coordinate $[FeL_2A_2]$ complexes (A = Cl, Br) [156]. The complexes formed with **26** are low spin, but complexes of **27** are high spin. For both ligands the bidentate coordination is via the azomethine nitrogen and the thione sulfur, and, like iron(III) complexes, the thiophene sulfur is not coordinated in iron(II) complexes.

4.3 Copper(II)

The copper(II) complexes of 3-ethoxy-2-oxobutyraldehyde bis(thiosemicarbazone) and related compounds are active in vivo agents [151, 158, 159]. The metal complexes of 2-heterocyclic thiosemicarbazones were evaluated for their cytotoxicities [160, 161]. Further studies have revealed that these ligand's iron and copper complexes are effective inhibitors of DNA synthesis at much lower concentrations than the free thiosemicarbazones without apparent cytotoxicity [127]. Although the iron(III) complex of 2-isoformylquinoline thiosemicarbazone, **21**, is considerably more active than free **21**, the copper(II) complex is only moderately more active [127].

The mechanism of antitumor action by these compounds still remains largely speculative although some excellent preliminary studies have appeared. For example, the copper(II) complex, CuL^+, was found to accumulate rapidly in cells [129] where it blocked the G/S interphase of the cell cycle, which is consistent with ribonucleotide reductase being affected [9]. While tracing the sequence of events leading to the accumulation of CuL^+, it was argued that its rapid interaction with the intracellular thiols may promote generation of toxic oxygen species such as superoxide and hydroxyl radicals (with concomitant oxidation of the thiols), and this might be responsible for the cellular damage [163]. Recent work, however, indicates that large amounts of hemoglobin and other radical scavengers are capable of detoxifying oxygen radicals at least for normal red cells [164].

4.3.1 2-Formylpyridine Thiosemicarbazones and S-Alkyldithiocarbazates

Among the first studies [126] of copper complexes involving a heterocyclic ring were those of 2-formylpyridine S-methyldithiocarbazate, **6**. Monomeric, green copper(II) complexes, [Cu(**6**-H)A] and [Cu(**6**-H)H_2O]A (A = Cl, Br, and NO_3), have broad absorption bands at about 16 000 cm^{-1}. 2-formylpyridine S-benzyldithiocarbazate, **29**, also forms green, monomeric complexes, [Cu(**29**-H)A], where A = Cl, Br, and NO_3 when prepared in ethanol-benzene solution [165]. Infrared assignments were used to establish coordination of **29**-H via the pyridyl nitrogen, azomethine nitrogen and thiol sulfur. The electronic spectra of all three complexes feature d–d bands in the 16 000 cm^{-1} region as dichloromethane solutions and nujol mulls.

In aqueous solution the formation constants of $Cu(\mathbf{5}\text{-H})^+$, (**5** is 2-formylpyridine thiosemicarbazone) have been studied in relation to the possible reaction of this ion with biological systems [166]. A pre-resonance Raman spectral study of the copper(II) complex of **5** in aqueous solutions of different pH's has been carried out [167]. The neutral ligand is bound in a tridentate manner at low pH via its pyridyl and azomethine nitrogens and thione sulfur. At higher pH's deprotonation at ^2N occurs, and both of the above nitrogens, as well as the thiol sulfur, are coordinated. Formation constants have also been determined for a series of 5-substituted-2-formylpyridine thiosemicarbazone copper(II) complexes in order to establish a relationship between complex stability and their ability to catalyze thiol oxidation in biological systems [159]. More recently [168], dimeric copper(II) complexes, [{Cu(**5**-H)A}$_2$], where A = OAc, Cl, ClO_4, and NO_3, have been isolated from a *N,N*-dimethylformamide-water mixture. A crystal study (Fig. 2) of the acetato complex shows that the dimer involves two acetato bridging ligands (using a single oxygen) with **5**-H coordinated in a planar fashion via the pyridyl nitrogen, azomethine nitrogen, and thiol sulfur. **5**-H and one of the bridging acetato oxygens provide an approximate plane and, a less strongly bound bridging oxygen gives a distorted square pyramidal geometry for each copper(II) ion. The powder ESR spectrum of the acetato dimer is rhombic with g-values of 2.20, 2.06, and 2.04. In frozen DMSO the dimer remains intact with g-values of 2.198 and 2.071, but in aqueous solution converts to [Cu(**5**-H)(H_2O)]$^+$. If the acetato dimer is dissolved in concentrated sulfuric acid, a sulfato dimer, [{Cu(**5**)SO_4}$_2$], involving the protonated thiosemicarbazone bonding as a tridentate ligand results. The bonding of **5** and **5**-H as tridentate ligands is structurally similar in the two complexes.

Spectroscopic studies [118] of [Cu(**5**)Br_2] suggest NNS coordination and a dimeric halogen bridged complex. This assignment is consistent with non-electrolytic behavior in DMF, but not with a magnetic moment at room temperature of 1.79 B.M. The shift of the thioamide **4** band from 815 cm^{-1} to 800 cm^{-1} on complexation is taken as evidence for coordination of the

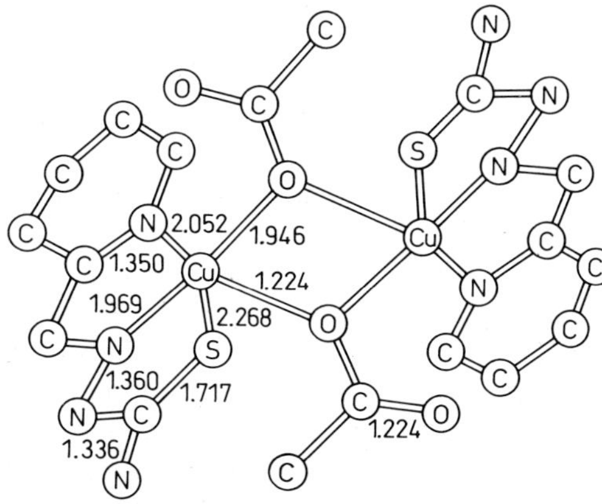

Fig. 2. Bis{u-acetato-2-formylpyridine thiosemicarbazonecopper(II)}, [Cu(**5**-H)OAc]$_2$

thione sulfur. The following complexes have recently been prepared [169]; [Cu(**5**)Cl_2] [Cu(**5**-H)Cl], [Cu(**5**-H)SCN], and [Cu(**5**-H)OAc]. The olive green [Cu(**5**)Cl_2] has a magnetic moment of 1.51 B.M., $g_{||} = 2.18$ and $g_{\perp} = 2.03$; the thioamide IV band is weakened, but shifted to higher energy. The complexes of **5**-H all have magnetic moments in the 1.8–1.9 B.M. range, involve NNS tridentate bonding and are considered distorted, planar, four-coordinate complexes.

Early studies [170] of copper(II) complexes of thiosemicarbazones were 2-formylpyridine ^2N-methylthiosemicarbazone, **30**, 6-methyl-2-formylpyridine ^2N-methylthiosemicarbazone, **31**, and 2-formylpyridine ^4N-dimethylthiosemicarbazone, **32**. With copper(II) chloride and bromide, monomeric complexes of stoichiometry [Cu(L)A_2] were isolated for each of these thiosemicarbazones. All six complexes had a band in the 14 000–15 000 cm^{-1} spectral region, but their stereochemistry was not specified.

Recently [171], a green copper(II) complex of 2-formylpyridine ^4N-phenylthiosemicarbazone, **33**, has been formulated as [Cu(**33**-H)$_2$]. The magnetic moment of 2.2 B.M. and observance of three different d–d bands (there appears to be an error in reporting the energy of the three bands) suggested the complex to be six-coordinate. However, the pyridyl nitrogen(s) were not considered coordinated making it difficult to understand bonding to the copper(II) center. 4-(*m*-aminophenyl)-2-formylpyridine, **7**, forms [Cu(**7**)Cl_2] which has bands at 13 330 and 10 100 cm^{-1} and is similar to the spectrum of [Cu(Me_6tren)Br]Br, a 5-coordinate trigonal bipyramid [172].

Although coordination of the heterocyclic nitrogen does not occur, two copper complexes of 3-hydroxy-5-hydroxymethyl-2-methyl-4-formylpyridine thiosemicarbazone, **34** have been isolated [173]. With copper(II) nitrate, the resulting [Cu(**34**-H)(NO_3)]·2H_2O involves loss of hydrogen from ^2N and transfer of the phenolic hydrogen to the pyridine nitrogen which results in an ONS tridentate ligand. When prepared from copper(II) acetate, [Cu(**34**-2H)]·2H_2O is formed on loss of both the ^2N and phenol protons and **34**-2H bonds as a ONS tridentate ligand. The complex is probably better formulated as [Cu(**34**-2H)H_2O]·H_2O. More recently [174], [Cu(**34**-H) (H_2O)]Cl·H_2O has been isolated from an aqueous solution at pH = 4. In this complex copper(II) coordinates in an approximate square planar environment involving a water molecule and the phenolic oxygen, azomethine nitrogen and sulfur of **34**-H. An elongated square pyramid is formed by a long interaction (i.e. 3.066 Å) between the copper(II) and a sulfur of a centrosymmetrically related center.

4.3.2 2-Acetylpyridine Thiosemicarbazones and S-Alkyldithiocarbazates

The first copper(II) complexes prepared with a 2-acetylpyridine thiosemicarbazone were those of the 3-azabicyclononyl-derivative, **4**, as well as the analogous selenosemicarbazone [175]. Monomeric [Cu(**4**-H) A] where A = Cl, Br, I, OAc and NO_3 were characterized by their magnetic and spectral measurements. A second report [128] on copper(II) complexes of **4** included [Cu(**4**) (**4**-H)]BF_4 as well as [Cu(**4**-H)NCS] and [Cu(**4**-H)N_3]. The spectral parameters change with the species completing the coordination sphere; the azido complex and [Cu(**4**)(**4**-H)]BF_4 have the highest d–d energies. The latter complex has **4** coordinated via its azomethine nitrogen. Similar copper(II) complexes were isolated with the analogous thiosemicarbazone prepared from 6-methyl-2-acetylpyridine, **15b**, and spectral studies indicated weaker bonding to copper(II) centers, presumably due to the methyl's steric effect [176]. For example, [Cu(**4**-H)Cl] has ν(CuCl) = 329 cm^{-1}, ν(d–d) = 17 400, 14 100 cm^{-1} and g_1 = 2.140, while [Cu(**15b**-H)Cl] has ν(CuCl) = 320 cm^{-1}, ν(d–d) = 16 390 cm^{-1} and g_1 = 2.165; all indicate stronger bonding for the former complex. **4** also has been shown to possess superior antileukemic activity [177].

Two copper(II) complexes of 2-acetylpyridine thiosemicarbazone, **8**, were included in a study of complexes of 2-formylpyridine thiosemicarbazone [169]. [Cu(**8**-H)OAc] has a magnetic moment consistent with a monomeric copper(II) center and both it and [Cu(**8**)Cl_2] have d_{x2-y2} ground state ESR spectra (Table 2). A d–d envelope and a magnetic moment of 1.68 B.M. have led others [178] to propose a distorted tetrahedral environment with metal-metal interaction for the brown complex, [Cu(**8**)Cl_2].

[Cu(**3a**)$_2$] $(BF_4)_2$, [Cu(**3a**-H)Cl], and [Cu(**3a**)Br_2], where **3a** is 2-acetylpyridine ^4N-methylthiosemicarbazone, have been prepared and studied spectroscopically [179]. The former complex involves planar bonding by pyridyl and azomethine nitrogen atoms of each ligand and likely involves axial coordination of the thione sulfur atoms since ν(d–d) = 15 670 cm^{-1}, which is at lower energy

Table 2. Electronic and ESR spectral values for solid copper(II) complexes of heterocyclic thiosemicarbazones

Compound	ν(d–d)		$g_{\|\|}$	$k_{\|\|}$	Ref.
From 2-formylpyridine					
[Cu(**5**)Cl_2]			2.18		[175]
[Cu(**5**-H)Cl]			2.17		[175]
From 2-acetylpyridine					
[Cu(**8**)Cl_2]			2.18		[175]
[Cu(**3a**)$_2$]$(BF_4)_2$	15.7		2.211		[186]
[Cu(**7a**-H)Cl]	15.4		2.170		[186]
[Cu(**4**)(**4**-H)]ClO_4	18.7	13.9	2.145	0.55	[132]
[Cu(**4**-H)Cl]	16.5		2.146		[181]
	17.0	14.1	2.140	0.54	[132]
[Cu(**9**-H)Cl]	19.4	17.3	2.152	0.63	[186]
[Cu(**12**-H)Cl]	18.1	17.1	2.256	0.81	[186]
[Cu(**10**)(**10**-H)]BF_4	16.6		2.134		[143]
[Cu(**13**-H)Cl]	16.2		2.159		[143]
[Cu(**11**-H)Cl]	16.4	14.0	2.145	0.52	[144]
	18.6		2.138		[189]
[Cu(**16**-H)Cl]	19.0	15.5	2.146	0.58	[187]
From 2-acetylpyridine N-oxide					
[Cu(**15**)(**15**-H)]ClO_4	15.8		2.278		[132]
[Cu(**25**-H)Cl]	13.9		2.135		[132]

than is generally found for CuN_4 chromophores. The difference in stoichiometry of the two halogen complexes is surprising, and it is possible that changing preparative methods could produce [Cu(**3a**)Cl_2] and/or [Cu(**3a**-H)Br].

Spectral results [180] for copper(II) complexes of 2-acetylpyridine ^4N-dimethylthiosemicarbazone, **9**, and 2-acetylpyridine 3-(4-methylpiperidine) thiosemicarbazone, **12**, are included in Table 2. While the complexes of **9**-H have spectral properties consistent with the other copper(II) complexes of Table 2, those of **12**-H, with the exception of [Cu(**12**-H)F], show considerably higher values of $g_{||}$, and consequently higher values of $k_{||}$. These values indicate that there is little in-plane π-bonding, possibly due to the bulkiness of the 4-methylpiperidine group.

Seven different copper(II) complexes [181] of 2-acetylpyridine ^4N-phenylthiosemicarbazone, **14**, all having the general formula, [Cu(**14**-H)A] have been prepared and characterized. Their spectral data are included in Table 2 and $g_{||}$ is similar to other copper(II) complexes of 2-acetylpyridine thiosemicarbazones. However, the d–d spectra all show two bands, suggesting planar stereochemistry; these bands are of higher energy than the analogous complexes of the bicyclononyl derivative, **4** [128, 175].

Copper (II) complexes prepared with **10** and **11** (^4N is included in a 6- or 7-membered ring [139, 140]) have deprotonated thiosemicarbazone ligands except for [Cu(**10**)Br_2] [139]. Also, [Cu(**10**)(**10**-H)]BF_4 was isolated, but a pure sample of the analogous complex was not isolated with **11**. However, [Cu(**11**)(**11**-H)]ClO_4, as well as [Cu(**11**-H)A] (A = Cl, Br, I, NCS, NO_3 and OAc), was recently reported to have **11**-H coordinated as a NNS tridentate ligand and **11** bonding via only the azomethine nitrogen [182]. The g values reported [140] for [Cu(**11**-H)A] complexes are the lowest for thiosemicarbazone complexes prepared from copper(II) halides suggesting strong planar bonding, and weak axial bonding which is likely due to steric bulk of **11**.

Two copper(II) complexes of 2-acetylpyridine ^4N-(2-thiazolyl)thiosemicarbazone, **35**, have been isolated from alcoholic solutions. [Cu(**35**)Cl_2] has ν(d–d) at 14 705 cm^{-1}, $g_{||} = 2.186$ and $A_{||} = 172$ G and considered [183] to be six-coordinate with **35** bonding via the pyridyl nitrogen, azomethine nitrogen, thione sulfur and thiazolyl nitrogen. [Cu(**35**-H)$_2$], which was isolated from a pH = 7.0 aqueous-ethanol solution, has ν(d–d) at 15 385 cm^{-1}, $g_{||} = 2.176$, and $A_{||} = 167$ G. Bonding in this six-coordinate complex is via the pyridyl nitrogen, azomethine nitrogen and thiol sulfur atoms. However, 2,6-diacetylpyridine ^4N-(2-thiazolyl)thiosemicarbazone formed [Cu(**36**)$_2Cl_2$] on stirring in ethanol-TMF [184]. The complex is considered to be octahedral, but with **36** bonding via the azomethine nitrogen and thione sulfur. The latter mode of coordination

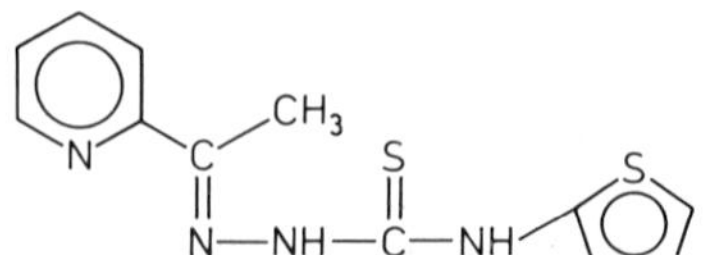

35

is surprising, since the band assigned to ν(C = S) shifts to higher energy in the complex's spectrum. [Cu(**36**-H)$_2$] was isolated from aqueous solution (pH = 7.0). The $g_{||}$ values are 2.294 and 2.331 for [Cu(**36**)$_2$Cl$_2$] and [Cu(**36**-H)$_2$], respectively, suggesting an absence of sulfur coordination. The d–d band maxima of about 13000 cm^{-1} are lower energy than most complexes listed in Table 2.

A distorted tetrahedral copper(II) center has been proposed [185] for [Cu(**37**)Cl$_2$] where **37** is 4-acctylpyridine thiosemicarbazone with the ring nitrogen not coordinated.

4.3.3 2-Picolyl- and 2-Lutidylketone Thiosemicarbazones and S-Alkyldithiocarbazates

Copper(II) complexes of 2,6-lutidylphenylketone thiosemicarbazone, **38**, have been prepared from copper(II) chloride and copper(II) bromide [186]. Similar to 2-pyridyl thiosemicarbazones, **38**-H coordinates via the ring nitrogen, the azomethine nitrogen and the thiol sulfur based on infrared spectral assignments. Magnetic susceptibilities and electron spin resonance spectra indicate dimeric complexes and both are formulated as [Cu(**38**-H)A]$_2$ with bridging sulfur atoms. The electronic spectra of both halide complexes show band maxima at 14500–14200 cm^{-1} with shoulders at 12100 cm^{-1}, which is consistent with a square pyramidal stereochemistry for a dimeric copper(II) center.

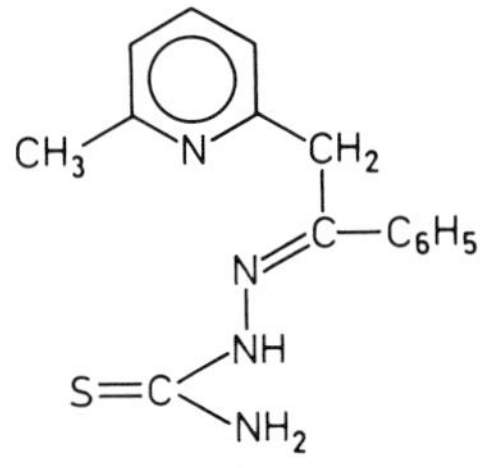

38

4.3.4 2-Formylquinoline and 1-Formylisoquinoline Thiosemicarbazones and S-Alkyldithiocarbazates

1-Formylisoquinoline S-methyldithiocarbazate, **19**, forms green [Cu(**19**-H)Cl], and its magnetic moment of 1.80 B.M. and ν(d–d) = 16000 cm^{-1} are cited as evidence for planar stereochemistry [146]. It and the analogous zinc complex showed the greatest antitumor activity of metal complexes tested.

Copper (II) complexes of stoichiometry [Cu(HL)A$_2$], where HL = **20** and **21** are 2-formylquinoline and 1-formylquinoline thiosemicarbazones, respectively, and A = Cl, Br, NO$_3$, NCS and NCSe, were prepared in aqueous ethanol [147].

All of these complexes have magnetic moments consistent with monomeric copper(II) centers and are various shades of green. Infrared spectra indicate both bond via the quinolyl nitrogen, azomethine nitrogen and thione sulfur. Further indication of five-coordinate trigonal bipyramidal stereochemistry is a broad band centered at 10 905 cm^{-1} which is split into two components at *ca.* 9800 and 12 400 cm^{-1}, as well as a rhombic powder ESR spectrum with a d_{z^2} ground state. More recently [187], [Cu(**21**)Cl_2] was again reported to be a trigonal bipyramid with d–d bands at 9260 and 12 500 cm^{-1}. This complex was shown to possess antitumor activity against P388 lymphocytic leukemica test system in mice.

4.3.5 2,6-Diacetylpyridine Bis(Thiosemicarbazones) and Bis(S-Alkyldithiocarbazates)

The pentadentate 2,6-diacetylpyridine bis(thiosemicarbazone), **28**, forms green, monomeric [Cu(**28**)Cl]Cl [148]. It is a distorted octahedral complex based on electronic and infrared spectra as well as 1:1 electrolytic behavior in methanol. Solution studies at various pH's show that the n–π* band of **28** shifts to lower energy upon loss of ^2N hydrogens [143]. This same shift was found for 6-acetyl-2-acetylpyridine thiosemicarbazone, and its copper(II) complexes and those of **28** were studied in solution, but no solids were isolated. Like iron(II) and iron(III), a large number of copper(II) complexes of 2,6-diacetylpyridine bis(^4N-azacyclic thiosemicarbazones) have been prepared [149]. All are 5-coordinate and have the ligand(s) bonding via the pyridyl nitrogen, two azomethine nitrogens, and two thiol sulfur atoms. The electronic and ESR spectra are reported to be consistent with a distorted trigonal bipyramidal arrangement of the pentadentate ligands. Few of these complexes possess significant activity against P388 tumor growth at the dosages tested. Also, the copper(II) complex of the related 2,6-diacetylpyridine bis(S-methyl dithiocarbazate), [Cu(**23**-2H)], has been prepared [150]. It has a magnetic moment consistent with a monomeric copper(II) center and its powder ESR spectrum has $g_{\|} = 2.12$ and $g_{\perp} = 2.01$.

4.3.6 2-Acetylpyridine N-Oxide Thiosemicarbazones and S-Alkyldithiocarbazates

Copper(II) complexes have been prepared with the 2-acetylpyridine *N*-oxide 3-azabicyclo[3.2.2.]nonylthiosemicarbazone, **25**, and bonding occurs via the pyridine *N*-oxide oxygen, azomethine nitrogen and thiol sulfur [128]. Based on electronic and ESR spectra, bonding to copper(II) of uninegative, tridentate **25**-H is considerably weaker than the related 2-acetylpyridine thiosemicarbazone, **4**-H. The other copper(II) complexes reported to date have been prepared from 2-acetylpyridine *N*-oxide ^4N-methylthiosemicarbazone, **24** [188], and ^4N-dimethylthiosemicarbazone, **34** [189]. Complexes of **24** are formed primarily as the neutral ligand and showed little tendency to lose the ^2N proton on complex formation. Both [Cu(**24**) Cl_2] and [Cu(**24**)Br_2] are green, have very low ν(d–d) maxima and high values of $g_{\parallel}$. In contrast, **34** forms [Cu(**34**-H)Cl] and [Cu(**34**-H)Br] which have considerably lower values of $g_{\parallel}$ than complexes of the pyridyl analog, **24**, suggesting a more planar arrangement of the donor atoms.

4.3.7 2-Formylpyrazole Thiosemicarbazones

5-Methyl-3-formylpyrazole thiosemicarbazone, **40**, forms monomeric copper(II) complexes having the general formula [Cu(**40**)A_2] where A = Cl, Br, NO_3, BF_4, ClO_4 and 1/2 SO_4 [190]. Infrared studies suggest tridentate bonding via the tertiary (iminic) nitrogen atom of the pyrazole ring, azomethine nitrogen and thione sulfur.

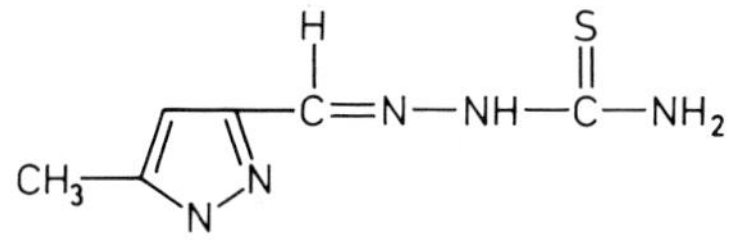

40

Copper(II) complexes of 4,5-dimethyl-3-formylpyrazole ^4N-phenylthiosemicarbazone, **41**, have been prepared by refluxing in ethanol [191]. [Cu(**41**-H)A], where A = Cl, Br, NO_3, NCS and N_3, were shown to involve bonding by ^2N(ring), azomethine nitrogen, and thiol sulfur atoms and all have magnetic moments in the range 1.8–1.9 B.M. The electronic and ESR spectral data for these complexes are comparable to that found for 2-acetylpyridine ^4N-phenylthiosemicarbazone, **14** [181].

4.3.8 2-Formyl- and 2-Acetylfuran Thiosemicarbazones and S-Alkyldithiocarbazates

A copper(II) complex [192] of 2-formylfuran *S*-benzyldithiocarbazate, [Cu(**42**-H)Cl], has a magnetic moment of 1.73 B.M, but, based on the complex's insolubility in non-polar and weakly polar solvents, a dimer structure is proposed. A dimer with two chloride bridges is more likely than one with two thiol bridges since the latter usually leads to strong antiferromagnetic interactions resulting in complete diamagnetism. The authors did not observe a band in the complex's infrared spectrum assignable to ν(CuO).

2-Formylfuran ^{4}N-phenylthiosemicarbazone, **43**, forms dimeric complexes, $[Cu(\mathbf{43}\text{-}H)A]_2$ where A = Cl, Br [193]. Both have magnetic moments below 0.8 B.M., ν(MA) at energies consistent with bridging halo ligands, and molecular weight determinations are consistent with dimeric structures for both. The infrared and electronic spectra indicate that **43**-H is coordinated as a bidentate ligand via the azomethine nitrogen and thiol sulfur with bridging halo ligands completing the planar arrangement of the copper(II) centers. The furan oxygen is not coordinated in these complexes.

Two copper(II) complexes of 4-(2-thiazolyl)-1-(2-acetylfuran)thiosemicarbazone, **44**, have been prepared and characterized [194]. Both $[Cu(\mathbf{44})_2Cl_2]$ and $[Cu(\mathbf{44}\text{-}H)_2]$ possess magnetic moments of about 1.1 indicating significant interaction between copper(II) centers. They are soluble in only DMSO and DMF. Infrared and conductivity studies suggest distorted octahedral coordination for $[Cu(\mathbf{44})_2Cl_2]$ with **44** bonding via the azomethine nitrogen and thione sulfur. For $[Cu(\mathbf{45}\text{-}H)_2]$, the thiazole nitrogen is also bound to the copper(II) based on the shift of ν(C = N) of the thiazole ring. It is suggested that these complexes possess two bridging thiosemicarbazone ligands: one bridging ligand provides a azomethine nitrogen to a copper(II) coordination sphere and the other a thione sulfur.

4.3.9 Other Polydentate Thiosemicarbazones

1-Benzoin-4-phenylthiosemicarbazone, although not a heterocyclic thiosemicarbazone, can function as an ONS tridentate ligand [195]. Its copper(II) complex, formed on loss of two hydrogen ions from the ligand, is dimeric or polymeric based on $\mu = 1.27$ B.M. and its lack of solubility in common solvents. Its activity against the fungi *Aspergillus niger* and *Candida tropicalis* and the bacteria *Escherichia coli* and *Staphyllococcus aureus* is 60% greater than the free thiosemicarbazone.

A copper(II) complexes of 5-phenylazo-3-methoxy salicylidene thiosemicarbazone has been shown to have promising growth inhibition activity against P388 lymphocytic leukemia cells sensitive and resistant to adriamycin [196]. The complex involves ON coordination of two deprotonated ligands and ν(CS) is reported to be unaltered in intensity and position in the complex from its position in the spectrum of the ligand. Inhibition studies with the uncomplexed thiosemicarbazone indicating an important role for the copper(II).

Salicylaldehyde and 5-bromosalicylaldehyde thiosemicarbazones coordinate as ONS donors and form 1:1 copper(II) and nickel(II) complexes which are dimeric with the oxygen donor bridging the two metal centers [197]. The metal complexes {copper(II)>nickel(II)} are more fungitoxic than the uncomplexed thiosemicarbazones against *Fusarium moniliforme*, *Drechslera orezae* and *Alternaria alternata*, with the latter fungus having the highest susceptibility.

4.4 Cobalt(III)

4.4.1 2-Formylpyridine Thiosemicarbazones and S-Alkyldithiocarbazates

Air oxidation of aqueous-alcoholic solutions of cobalt(II) salts and **5** yield diamagnetic [Co (**5**-H)$_2$]A where A = Cl, Br, I, NO_3 or ClO_4 [156a]. A representative complex was found by crystallography to be the *mer*-isomer (Fig. 3) [156b]. The same complexes were isolated regardless of preparative mole ratios of **5** to cobalt(II) salt [198]. From acidified aqueous ethanol solutions, [Co(**5**-H)$_2$]A (A = Cl, NCS, NO_3, N_3, and CN) have been isolated with magnetic moments of 3.86–3.93 B.M. [199]. Conductivity and spectral measurements suggest association by the anions in both the solid state and DMF. More recently [118], diamagnetic [Co(**5**-H)$_2$]BF_4 was isolated from boiling ethanol and spectroscopic methods indicate that bonding is via the pyridyl nitrogen, azomethine nitrogen and thiol sulfur.

A warmed alcoholic solution of cobalt(II) nitrate and 2-formylpyridine S-methyldithiocarbazate, **6**, yielded diamagnetic [Co(**6**-H)$_2$]NO_3 [126]. However, cobalt(II) chloride, bromide and thiocyanate yielded complexes with cobalt(III) cations and cobalt(II) anions, [Co(**9**-H)$_2$]$_2$ [CoA_4].

4.4.2 2-Acetylpyridine Thiosemicarbazones and S-Alkyldithiocarbazates

Preparations of heterocyclic thiosemicarbazone complexes with cobalt(II) salts of weakly coordinating anions (e.g., perchlorate and tetrafluoroborate) often result in cobalt(III) complexes due to air oxidation. The first 2-acetylpyridine thiosemicarbazone cobalt(III) complex was the diamagnetic [Co(**15b**-H)$_2$]ClO_4

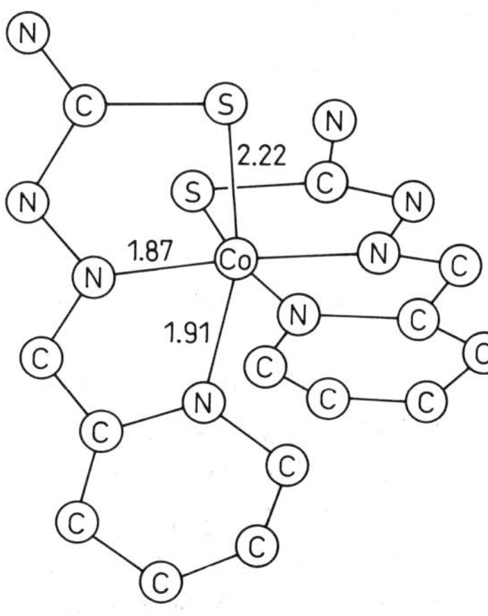

Fig. 3. Bis(2-formylpyridine thiosemicarbazone)cobalt(III) ion, [Co(**5**-H)$_2$]$^+$

Table 3. Ligand field parameters for cobalt(III) complexes of heterocyclic thiosemicarbazones[a]

Compound	B	C	Dq	Ref.
From 2-acetylpyridine				
$[Co(\mathbf{3a}\text{-}H)_2]^+$	550	4680	2490	[179]
$[Co(\mathbf{3b}\text{-}H)_2]^+$	778	4700	2285	[b]
$[Co(\mathbf{13}\text{-}H)_2]^+$	594	4720	2260	[141]
$[Co(\mathbf{9}\text{-}H)_2]^+$	667	4500	2205	[137]
$[Co(L4DE)_2]^{+c}$	582	4120	1990	[b]
$[Co(L4DP)_2]^{+d}$	505	4380	2075	[b]
$[Co(\mathbf{10}\text{-}H)_2]^+$	717	4560	2290	[139]
$[Co(\mathbf{11}\text{-}H)_2]^+$	740	4760	2330	[140]
From 2-formylpyrazole				
$[Co(\mathbf{40}\text{-}H)_2]^+$	588	4615	2385	[190]
$[Co(\mathbf{45}\text{-}H)_2]^+$	580	4810	2470	[191]
From 2-acetylpyridine				
$[Co(\mathbf{24}\text{-}H)_2]^+$	500	3700	2200	[188]
$[Co(\mathbf{39}\text{-}H)_2]^+$	595	3825	2235	[189]

[a] Primarily BF_4^- salts
[b] D.X. West et al., unpublished results
[c] L4DE = anion of 2-acetylpyridine ^4N-diethylthiosemicarbazone
[d] L4DP = anion of 2-acetylpyridine ^4N-dipropylthiosemicarbazone

[176]. An intense peak at m/e = 689, consistent with $[Co(\mathbf{15b}\text{-}H)_2^+]$, was observed in the positive ion plasma desorption mass spectrum.

Cobalt(III) complexes of the general formula $[CoL_2]BF_4$ have been prepared for **3a** [179], **9** [137], **10** [139], **11** [140], and **12** [141] and their values of Dq are 2200–2500 cm^{-1} and values of the covalency parameter, B, are 550–740 cm^{-1}. The strongest field and most covalent bonding is provided by **11**, 2-acetylpyridine ^3C-hexamethyleneiminethioemicarbazone, which has the largest alkyl function attached at ^4N.

4.4.3 2-Formylquinoline and 1-Formylquinoline Thiosemicarbazones and S-Alkyldithiocarbazates

The S-methyldithiocarbazates of both 2-formylquinoline, **18**, and 1-formylisoquinoline, **19**, when mixed with cobalt(II) salts in ethanol-dimethoxypropane, yield $[Co(L\text{-}H)_2]NO_3$ and $[Co(L\text{-}H)_2]_2[CoA_4]$, where A = Cl, Br, I, NCS, or NCSe [145]. Magnetic and spectral data support these stoichiometries and bonding is via the ring nitrogen, azomethine nitrogen, and thiol sulfur. More recently [146], $[Co(\mathbf{19}\text{-}H)_2]_2[CoCl_4]$ has been shown to have d–d bands consistent with the presence of both cobalt centers.

4.4.4 2,6-Diacetylpyridine Bis(Thiosemicarbazones) and Bis(S-Alkyldithiocarbazates)

There have been no reports of cobalt(III) complexes of these potentially pentadentate ligands.

4.4.5 2-Acetylpyridine N-Oxide Thiosemicarbazones and S-Alkyldithiocarbazates

Brown [Co(**24**-H)$_2$]A (A = ClO_4, NO_3) and [Co(**24**) (**24**-H)][CoA_4] (A = Cl, Br) have been isolated [188] and the cobalt(III) centers have B and Dq in the range 430–580 cm^{-1} and 2070–2320 cm^{-1}, respectively. These complexes have greater covalency and a weaker ligand field than the analogous 2-acetylpyridine ^4N-alkylthiosemicarbazone complexes. Plasma desorption mass spectra are supportive of the stoichiometries shown above. More recently the ^4N-dimethyl-derivative, [Co(**34**-H)$_2$]ClO_4 has been prepared and characterized [189]. **34**-H provides a similar ligand field to that of **24**-H, but has greater covalency.

4.4.6 2-Formylpyrazole Thiosemicarbazones

Aqueous ethanol solutions of 5-methyl-3-formylpyrazole thiosemicarbazone, **40**, yielded diamagnetic [Co(**40**-H)$_2$]A (A = Cl, Br, I, BF_4, ClO_4, NO_3, and 1/2SO_4) upon bubbling oxygen gas through mixtures containing the various cobalt(II) salts [190]. Bonding is via the ring nitrogen, ^2N, azomethine nitrogen and thiol sulfur. Values of 2500 and 500 cm^{-1} were found for Dq and B, respectively, similar to those of the 2-acetylpyridine thiosemicarbazones [137, 139–141, 179].

5-Methyl-3-formylpyrazole 4-phenylthiosemicarbazone, **45**, also coordinates as a deprotonated NNS ligand to form spin-paired cobalt(III) complexes [200]. Values of Dq and B are 2500 and 550 cm^{-1}, respectively, indicating less covalency than the complexes of **40**. Analogous complexes have been isolated with the related **41** having values of Dq and B of 2460 and 580 cm^{-1}, respectively, for [Co(**41**-H)$_2$$^{2+}$] [191].

4.4.7 2-Formyl- and 2-Acetylfuran Thiosemicarbazones and S-Alkyldithiocarbazates

2-Formylfuran *S*-benzyldithiocarbazate, **42**, yielded diamagnetic [Co(**42**-H)$_3$] on refluxing in ethanol [192]. Coordination is via the thione sulfur and azomethine nitrogen, but not the furan oxygen. 2-Formylfuran ^4N-phenylthiosemicarbazone, **43**, formed diamagnetic [Co(**43**-H)$_3$]·2H_2O on reflux of a 3:1 ligand to cobalt(II) chloride mixture in ethanol and addition of water after cooling [193]. Spectral studies indicate coordination via the azomethine nitrogen and thiol sulfur and values of Dq and B were calculated to be 2100 and 542 cm^{-1}, respectively.

4.4.8 2-Formyl- and 2-Acetylthiophene Thiosemicarbazones

2-Formylthiophene thiosemicarbazone, **26**, forms [Co(**26**)$_2$A$_2$]A and [Co(**26**)A$_3$(H_2O)] (A = Cl, Br, I, NO_3, and OAc), by varying the preparative ligand to cobalt(II) salt mole ratio in hot ethanol [201]. Addition of sodium hydroxide allows isolation of [Co(**29**-H)$_3$]. All complexes are weakly paramagnetic indicating spin paired complexes. Spectroscopic studies showed bonding via the azomethine nitrogen and thiol sulfur, but not the thiophene sulfur.

4.5 Cobalt(II)

4.5.1 2-Formylpyridine Thiosemicarbazones and S-Alkyldithiocarbazates

Impure cobalt(II) complexes have been isolated from deoxygenated alcoholic solutions of cobalt(II) salts and 2-formylpyridine thiosemicarbazone [156]. 2-formylpyridine ^{4}*N*-phenylthiosemicarbazone, **33**, forms [Co(**33**-H)$_2$(H_2O)$_2$] and spectroscopic evidence suggests that the ring nitrogen is not coordinated [171]. The magnetic moment is 4.90 B.M., Dq = 1170 cm^{-1} and B = 874 cm^{-1}, the latter quantities being somewhat higher than would be expected for a [Co(NS)$_2$O$_2$] center [194]. 4-(*m*-aminophenyl)-2-formylpyridine thiosemicarbazone, **7**, forms [Co(**7**)Cl_2] which is brown, has a magnetic moment of 4.43 B.M., and is a non-electrolyte [172]. Coordination is NNS, and the electronic spectrum is consistent with C_{3v} symmetry.

Although coordination of the heterocyclic nitrogen does not occur, two cobalt(II) complexes of 3-hydroxy-5-hydroxymethyl-2-methyl-4-formylpyridine have been isolated with stereochemistry [Co(**34**-H)A]·2H_2O (A = NO_3, OAc) [173]. For both complexes coordination is ONS (deprotonated phenolic oxygen), but magnetic or electronic spectral data are not included.

4.5.2 2-Acetylpyridine Thiosemicarbazones and S-Alkyldithiocarbazates

2-Acetylpyridine thiosemicarbazone forms [Co(**8**)$_2$$Cl_2$], which is isolated from hot ethanol [178]. Based on infrared spectra the pyridyl nitrogen is coordinated and bonding is NS with two chlorines bringing the coordination number to six. The complex is a non-electrolyte in DMF, has a magnetic moment of 4.13 B.M., and the electronic spectrum has bands at about 8160 and 17860 cm^{-1}, consistent with octahedral stereochemistry.

From hot ethanol 2-acetylpyridine ^4N-alkyl- [141, 179] and ^4N-dialkylthiosemicarbazones [134] generally give cobalt(III) complexes of stoichiometry [Co(**50**-H)$_2$]A (A = ClO_4, BF_4). However, cold ethanol allowed isolation of red brown [Co(**9**)$_2$]$(BF_4)_2$ having a magnetic moment and d–d spectrum consistent with a low spin, planar cobalt(II) complex and coordination proposed to be NS (azomethine nitrogen) [137]. When starting with cobalt(II) halides, tetrahedral [Co(**50**-H)A] complexes have been isolated for **9** [137], **11** [140] and **13** [141] and there appears to be a greater tendency for this stoichiometry and stereochemistry to occur with the larger bromo ligand (e.g. [Co(**13**-H)Br] and [Co(**13**)Cl_2] were isolated from a refluxed ethanol solution). While the deprotonated complexes coordinate NNS, [Co(**13**)Cl_2] was found to be tetrahedral and assumed to have **13** bound through the two nitrogens [141]. When the pyridyl nitrogen is unable to be a part of a chelate ring such as with 4-acetylpyridine thiosemicarbazone, **37**, [Co(**37**)$_2Cl_2$] was found to be too unstable for a magnetic moment determination [185]. The complex is considered to be octahedral with NS (azomethine) coordination.

4.5.3 2-Formylquinoline and 1-Formylquinoline Thiosemicarbazones and S-Alkyldithiocarbazates

Ethanol-dimethoxypropane solutions of either 1-formylisoquinoline or 2-formylquinoline thiosemicarbazone and cobalt(II) salts yield [Co(L)A_2] complexes where A = Cl, Br, I, NO_3, NCS, or NCSe [147]. All are non-electrolytes, have magnetic moments of 4.30–4.70 B.M. and are five coordinate with approximate trigonal bipyramidal stereochemistry involving NNS coordination based on electronic and infrared spectra. [Co(**21**-H)$_2$]·$2H_2O$ was isolated from a cold methanolic solution of cobalt(II) chloride and 1-formylisoquinoline thiosemicarbazone [187]. Infrared spectral studies show NNS coordination; the electronic spectral bands fit a distorted octahedral symmetry, and the magnetic moment is 4.48 B.M.

4-Methyl-5-amino-1-formylisoquinoline thiosemicarbazone, **22**, also yields cobalt(II) complexes from unheated methanol solution [202]. However, due to this ligand's added steric requirements, a complex, [Co(**22**)Cl_2], with one ligand per metal ion center is formed. This brown solid has a magnetic moment of 4.42 B.M., is a non-electrolyte, has coordination of a neutral NNS ligand, and the electronic spectrum indicates approximate trigonal bipyramidal stereochemistry.

4.5.4 2,6-Diacetylpyridine Bis(Thiosemicarbazones) and Bis(Dialkyldithiocarbazates)

When a methanol solution containing a 1:1 molar ratio of cobalt(II) chloride to 2,6-diacetylpyridine bis(thiosemicarbazone), **28**, is refluxed, [Co(**28**)Cl]Cl is formed [147]. Spectral studies indicate bonding as a SNNNS pentadentate ligand with the cobalt(II) center having distorted octahedral symmetry. Solution studies of cobalt(II) complexes of **28**, as well as 6-acetyl-2-acetylpyridine thiosemicarbazone, have been carried out, but no solids were isolated [143]. A series of 5-coordinate [Co(L-2H)] complexes have been prepared by stirring methanolic solutions of cobalt(II) chloride and various 2,6-diacetylpyridine bis(^{4}*N*-azacyclic thiosemicarbazones) [153]. These pentadentate ligands yield complexes with five d–d bands including a splitting of the $^4A'_2(F) \rightarrow {}^4E'(F)$ transition indicating an intermediate configuration (C_{2v}).

2,6-Diacetylpyridine bis(S-methylisothiosemicarbazone), **46**, formed [Co(**46**-H)I_2]$\cdot H_2O$ from a mixture of cobalt(II) acetate, **46**, and hydrogen iodide in heated ethanol [203]. The complex is 7-coordinate with **46** functioning as a NNNNN pentadentate ligand.

4.5.5 2-Acetylpyridine N-Oxide Thiosemicarbazones and S-Alkyldithiocarbazates

To date, the only reported cobalt(II) complexes involve 2-acetylpyridine *N*-oxide ^{4}*N*-dimethylthiosemicarbazone, **34** [189]. [Co(**39**-H)A] (A = Cl, Br) have magnetic moments suggestive of flattened tetrahedra, but calculations based on T_d symmetry yield values of Dq of about 330 cm^{-1} and B values of 743 and 707 cm^{-1} for the chloro and bromo complexes, respectively.

4.5.6 2-Formyl- and 2-Acetylfuran Thiosemicarbazones and S-Alkyldithiocarbazates

Cobalt(II) complexes of 2-formylfuran thiosemicarbazone, **47**, have been prepared from aqueous methanolic solutions with a variety of cobalt(II) salts [204]. A molar ratio of 2:1 ligand-to-metal salt yielded [Co(**47**)$_2$$A_2$] (A = Cl, Br, I, NO_3, and SCN), and complexes of this stoichiometry are octahedral with NS coordination, but not the furan oxygen. When sodium hydroxide is added to the preparative mixture [Co(**47**-H)$_2$] is isolated, and this complex has magnetic and spectral properties consistent with octahedral symmetry indicating coordination via the furan oxygen. A molar ratio of 1:1 yields [Co(**47**)A_2(H_2O)$_2$] (A = Cl, Br, I, NO_3 and SCN) whose octahedral coordination is completed by two water molecules. In the presence of sodium hydroxide, complexes of stoichiometry [Co(**47**-H)A] (A = Cl, Br, I, NO_3 and NCS) are formed which have magnetic moments in the range 2.15–2.70 B.M. prompting the authors to propose dimeric (A bridges) complexes with planar cobalt(II) centers. Diamagnetic complexes are isolated from the same preparative method when cobalt(II) acetate and cobalt(II) sulfate are used. The resulting [Co(**47**-H)SO_4]$\cdot 2H_2O$ and [Co(**47**-H)(OAc)$_2$]$\cdot H_2O$ are cobalt(III) complexes based on their diamagnetism and d–d bands at 17000 and 25000 cm^{-1}.

Cobalt(II) complexes of 4-(2-thiazolyl)-1-(2-acetylfuran) thiosemicarbazone, **44**, were isolated having stoichiometries of [Co(**44**)$_2$Cl$_2$] and [Co(**44**-H)$_2$] [194]. The former complex is isolated from neutral solution and is octahedral with each **44** bonding NS. The latter complex is formed in basic media, involves a tetrahedral cobalt(II) center and has NS bonding based on infrared studies.

4.5.7 Other Polydentate Thiosemicarbazones

Pale brown, high spin octahedral [Co(HBPS)$_2$], H$_2$BPS = 1-benzoin-4-phenylthiosemicarbazone, has Dq = 736 cm^{-1}, B = 717 cm^{-1} and $\beta = 0.74$ and is particularly effective against *C. tropicalis* [195]. [Co(HPBS)$_2$] is also more active against *A. niger*, *E. coli* and *S. aureus* than the uncomplexed H$_2$PBS, though not as active as [Cu(PBS)] against any of the latter three.

[Co(oAPT)$_2$] has Dq = 777 cm^{-1}, B = 850 cm^{-1} and $\beta = 0.88$ and is more than twice as active as HoAPT {1-(*o*-aminoacetophenone)-4-phenylthiosemicarbazone} against *A. niger*, comparable to HoAPT against *S. aureus* and *E. coli*, and considerably less active against *C. tropicalis* [205].

4.6 Nickel(II)

4.6.1 2-Formylpyridine Thiosemicarbazones and S-Alkyldithiocarbazates

With 2-formylpyridine thiosemicarbazone, **5**, octahedral, paramagnetic nickel(II) complexes of general formula [Ni(**5**)$_2$A$_2$], where A = Cl, Br, NO$_3$ or ClO$_4$, have been prepared in slightly acidic solution [153]. Also, the octahedral [Ni(**5**-H)$_2$] was formed in the presence of a base in hot alcoholic solution. However, more recently complexes of stoichiometry [Ni(**5**)$_2$A$_2$] and [Ni(**5**)A$_2$] along with [Ni(**5**-H)A] and [Ni(**5**-H)$_2$] where A = Cl, Br, NCS etc., have been reported [198]. Coordination by both **5** and **5**-H is considered to be tridentate NNS. The [Ni(**5**-H)A] complexes are diamagnetic and planar, but the rest of the complexes studied were paramagnetic, distorted octahedral centers. [Ni(**5**)$_2$](NO$_3$)$_2\cdot$EtOH was recently isolated from hot ethanol solution, and spectroscopic evidence indicates a distorted octahedral structure {$^3A_{1g} \rightarrow {}^3T_{2g}(10Dq) = \nu_1 = 11\,910$ cm^{-1}; $\rightarrow {}^3T_{1g} = \nu_2 = 18\,180$ cm^{-1}; and $\rightarrow {}^3T_{1g}(P) = \nu_3 = 25\,000$ cm^{-1}} with a *cis* arrangement of the two tridentate ligands [118]. On refluxing (2–3 h) an aqueous-ethanol mixture of nickel(II) acetate and **5**, a violet, diamagnetic, planar [Ni(**5**-H)$_2$] complex was isolated and the bonding of **5**-H is proposed to be NS (azomethine nitrogen), but not the pyridyl nitrogen [171].

The same preparative method [171] involving the ^4N-phenylderivative, **33**, yielded [Ni(**33**-H)$_2$(H$_2$O)$_2$] and TGA studies showed no mass loss up to 200 °C indicating the water to be coordinated. The magnetic moment is 3.50 B.M. and the complex has bands at 10 204, 17 241, and 26 315 cm^{-1}, assignable to ν_1, ν_2 and ν_3, respectively, for an approximate octahedron. The pyridyl nitrogen is again proposed to be uncoordinated. The more sterically demanding 4-(*m*-aminophenyl)-2-formylpyridine thiosemicarbazone, **7**, forms high spin [Ni(**7**)Cl_2] with NNS bonding [172].

Both 2-formylpyridine and 6-methyl-2-formylpyridine ^2N-methylthiosemicarbazones, **30** and **31**, respectively, form high spin $NiLA_2$ (A = Cl, Br, NCS) and $[Ni(L)_2]A_2$ (A = ClO_4, NO_3) [170]. The former complexes are 5- or 6-coordinate with bridging halogen or psuedohalogen ligands while the latter are octahedral with NNS coordination.

2-Methyl-3-hydroxy-4-formyl-5-hydroxymethylpyridine thiosemicarbazone, **34**, does not bond via the pyridyl nitrogen in [Ni(**34**-H)NO_3]·$2H_2O$ and [Ni(**34**-2H)]·$2H_2O$ [173].

With 2-formylpyridine S-methyldithiocarbazate, planar, diamagnetic nickel(II) complexes with stoichiometry [Ni(**6**-H)A], where A = Cl, Br, I, and NCS, have been isolated [126]. Also, paramagnetic, octahedral [Ni(**6**-H)$_2$] has been prepared and spectrally characterized. These complexes have NNS coordination as does the related S-benzyldithiocarbazate, **24**, in [Ni(**31**-H)A], A = Cl, Br [165].

4.6.2 2-Acetylpyridine Thiosemicarbazones and S-Methyldithiocarbazates

Diamagnetic, planar [Ni(**4**-H)A] and the seleno analog of **4** were included in a study of copper(II) complexes [175]. Additional nickel(II) complexes of the azabicyclononyl thiosemicarbazone having the following stoichiometries were studied: planar [Ni(**4**-H)A], where A = Cl, Br, NCS, NCSe, N_3, and NO_2; octahedral [Ni(**4**)$_2$(H_2O)$_2$]A_2, where A = BF_4 and I; and octahedral [N(**4**)$_2$(NO_3)$_2$] and [Ni(**4**-H)$_2$] [125]. Also, the selenosemicarbazone derivative of [Ni(**4**-H)Cl] was isolated. A crystal structure of **4** (Fig. 1) indicated it existed in the tautomeric form having the hydrogen attached to the azomethine nitrogen, ^1N, with hydrogen bonding to the pyridyl nitrogen and thiol sulfur atoms. The structure (Fig. 4) of [Ni(**4**-H)NCS] indicated the complex to be nearly planar with NNS coordination and **4**-H has essentially the same stereochemistry as the free **4**. Electronic spectra results indicate that **4**-H has a stronger ligand field than its selenium analog, and the ring substituted (i.e. 6-methyl-) **15b**-H [176] ligand. The diamagnetic nickel(II) complexes gave NMR spectra similar to the uncoordinated **4**, the most noticeable difference being the collapse of the triplet methyl- (acetyl) at 2.6, 2.5, and 2.4 ppm to a singlet at about 2.2 ppm [125].

In contrast to **4**, 2-acetylpyridine ^4N-methylthiosemicarbazone, **3a**, formed yellow-green, paramagnetic, octahedral [Ni(**3a**)$_2A_2$] complexes with nickel(II) chloride and bromide. The neutral form of the ligand was proposed to be NN coordinated [180]. Brown, paramagnetic [Ni(**3a**-H)$_2$] was prepared from nickel(II) acetate with NNS coordination, and its d–d spectrum and ligand field

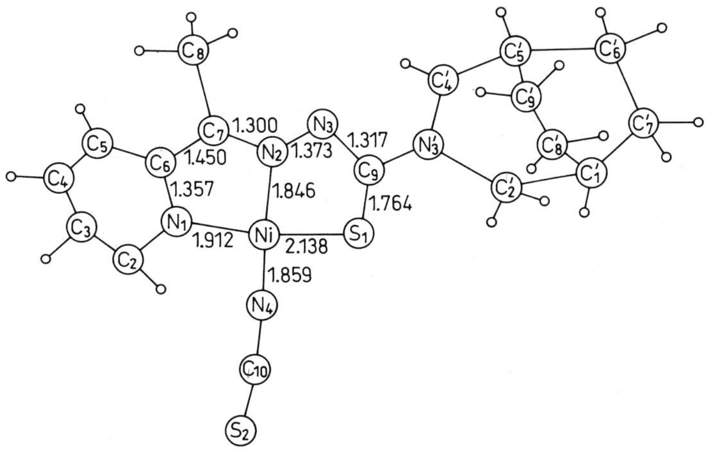

Fig. 4. 2-Acetylpyridine azabicyclo[3.2.2.]nonylthiosemicarbazonethiocyanatonickel(II), [Ni(**4**-H)NCS]

parameters were considerably different than the spectra of the NN coordinated complexes. The ^4N-cyclohexylthiosemicarbazones formed brown [Ni(**13**)$_2$Cl$_2$], but 4-coordinate [Ni(**13**-H)Br], which is also brown and has a magnetic moment consistent with tetrahedral symmetry [141]. However, the ^4N-phenylthiosemicarbazone forms [Ni(**14**-H)A] (A = Cl, Br, OAc, and NO_3) with red [Ni(**14**-H)Cl] being diamagnetic and planar, but the other three complexes, which are various shades of brown, are paramagnetic and tetrahedral [206]. Thermal decomposition of these complexes leaves NiO after several decomposition steps.

Although thiosemicarbazones which have a ^4N-hydrogen seem to yield a variety of nickel(II) centers depending on the nature of the alkyl group attached at ^4N, the ^4N-dialkyl- [137] and ^3C-azabicyclothiosemicarbazones [139, 140] studied to date all yield brown, diamagnetic [Ni(L-H)Cl] complexes. With the exception of [Ni(**10**)$_2$Br]Br [139], the bromide salts have also yielded diamagnetic [Ni(L-H)Br] complexes.

4.6.3 2-Picolylketone Thiosemicarbazone and S-Alkyldithiocarbazates

Diamagnetic nickel(II) complexes of the S-methyldithiocarbazate of 2-picolylphenylketone, **16**, having the general formula [Ni(**16**-H)A]$_2$ (A = Cl, Br, and I) have been isolated from boiling ethanol [144]. When 2,6-lutidylphenylketone S-methyldithiocarbazate, **17**, is the ligand, diamagnetic complexes of general formula [Ni(**17**-H)$_2$]$_2$A$_2$ were isolated. The electronic spectra show the **16**-H complexes are dimeric with 5-coordinate nickel(II) centers involving two sulfur bridging atoms. In pyridine solution the spectrum is altered by the formation of monomeric, planar [Ni(**16**-H)A] (no adducts could be isolated). The dimeric complexes of **17**-H have two sulfur bridging donor atoms and both nickel(II) centers are considered to be planar. Again, no adduct formation occurs in pyridine.

2-Picolylphenylketone *S*-benzyldithiocarbazate, **48**, yielded paramagnetic [{Ni(**48**-H)A}$_2$] (A = Cl, Br) and diamagnetic [Ni(**48**-H)I] [207]. All three compounds are non-electrolytes and the iodo complex is planar while the other two complexes involve sulfur bridging atoms and five-coordinate nickel(II) centers. All three complexes can be converted to monomeric, octahedral complexes by addition of pyridine, 2-picoline or quinoline.

2,6-Lutidylphenylketone thiosemicarbazone, **38**, and nickel(II) chloride, bromide, or iodide yield [Ni(**38**-H)$_2$]A$_2$ with NNS coordination based on infrared spectra [187]. All three complexes are electrolytes and the cations are dimeric with sulfur atoms acting as bridges; interdimer interactions via sulfur are present. The electronic spectra suggest a nickel(II) stereochemistry other than tetrahedral or octahedral for these paramagnetic complexes.

4.6.4 2-Formylquinoline and 1-Formylisoquinoline Thiosemicarbazones and S-Alkyldithiocarbazates

One of the few crystal structures solved for a metal complex of a heterocyclic thiosemicarbazone is bis(1-formylisoquinoline thiosemicarbazone) nickel(II) monohydrate, [Ni(**21**-H)$_2$] (Fig. 5) [208]. The nickel(II) center was found to be NNS coordinated by two approximately planar thiosemicarbazone ligands.

More recently [147], five-coordinate, paramagnetic complexes of stoichiometry [NiLA$_2$] (A = Cl, Br, I, NO$_3$, NCS, and NCSe) have been prepared from both 2-formylquinoline and 1-formylisoquinoline thiosemicarbazone, **20** and **21**, respectively. Bonding of the neutral ligands is considered to be NNS.

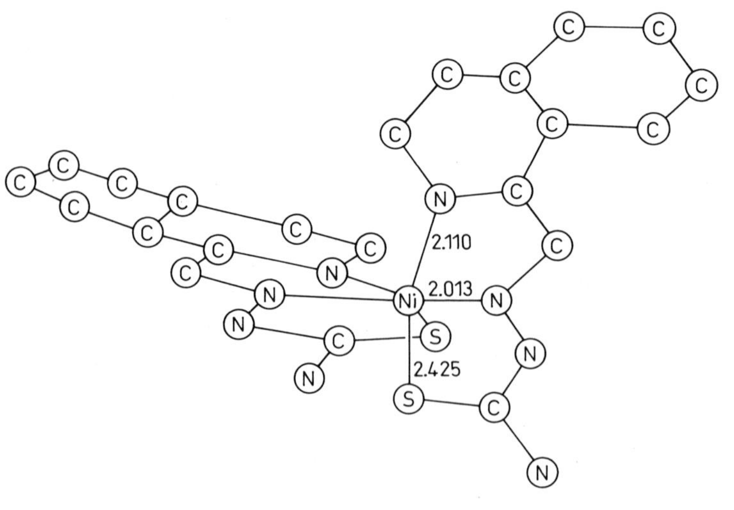

Fig. 5. Bis(1-formylisoquinoline thiosemicarbazone)nickel(II), [Ni(**21**-H)$_2$]

The electronic spectra and magnetic susceptibility of [Ni(**21**)Cl_2] were found to be consistent with a five-coordinate high spin complex. This complex has greater antitumor activity against P388 lymphocytic leukemia test system in mice than cobalt(II), copper(II), zinc(II) and platinum(II) complexes of **21** [187].

Although 4-methyl-5-amino-1-formylisoquinoline thiosemicarbazone, **22**, also forms [Ni(**22**)Cl_2] with chemical properties similar to [Ni(**21**)Cl_2], it, along with other metal complexes of **22**, showed no activity against P388 lymphocytic leukemia test system in mice [202].

Complexes of stoichiometry [Ni(L-H)A], where A = Cl, Br, I, NO_3, NCS, and NCSe, have been prepared with the S-methyldithiocarbazates of both 2-formylquinoline and 1-formylisoquinoline [145]. All complexes involve planar, diamagnetic nickel(II) and are NNS coordinated. In addition, each S-methyldithiocarbazate forms a paramagnetic, octahedral [Ni(L-H)$_2$] complex with Dq about 1120 cm^{-1}. Both [Ni(**19**-H)$_2$] and [Ni(**19**-H)Cl], the former a paramagnetic, octahedral complex and the latter a diamagnetic, planar species, possess significant activity against the P388 lymphocytic leukemia test system in mice [146].

4.6.5 2,6-Diacetylpyridine Bis(Thiosemicarbazones) and Bis(S-Alkyldithiocarbazates)

Solution studies of nickel(II) complexes of 2,6-diacetylpyridine bis(thiosemicarbazone), **28**, have been carried out [147]. From a refluxed methanolic solution containing **28** and nickel(II) chloride, [Ni(**28**)Cl_2] was isolated which behaved as a 1:1 electrolyte in methanol [152]. Based on spectral information it is proposed that the complex is a distorted octahedron with a SNNNS pentadentate ligand.

A series of paramagnetic [NiL-2H] complexes have been isolated in which L = 2,6-diacetylpyridine bis(azacyclothiosemicarbazones) with Ni(II) assuming a distorted five-coordinate structure with d–d bands at about 7250, 10 510, 12 500, 14 400, 19 200 and 20 500 cm^{-1} [153]. None of these Ni(II) complexes were reported to have activity against the P388 lymphocytic leukemia test system in mice.

2,6-Diacetylpyridine bis(S-methylisothiosemicarbazone), **46**, yields five-coordinate [Ni(**46**-H)]X (X = I, NCS) and six-coordinate [Ni(**46**-2H)] [203]. The latter compound is thought to be dimeric or polymeric with octahedral nickel(II) centers.

4.6.6 2-Acetylpyridine N-Oxide Thiosemicarbazones and S-Alkyldithiocarbazates

With 2-acetylpyridine *N*-oxide ^4N-methylthiosemicarbazone, neutral, tridentate coordination occurs in [Ni(**24**)$_2$](ClO_4)$_2$, but [Ni(**24**)$_2$(NO_3)$_2$] is isolated from nickel(II) nitrate [188]. Both complexes have Dq values of over 1000 cm^{-1}, but the former complex has a considerably lower value (i.e. 594 cm^{-1}) for the Racah parameter, B, indicating sulfur coordination. When the nickel(II) salt is a chloride or bromide, yellow [Ni(**24**)X_2] complexes are formed, and both have magnetic and spectral properties consistent with tetrahedral stereochemistry. While **24** coordinates primarily as the neutral ligand, the related ^4N-dimethylthiosemicarbazone, **34**, bonds as the anion to form brown complexes of stoichiometry [Ni(**34**-H)A] (A = Cl, Br) [189]. The magnetic moments suggest a tetrahedral stereochemistry, but unlike the related cobalt(II) complexes, calculation of ligand field parameters does not give reasonable values. The nickel(II) complexes prepared with the analogous 2-acetylpyridine ^4N-dimethylthiosemicarbazone, **9**, have magnetic moments less than 1.0 B.M. suggesting them to be more planar, consistent with a 5,5-chelating ring system compared to the 6,5-system of **34**-H. Although thione sulfur coordination often occurs in complexes of thiosemicarbazones, [Ni(**34**)$_3$]$(ClO_4)_2$ has ON coordination based on Dq = 980 cm^{-1} and B = 832 cm^{-1} [189].

4.6.7 2-Formylpyrazole Thiosemicarbazones

5-Methyl-3-formylpyrazole thiosemicarbazone, **40**, yielded [Ni(**40**)$_2$]A_2 (A = Cl, Br, NO_3, BF_4, ClO_4 and 1/2 SO_4) from weakly acidic solution [190]. The molecule **40** bonds as a NNS tridentate ligand and all complexes behave as 1:2 electrolytes. Their electronic spectra are consistent with an octahedral stereochemistry and Dq values range from 1111–1190 cm^{-1} and B values from 629 to 741 cm^{-1}.

5-Methyl-3-formylpyrazole ^4N-phenylthiosemicarbazone, **45**, coordinates as a deprotonated NNS tridentate ligand to form the diamagnetic, spin-paired [Ni(**45**-H)NCS] complex [200]. However, the complexes formed with other nickel(II) salts have stoichiometry [Ni(**45**)$_2$]A_2, where A = Cl, Br, NO_3, ClO_4, and BF_4. These octahedral complexes NNS coordinated neutral ligands have Dq values between 1086 and 1150 cm^{-1}.

4.6.8 2-Formyl- and 2-Acetylfuran Thiosemicarbazones and S-Alkyldithiocarbazates

2-formylfuran thiosemicarbazone, **47**, ^4N-methylthiosemicarbazone, **49**, and ^4N-phenylthiosemicarbazone, **43**, all yield [NiL_2] complexes from heated aqueous alcohol solutions brought to above pH = 7 with ammonia [209]. All complexes are four-coordinate, planar species which do not have the furan oxygen coordinated. [Ni(**43**-H)$_2$] was also isolated from anhydrous ethanol with sodium acetate added [193].

Nickel(II) complexes, [Ni(**44**)$_2Cl_2$] and [Ni(**44**-H)$_2$], have been isolated from 2-acetylfuran ^4N-(2-thiazolyl)thiosemicarbazone, **44** [194]. The former complex is paramagnetic and has d–d bands at 9805, 15 875 and 23 810 cm^{-1}

suggesting an approximate octahedral stereochemistry while the latter is diamagnetic and has bands at 16 530 and 23 250 cm^{-1} with a spin-forbidden band at 10 255 cm^{-1}. Both **44** and **44**-H bond to Ni(II) as NS ligands without coordination of hetero atoms from either ring. Diamagnetic, planar [Ni(**42**-H)$_2$] has been prepared from 2-formylfuran *S*-benzyldithiocarbazate [192].

2-Formylfuran *S*-methylthiosemicarbazone, **50**, forms [Ni(**50**)$_2$(H_2O)$_2$](ClO_4)$_2$ and [Ni(**50**)$_2$](ClO_4)$_2$], both of which are paramagnetic with Dq = 980 and 910 cm^{-1}, respectively [210]. Both have B about 950 cm^{-1}, which is consistent with NiN_4O_2 coordination. A single crystal determination shows coplanar coordination of azomethine and ^4N-thioamide nitrogens from each of the *S*-methylthiosemicarbazones with two aquo ligands occupying apical positions in the distorted octahedron.

4.6.9 2-Formyl- and 2-Acetylthiophene Thiosemicarbazones

2-Formylthiophene thiosemicarbazone, **26**, forms nickel(II) complexes of stoichiometry of [Ni(**26**)$_2$]A_2, [Ni(**26**)A_2], Ni(**26**-H)A, and [Ni(**26**-H)$_2$] where A = Cl, Br, I, NCS etc. [211]. Except when A = NCS in the first two types of complexes, the rest are diamagnetic and planar. The thiosemicarbazone moiety bonds NS, but there is no indication whether the Ni(**26**-H)A complexes are dimeric, or the thiophene sulfur completes the coordination sphere, which is less likely.

2-Formylthiophene thiosemicarbazone, **26**, as well as the ^4N-methylthiosemicarbazone, and ^4N-phenylthiosemicarbazone, each yield complexes of stoichiometry [Ni(**26**-H)$_2$] from heated aqueous alcohol solutions brought to above pH = 7 with ammonia [209]. All complexes are four-coordinate, diamagnetic and the thiophene sulfur does not bond to the nickel(II) center.

4.6.10 Other Polydentate Thiosemicarbazones

Octahedral [Ni(HBPS)], where HBPS is 1-benzoin-4-phenylthiosemicarbazone, has Dq = 856 cm^{-1} and B = 732 cm^{-1} similar to the heterocyclic thiosemicarbazones [198]. Its activity against fungi and bacteria is about the same as the free ligand. [Ni(oAPT)Cl] · $3H_2O$ is considered to have both square planar and octahedral nickel(II) centers present [205]. This nickel(II) complex is less active against *C. tropicalis*, somewhat more active against *S. aureus* and *E. coli*, and more active against *A. niger* than HoAPT {1-(*o*-aminoacetophenone)-4-phenylthiosemicarbazone}.

Nickel(II) complexes of *p*-anisaldehyde thiosemicarbazone, [Ni(HL)$_2$$X_2$] with X = Cl [212], Br [213], showed greater activity than the corresponding iron(II), manganese(II) and cobalt(II) complexes against *Alternaria* (Sp.), *Paecilomyces* (Sp.) and *Pestalotia* (Sp.). All complexes were more active than the uncomplexed thiosemicarbazone.

4.7 *Other Metal Ions*

Recently mononuclear ruthenium(II) and binuclear ruthenium(III) complexes of **5** and its ^4N-phenyl derivative have been isolated [214]. Although many of the complexes were not soluble in the culture medium, both ruthenium(II) and ruthenium(III) complexes showed considerably more activity against *E. coli* than the uncomplexed thiosemicarbazones. Similar results were found for the oxovanadium(IV) complexes of the same ligands; $[VO(L)_2Br]Br$, where L is the deprotonated 2-formylpyridine ^4N-phenylthiosemicarbazone, was the most potent inhibitory activity [215].

Six-coordinated dioxomolybdenum(VI) complexes of general formula $[MoO_2(L)_2]$, where L is the anion of a heterocyclic thiosemicarbazone have been prepared and spectrally characterized [216]. These complexes were marginally more active than the uncomplexed thiosemicarbazones against *Aspergillus flaves*, *Aspergillus niger* and *Alternaria alternata.*

5 Conclusions and Suggestions for Further Studies

The biological activity of the metal complexes is often greater than the uncomplexed ligand. Testing to date suggests that the copper(II) complexes have the greatest activity with the nickel(II) complexes next. Recent studies in our laboratories indicate that coordination by a heterocyclic atom, as well as the thiosemicarbazone moiety, is important to antifungal activity. Also, the nature of the group(s) attached to ^4N seems to affect the complexes' and thiosemicarbazone's activity. More studies involving structural variation of ^4N substituents as well as modification of the heterocyclic ring may prove beneficial to the enhancement of biological activity of metal complexes. A comparison of the biological activity of complexes of the same ^4N-substituted thiosemicarbazones prepared from different heterocycles may be of interest.

Acknowledgement: The authors thank Ms. Cinda S. Carlson, Mr. Rajeev C. Chikate, and Professor Anthony E. Liberta for their helpful comments concerning this manuscript.

6 References

1. e.g. Blanz EJ Jr, French FA (1968) Cancer Res 28: 2419
2. e.g. Logan JC, Fox MP, Morgan JH, Makohon AM, Pfau CJ (1975) J Gen Virol, 28: 271
3. Dobek AS, Klayman DL, Dickson EJ Jr, Scovill JP, Tramont EC, Antimicrob Agents Chemother 18: 27 (1980) and references therein

4. Klayman DL, Bartosevich JE, Griffin TS, Mason CJ, Scovill JP (1979) J Med Chem 22: 855 and references therein
5. Mittal SP, Sharma SK, Singh RV, Tandon JP (1981) Curr Sci 50: 483 and references therein
6. French FA, Blanz EJ Jr, Do Amaral JR, French DA (1970) J Med Chem 13: 1117
7. Cory JG, Fleischer AE (1979) Cancer Res 39: 4600
8. Sartorelli AC, Agrawal KC, Tsiftsoglou AS, Moore EC (1977) Advances in Enzyme Regulation 15: 117
9. Saryan LA, Ankel E, Krishnamurti C, Antholine W, Petering DH (1981) Biochem Pharmacol 30: 1595
10. Karon M, Benedict WF (1972) Science 178: 62
11. Livingstone SE (1965) Quart Rev Chem Soc, 19: 386
12. Lindoy LF (1969) Coord Chem Rev 4: 41
13. Ali MA, Livingstone SE (1974) Coord Chem Rev 13: 101
14. Wasson JR, Woltermann GM, Stoklosa HJ (1973) Transition metal dithio- and diselenophosphate complexes in inorganic chemistry, Topics in Current Chemistry, Springer, Berlin Heidelberg New York, 35: 65
15. Thorn GD, Ludwig RA (1962) The dithiocarbamates and related compounds, Elsevier, Amsterdam
16. Coucovanis D (1970) Prog Inorg Chem 11: 233
17. Gray, HB (1968) Transition Met Chem 1: 240 (1970); Jorgensen CK (1968) Inorg Chim Acta Rev, 2: 65; and McCleverty JA (1968) Prog Inorg Chem 10: 49 (1968)
18. Cox M, Darken J (1971) Coord Chem Rev 7: 29 and Livingstone SE (1971) Coord Chem Rev 7: 59
19. Singh RB, Garg BS, Singh RP (1978) Talanta 25: 619
20. Campbell MJM (1975) Coord Chem Rev 15: 279
21. Padhye S, Kauffman GB (1985) Coord Chem Rev 63: 127
22. Agrawal KC, Sartorelli AC (1968) J Pharm Sci 57: 1948
23. Hughes MN (1981) The inorganic chemistry of biological processes, John Wiley, Chichester
24. Petering DH (1973) In: Sigel H (ed) Metal ions in biological systems, vol 11, Marcel Dekker, New York
25. Perrin DD (1979) In: Topics in Current Chemistry, vol 000, Springer, Berlin Heidelberg New York
26. e.g. Agrawal KC, Sartorelli AC (1978) In: Ellis GP, West GB (eds) Progress in medicinal chemistry, Elsevier, North Holland Biomedical Press 15: 321; DeMilo AB, Redfern RE, Borkovec AB (1983) J Agric Food Chem 31: 713; Scovill JP, Klayman DL, Lambros C, Childs GE, Notsch JD (1984) J Med Chem 27: 87
27. Brockman RW, Sidwell RW, Arnett G, Shaddix S (1970) Proc Soc Expt Biol Med 133: 609
28. Domagk G, Behnich R, Mietzsch F, Schmidt H (1946) Naturwissenschaften 33: 315
29. Domagk G (1947) Zentralbl Gynaekol 69: 833
30. Hoggarth E, Martin A, Storey M, Young E (1949) Br J Pharmacol 4: 248
31. Donovick R, Pansy F, Stryker G, Bernstein J (1950) J Bacteriol 59: 667
32. Hamre D, Bernstein J, Donovick R (1950) J Bacteriol 59: 675
33. Berger A (1970) "Medicinal Chemistry", 3rd edn, Vol II, Wiley-Interscience, p 429
34. Miller A, Fox W, Tall R (1966) Tubercule 47: 33
35. Ctron K (1972) Br Med J 1: 426
36. Briggs I, Rochester W, Shennan D, Redell R, Fox W, Heffernan J, Miller A, Nann A, Stott H, Toll R (1968) Tubercule 49: 48
37. East Africa/British Medical Research Council (1972) Lancet 1: 1079
38. Drugs of the Future (1982) VII, 38
39. Hooper M, Purohit MG (1983) in Prog Med Chem, ed Ellis GP, West GB, Vol 20, Elsevier, North Holland, 1983, p 1
40. Rees RJW (1987) Int J Lepr 55: 11
41. Collins FM, Klayman DL, Morrison NE (1982) J Gen Microbiol 128: 1349
42. Syndell JK (1951) Int J Lepr 49: 90
43. Klayman DL, Scovill JP, Bartosevich JF, Mason CJ (1979) Eur J Med Chem 14: 317
44. Schaper KJ, Seydel JK, Rosenfeld M, Kazda J (1986) Lepr Rev 57: Supplement 3, 254
45. Wiles DM, Supruchunk T (1969) J Med Chem 12: 526
46. Prescott B, Lones GW, Peacock CL, Caldes G (1969) Antimicrob Agents Chemother 275
47. Douros JD, Brokl M, Krest A (1973) Antimicrob Agents Chemother 78: 124
48. Saxena A, Kochar JK, Tandon JP (1981) J Antibact Antifungal Agents 9: 435

49. Giri S, Nizamuddin, Srivastava UC (1983) Agricultural Biol and Chem 47: 103
50. Tayal HD, Gupta AK, Gupta DR (1983) J Antibact 11: 99
51. Pandeya SN (1985) Himalayan Chem Pharm Bull 2: 1
52. Parwana HK, Singh G, Talwar P (1985) Inorg Chim Acta 108: 87
53. Hamre D, Brownlee K, Donovick R (1950) J Immunol 67: 305
54. Hamre D, Bernstein J, Donovick R (1950) Proc Soc Expt Biol Med 73: 275
55. Thompson R, Minton S Jr, Officer J, Hitchings G (1953) J Immunol 70: 229
56. Thompson R, Price M, Minton S (1951) Proc Soc Exp Biol Med 78: 11
57. Thompson R, Davis J, Russell P, Hitchings G (1953) Proc Soc Exp Biol Med 84: 496
58. Bauer D (1955) Br J Exp Pathol 36: 105
59. Bauer D, Sadler P (1961) Nature (London) 190: 1167
60. Bauer D (1963) Br J Exp Pathol 44: 233
61. Bauer D, Sadler P (1959) Nature (London) 184: 1496
62. Bauer D, Sadler P (1960) Br J Pharmacol 15: 101
63. Bauer D, Dumbell K, Fox-Hulme P, Sadler P (1962) Bull WHO 26: 772
64. Bauer D, Sadler P (1960) Lancet 1: 1110
65. Bauer D, St Vincent L, Kempe C, Downie A (1963) Lancet 2, 494
66. Bauer D (1965) Ann N Y Acad Sci 130: 110
67. Bauer D, St Vincent L, Kempe C, Young P, Downie A (1969) Am J Epidemial 90: 130
68. Keiner G, Fatima N, Russell P, Haase A, Ahmad N, Mohammed N, Thomas D, Mack T, Kahan M, Knatterud R, Anthony R, McCrumb F (1971) Am J Epidemial 94: 435.
69. Levinson W (1972) ed Corter WA, In Selective Inhibitors of Viral Functions, CRS Press, p 213
70. Hutfield D, Csonka G (1964) Lancet 1: 329
71. Underwood G, Nichol F (1971) Appl Microbiol 22: 558
72. Munro T, Sabena L (1970) J Gen Virol 7: 55
73. Liberman M, Pascale A, Schafer T, Came P (1972) Antimicrob Agents Chemother 1: 143
74. Shipman C Jr, Smith SH, Drach JC, Klayman DL (1981) Antimicrob Agents Chemother 19: 682
75. Shipman C Jr, Smith SH, Drach JC, Klayman DL (1986) Antiviral Res 6: 197
76. Crumpacker C (1984) In: Rapp E F, Herpesvirus, Alan R Liss, Inc, NU p 687
77. Turk SR, Shipman C Jr, Drach JC (1986) Biochem Pharmacology 35: 1539
78. Bauer D, Apostolov K (1966) Science 154: 796
79. Pearson G, Zimmerman E (1969) Virology 38: 641
80. Gladych J, Hunt J, Jack D, Haff R, Boyle J, Stewart R, Ferlauto R (1969) Nature (London) 221: 286
81. Levinson W, Woodson B, Jackson J (1971) Nat New Biol 232: 116
82. Levinson W, Faras A, Woodson B, Jackson J, Bishop JM (1973) Proc Nat Acad Sci USA 70: 164
83. Kaska WC, Carrano C, Michalowski J, Jackson J, Levinson (1978) Bioinorg Chem 8: 225
84. Agrawal K, Booth B, Michaud R, Moore E, Sartorelli A (1974) Biochem Pharm 23: 2421
85. Klayman DL, Scovill JP, Bartosevich JF, Mason CJ (1979) 22: 1367
86. Christenson B, Rodriguez JR, Gorbea HF, Ramirez-Ronda CH (1985) Antimicrob Agents and Chemother 27: 570
87. Emery CE, Stancato FA, Brown RE, Prichard DA, Wolfe AD (1983) Life Sci 33: 1285
88. Dodd RH, Ouannes C, Robert-Gero M, Potier P (1989) J Med Chem 32: 1272
89. French FA, Freedlander R (1960) Can Res 21: 505
90. Petering HG, Van Giessen GJ (1966) In: Peisach J, Aisen P, Blumberg W (eds) 'The Biochemistry of Copper', Academic Press, NY p 197–210
91. Petering DH, Petering HG (1975) In Sartorelli AC, Johns DG (eds) Handbook of Experimental Pharmacology 38/2: Springer Verlag, NY, p 841
92. Regeslson W, Holland JF, Talley RW (1967) Cancer Chemother Rep 51: 171
93. Wilkelman DA, Bermke Y, Petering DH (1974) Bioinorg Chem 3: 261
94. Petering DH, Antholine WE, Saryan LA (1984) In Ottenbrite RM, Butter GB (eds) Metal Complexes as Antitumor Agents in Anticancer and Interferron Agents, Marcel Dekker, NY, p 203
95. Kraker A, Krezoski S, Schneider J, Minkel D, Petering DH (1985) J Biol Chem 260: 13710
96. Antholine WE, Basosi R, Hyde JS, Petering DH (1987) In Sorenson JR (ed) Biology of Copper Complexes, Humana, Clifton, NJ
97. Subczynski WK, Antholine WE, Hyde JS, Petering DH (1987) J Am Chem Soc 109: 46
98. Brockman RW, Thomson JR, Bell MJ, Skipper HE (1956) Cancer Res 21: 349

99. French FA, Freedlander BL (1958) Cancer Res 18: 1290
100. French FA, Freedlander BL, Blanz EJ Jr (1961) Cancer Res 21: 349
101. French FA, Blanz EJ Jr (1965) Cancer Res 25: 1454
102. French FA, Blanz EJ Jr (1966) Cancer Res 26: 1638
103. French FA, Blanz EJ Jr (1966) J Med Chem 9: 585
104. Agrawal KC, Sartorelli AC (1978) In Ellis GP, West GB (eds) Progress in Medicinal Chemistry, Vol 15, Elsevier, p 321
105. Petering DH (1973) In Sigel H (ed) Metal Ions in Biological Systems, Vol 11, Marcel Dekker, p 198
106. Sartorelli AC (1967) Biochem Biophys Res Comm 27: 26
107. Moore EC, Zedeck MS, Agrawal KC, Sartorelli AC (1970) Biochemistry 9: 4492
108. Sartorelli AC, Agrawal KC, Moore EC (1971) Biochem Pharmacol 20: 3119
109. Agrawal KC, Lin AJ, Booth BA, Wheaton JR, Sartorelli AC (1974) J Med Chem 17: 631
110. Agrawal KC, Booth BA, Dennuzzo SM, Sartorelli AC (1975) J Med Chem 17: 368
111. West DX, Paulson CA, unpublished results
112. Klayman DL, Lin AJ (1984) Org Prep Proc Int 16: 79
113. Domiano P, Gasparii Fava G, Nardelli M, Sgarabotto P (1969) Acta Crystallogr 25B: 343; Andretti GD, Domiano P, Gasparii Fava G, Nardelli M, Sgarabotto P (1970) Acta Crystallogr 26B: 1005
114. Gerbeleu NV, Revenko MD, Leovac VM (1977) Russ J Inorg Chem 22: 1009
115. Tui F, Turta KI, Gerbeleu NV (1977) Russ J Inorg Chem 22: 1497
116. West DX, Huffman DL (1989) Transition Met Chem 14: 190
117. West DX, Ahrweiler PM, Ertem G, Scovill JP, Klayman DL, Flippen-Anderson JL, Gilardi R, George C, Pannell LK (1985) Transition Met Chem 10: 264
118. Parwana HK, Singh G (1987) Ind J Chem 26A: 581
119. Timken MD, Wilson SR, Hendrickson DN (1985) Inorg Chem 24: 3450
120. West DX, Makeever RM, Scovill JP, Klayman DL (1984) Polyhedron 3: 947
121. Sonawame PB, Chikate RC, Padhye S, Doedens RJ, unpublished results
122. Ablov AV, Gerbeleu NV (1964) Russ J Inorg Chem 9: 1260
123. Antonini I, Claudia F, Franchetti P, Grifantini M, Martelli S (1977) J Med Chem 20: 447
124. Antonini I, Claudia F, Franchetti P, Grifantini M, Martelli S (1977) Eur J Med Chem 14: 89; Lemke TL, Shek TW, Cates LA, Smith LK (1977) J Med Chem 20: 1351; Khetan SK (1973) Chem Ind (London) 183
125. West DX, Scovill JP, Silverton J, Bavoso A (1986) Transition Met Chem 11: 123
126. Ali MA, Livingstone SE, Phillips DJ (1971) Inorg Chim Acta 5: 493
127. Spingarn NE, Sartorelli AC (1977) J Med Chem 22: 1314
128. Saryan LA, Ankel E, Krishnamurti C, Petering DH, Elford H (1979) J Med Chem 22: 1218
129. Antholine WE, Knight JM, Petering DH (1976) J Med Chem 19: 339
130. Beraldo H, Tosi L (1983) Inorg Chim Acta 75: 249
131. Raina H, Srivastava TS (1982) Inorg Chim Acta 67: 83
132. Mohan M, Kumar M, Kumar A, Madhuranath PH, Jha NK (1988) J Inorg Biochem 32: 239
133. Oxford JS, Perrin DD (1974) Gen Virol, 23: 59
134. Raina R, Srivastava TS (1983) Ind J Chem, 22A: 701
135. Bhoon YK, Mitra S, Scovill JP, Klayman DL (1982) Transition Met Chem 7: 264
136. Klayman DL, Scovill JP, Franchino CF (1982) J Med Chem 25: 1261
137. West DX, Lewis N (1988) Transition Met Chem 13: 277
138. Garg BS, Kurup MRP, Jain SK, Bhoon YK (1988) Transition Met Chem 13: 247
139. West DX, Galloway DS, Case DA (1988) Transition Met Chem 13: 415
140. West DX, Galloway DS (1988) Transition Met Chem 13: 410
141. West DX, Profilet RD, Hines JL (1988) Transition Met Chem 13: 467
142. Jain SK, Garg BS, Bhoon YK (1986) Transition Met Chem 11: 89
143. Martinez-Martinez MP, Garcia-Vargas M, Munoz-Leyva JA (1986) Spectrochim Acta 42A: 701
144. El-Dissouky A, El-Sonbati AZ (1986) Synth React Inorg Met-org Chem 16: 1035
145. Mohan M, Kumar M (1984) Synth React Inorg Met-org Chem 14: 615
146. Mohan M, Kumar M, Kumar A, Madhuranath PH, Jha NK (1988) J Inorg Biochem 33: 121
147. Mohan M, Marmohan (1982) Synth React Inorg Met-org Chem 12: 761
148. DeConti RC, Toftness BR, Agrawal KC, Tomchick R, Mead JAR, Bertino JR, Sartorelli AC, Creasy WA (1972) Cancer Res 32: 455
149. French FA, Lewis AE, Sheena AH, Blanz EJ Jr, (1965) Fed Proc Fed Am Soc Exp Biol 24: 402

150. Krakoff IH, Etcubanas E, Man C, Mayer K, Bethune V, Burchenal JH (1974) Cancer Chemother Res 58: 207
151. Petering HB, Bustrik HH, Crim JA (1967) Cancer Res 27: 1115
152. Mohan M, Sharma P, Kumar M, Jha NK (1986) Inorg Chim Acta 125: 9
153. Mohan M, Agarawal A, Jha NK (1988) J Inorg Biochem 34: 41
154. Majumbder SMMH, Ali MA, Smith FE, Mridha MAU (1988) Polyhedron 7: 2183
155. Burman S, Sathyanarayana DN (1981) Ind J Chem 20A: 57
156. Ablov AV, Belichuk NI (1969) Russ J Inorg Chem 14: 93; Tarkhova TN, Akafova KN, Belov NV (1974) Sov Phys Crystallogr 19: 39
157. Knight JM, Whelan H, Petering DH (1979) J Inorg Biochem 11: 327
158. Miertus S, Filipovic P (1982) Eur J Med Chem-Chim Ther 17: 145
159. Crim JA, Petering HG (1967) Cancer Res 27: 1278
160. Van Giessen GJ, Crim JA, Petering DH, Petering HG (1973) J Natl Cancer Inst 51: 139
161. Agrawal KC, Booth BA, Moore EC, Sartorelli AC (1973) Proc Am Assoc Cancer Inst 51: 139
162. Antholine WE, Gunn P, Hopwood LE (1981) Int J Radiat Oncol Biol 7: 491
163. Antholine WE, Taketa F (1984) J Inorg Biochem 20: 69
164. Antholine WE, Knight J, Whelan H, Petering DH (1977) Mol Pharmacol 13: 89
165. Ali MA, Tarafdar MTH (1977) J Inorg Nucl Chem 39: 1785
166. Antholine WE, Knight JM, Petering DH (1977) Inorg Chem 16: 569
167. Beraldo H, Tosi L (1986) Inorg Chim Acta 125: 173
168. Bingham AG, Bogge H, Muller A, Ainscough EW, Brodie AM (1987) J Chem Soc Dalton 493
169. Bell CF, Lott KAK, Hearn N (1987) Polyhedron 6: 39
170. Malik MA, Phillips DJ (1974) J Inorg Nucl Chem 36: 2229
171. Reddy PSN, Agarwala BV (1987) Synth React Inorg Met-org Chem 17: 585
172. Mohan M, Sharma P, Jha NK, (1985) Inorg Chim Acta 107: 91
173. Belichi MF, Gasparri GF, Leporati E, Pelizzi C, Tarasconi P, Tosi G (1986) J Chem Soc, Dalton, 2455
174. Ferrari MB, Fava GG, Pellizzi C, Tarasconi P, Tosi G (1987) J Chem Soc, Dalton, 227
175. Bhoon YK, Scovill JP, Klayman DL (1984) Ind J Chem 22A: 267 and Spectrochim Acta 40A: 691
176. West DX, Ertem G, Makeever RM, Scovill JP, Klayman DL (1985) Transition Met Chem 10: 41
177. Scovill JP, Klayman DL, Franchino CL (1982) J Med Chem 25: 1261
178. Perez-Dubois P, Souza P, Masaguer JR, Arquero A (1987) Transition Met Chem 12: 200
179. West DX, Lewis NC (1987) Transition Met Chem 12: 365
180. Jain SK, Garg BS, Bhoon YK (1986) Spectrochim Acta 42A: 959
181. Jain SK, Garg BS, Bhoon YK, Klayman DL, Scovill JP (1985) Spectrochim Acta 41A: 407
182. Garg BS, Kurup MRP, Jain SK, Bhoon YK (1988) Transition Met Chem 13: 309
183. Singh B, Srivastava U (1988) Transition Met Chem 13: 205
184. Singh B, Srivastava U (1988) Transition Met Chem 13: 241
185. Abou-Sekkina MM, El-Dissouky A, El-Sonbati AZ (1983) Acta Chim Hung 114: 29
186. Souza P, Sanchez-Kaiser F, Masaguer JR, Arquero A (1987) Transition Met Chem 12: 128
187. Mohan M, Sharma P, Jha NK (1985) Inorg Chim Acta 106: 117
188. West DX, Makeever RM, Ertem G, Scovill JP, Pannell LK (1987) Transition Met Chem 11: 131
189. West DX, Pannell LK (1989) Transition Met Chem 14: 457
190. Saha N, Mukherjee N (1984) Polyhedron 3: 1135
191. El-Dissouky A (1987) Spectrochim Acta 43A: 1177
192. Ali MA, Bose RN (1984) Polyhedron 3: 517
193. Ibrahim KI, El-Asmy AA, Bekheit MM, Mostafa MM (1985) Synth React Inorg Met-org Chem 15: 1247
194. Singh B, Srivastava U (1988) Synth React Inorg Met-org Chem 18: 515
195. Patil BG, Havinale BR, Shallom JM, Chitnis MP (1989) J Inorg Biochem 36: 107
196. Akbar Ali M, Kabir MH, Nazimuddin M, Majumbder SMMH, Tarafder MTH, Khair MA (1988) Ind J Chem 27A: 1064
197. El Asmy A, Shaibi YM, Babaqi AS, Mounir M, Ashour SA (1988) Transition Met Chem 13: 332
198. Iskander MF, El-Sayed L, Zaki KI (1979) Transition Met Chem 4: 225
199. Bamgboye TT, Bamgboye OA (1985) Inorg Chim Acta 105: 223
200. Saha N, Mukherjee N (1984) Synth React Inorg Met-org Chem 14: 1151
201. Mukkanti K, Pandeya KB, Singh RP (1985) Synth React Inorg Met-org Chem 15: 613

202. Mohan M, Sharma P, Jha NK (1985) Inorg Chim Acta 106: 197
203. Leovac VM, Cesljevic VI (1987) Transition Met Chem 12: 504
204. Mukkanti K, Pandeya KB, Singh RP (1986) Synth React Inorg Met-org Chem 16: 229
205. El Asmy A, Shaibi YM, Shedaiwa IM, Khattab MA (1988) Synth React Inorg Met-org Chem 18: 331
206. Jain SK, Garg BS, Bhoon YK (1987) Transition Met Chem 12: 13
207. El-Dissouky A, Abou-Sekkina MM, El-Kersh M, El-Sonbati AZ (1984) Transition Met Chem 9: 372
208. Mathew M, Palenik GJ (1969) J Am Chem Soc 91: 6310
209. Mukkanti K, Pandeya KB, Singh RP (1982) Indian J Chem 21A: 641
210. Levac VM, Divjakovic V, Fazlic R, Engel P (1987) Transition Met Chem 12: 73
211. Umapathy P, Budhkar AP, Dorai CS (1986) J Indian Chem Soc 63: 714
212. Thimmaiah KN, Lloyd WD, Chandrappa GT (1985) Inorg Chim Acta 106: 81
213. Bamgboye TT, Bamgboye OA (1987) Inorg Chim Acta 133: 247
214. Chattopadhyay SK, Ghosh S (1989) Inorg Chim Acta 163: 245
215. Maiti A, Guhu AK, Ghosh S (1988) J Inorg Biochem 33: 57
216. Kanoongo N, Singh R, Tandon JP (1989) Bull Chem Soc Japan 62: 1385

7 Appendix: Symbols for Organic Molecules

1 *p*-acetamidobenzaldehyde thiosemicarbazone (thiacetazone)
2 isatin-3-thiosemicarbazone
3 2-acetylpyridine **a)** ^4N-methylthiosemicarbazone **b)** ^4N-ethylthiosemicarbazone
4 2-acetylpyridine azabicyclo[3.2.2.]nonylthiosemicarbazone
5 2-formylpyridine thiosemicarbazone
6 2-formylpyridine S-methyldithiocarbazate
7 4-(*m*-aminophenyl)-2-formylpyridine thiosemicarbazone
8 2-acetylpyridine thiosemicarbazone
9 2-acetylpyridine ^4N-dimethylthiosemicarbazone
10 2-acetylpyridine 3-piperidinylthiosemicarbazone
11 2-acetylpyridine 3-hexamethyleneiminethiosemicarbazone
12 2-acetylpyridine 3(4-methylpiperidinyl)thiosemicarbazone
13 2-acetylpyridine ^4N-cyclohexylthiosemicarbazone
14 2-acetylpyridine ^4N-phenylthiosemicarbazone
15 6-methyl-2-acetylpyridine **a)** ^4N-methylthiosemicarbazone **b)** azabicyclo[3.2.2.]nonylthiosemicarbazone
16 2-picolylphenylketone S-methyldithiocarbazate
17 2,6-lutidylphenylketone S-methyldithiocarbazate
18 2-formylquinoline S-methyldithiocarbazate
19 1-formylisoquinoline S-methyldithiocarbazate
20 2-formylquinoline thiosemicarbazone
21 1-formylisoquinoline thiosemicarbazone
22 4-methyl-5-amino-1-formylisoquinoline thiosemicarbazone
23 2,6-diacetylpyridine bis(S-methyldithiocarbazate)

24 2-acetylpyridine *N*-oxide ^4N-methylthiosemicarbazone
25 2-acetylpyridine *N*-oxide 3-azabicyclo[3.2.2.]nonylthiosemicarbazone
26 2-formylthiophene thiosemicarbazone
27 2-acetylthiophene thiosemicarbazone
28 2,6-diacetylpyridine bis(thiosemicarbazone)
29 2-formylpyridine S-benzyldithiocarbazate
30 2-formylpyridine ^2N-methylthiosemicarbazone
31 6-methyl-2-formylpyridine ^2N-methylthiosemicarbazone
32 2-formylpyridine ^4N-dimethylthiosemicarbazone
33 2-formylpyridine ^4N-phenylthiosemicarbazone
34 3-hydroxy-5-hydroxymethyl-2-methyl-4-formylpyridine thiosemicarbazone
35 2-acetylpyridine ^4N-(2-thiazolyl)thiosemicarbazone
36 2,6-diacetylpyridine ^4N-(2-thiazolyl)thiosemicarbazone
37 4-acetylpyridine thiosemicarbazone
38 2,6-lutidylphenylketone thiosemicarbazone
39 2-acetylpyridine *N*-oxide ^4N-dimethylthiosemicarbazone
40 5-methyl-3-formylpyrazole thiosemicarbazone
41 4,5-dimethyl-3-formylpyrazole ^4N-phenylthiosemicarbazone
42 2-formylfuran S-benzyldithiocarbazate
43 2-formylfuran ^4N-phenylthiosemicarbazone
44 2-acetylfuran ^4N-(2-thiazolyl)thiosemicarbazone
45 5-methyl-3-formylpyrazole ^4N-phenylthiosemicarbazone
46 2,6-diacetylpyridine bis(S-methylisothiosemicarbazone)
47 2-formylfuran thiosemicarbazone
48 2-picolylphenylketone S-benzyldithiocarbazate
49 2-formylfuran ^4N-methylthiosemicarbazone
50 2-formylfuran S-methylthiosemicarbazone

Nature and Dynamics of the Spin-State Interconversion in Metal Complexes

Edgar König

Institut für Physikalische und Theoretische Chemie, University of Erlangen-Nürnberg, D-8520 Erlangen, Germany

Spin-state transitions in metal complexes which are driven by a change of temperature T or pressure p are always associated with a considerable reorganization of molecular geometry. The change involves metal-ligand bond lengths R, bond angles, and a variation of ligand orientation. In particular, the elongation ΔR by up to $\sim 10\%$ occurring in the course of the LS $\rightarrow$ HS conversion produces an expansion of molecular volume $\Delta V \simeq 25$ Å^3 per metal atom. The average crystal structure for given values of T and p is reproduced by the fractional occupancy of the individual structures of the high-spin (HS) and low-spin (LS) isomer. The transitions are reasonably well described by a number of theoretical models which are equivalent to the Bragg and Williams approximation of the Ising model. The dynamics of the spin-state transitions in solution, based on measurements by ultrasonic and photo-perturbation techniques, is in general rapid with rate constants between 4×10^5 and 3×10^8 s^{-1}. Similar results are obtained for the spin conversion in solid complexes where the line shape analysis of Mössbauer spectra based on the theory of Blume and Tjon is applied. The dynamics may be rationalized employing one-dimensional cross sections through Gibbs free-energy surfaces $G = G(R)$, an alternative being the comparison of the results with quantum-mechanical calculations for a radiationless non-adiabatic multiphonon process.

1 Introduction 53

2 The Variation of Molecular Volume and Some Theoretical Implications 59

3 Methods and Basic Relations for Relaxation Measurements of Spin-State Equilibria 68

4 Relaxation of Spin-State Equilibria of Metal Complexes in Solution 74
4.1 Complexes of Iron(II) 74
4.2 Complexes of Iron(III) 79
4.3 Complexes of Cobalt(II) 82

5 Consequences with Respect to the Energetic and Molecular Properties of the Activated Complex from Absolute Reaction Rate Theory 83

6 Description in Terms of a Radiationless Nonadiabatic Multiphonon Process 93
6.1 Outline of the Method 94
6.2 Calculation of Electronic Matrix Elements 97
6.3 Calculation of Rate Constants 99

7 Relaxation of Thermally Quenched or Optically Excited High-Spin States Resulting from the Spin-State Transition in Solid Metal Complexes 101
7.1 Survey of the Available Results 101
7.2 Studies on Individual Systems 103

8 Rates of Spin-State Transition for Solid Metal Complexes Based on the Line Shape Analysis of Mössbauer Spectra 107
8.1 Lifetime Limits and the Rate of Spin-State Conversion 107
8.2 The Relaxation Theory of Blume and Tjon 109

Structure and Bonding 76

8.3 A Simple Example of Line Shape Analysis: Spin-State Relaxation in the Nitrosyl Complex Fe(J-mph)NO 115
8.4 Individual Spin-State Interconversion Rate Studies 119

9 The Molecular Reorganization Associated with Spin-State Equilibria in Solution and the Structural Changes Accompanying Spin-State Transitions in Solid Metal Complexes 134

10 Conclusions 145

11 References 148

1 Introduction

First-row transition-metal complexes having electronic configurations $3d^4$, $3d^5$, $3d^6$, and $3d^7$ can exist, in octahedral symmetry, in the high-spin (HS) or low-spin (LS) ground state depending on the strength of the ligand field. The critical factor is the relation between the cubic ligand field splitting energy $\Delta = 10$ Dq and the mean spin-pairing energy Π. Thus, HS states are formed if $\Delta < \Pi$ (weak fields), whereas LS states arise for $\Delta > \Pi$ (strong fields). In the particular case where the energy difference $|\Delta - \Pi|$ becomes of the order of the thermal energy $k_B T$, the spin states HS and LS may be interconvertible (HS $\leftrightarrow$ LS transition in the solid complex or HS $\rightleftharpoons$ LS equilibrium in solution) subject to a change of an external variable such as temperature T, pressure p or concentration of a solvent x. It should be noted that spin-state changes have been also observed for the configuration $3d^8$ of nickel(II). However, these transitions involve either the conformational change between paramagnetic tetrahedral and diamagnetic planar isomers in four-coordinate d^8 complexes [1, 2] or the change between paramagnetic five- or six-coordinate complexes and diamagnetic planar four-coordinate d^8 species [3]. A small number of examples has been reported even for $4d$ element complexes such as those of molybdenum(II) [4] and for clusters such as Nb_6I_{11} and HNb_6I_{11} [5].

Most complexes showing spin-state transitions are in fact of low symmetry. In order to describe their electronic structure it is convenient to employ term symbols appropriate to cubic symmetry and this practice will be followed below. The most common transition-metal ions for which spin-state transitions have been observed are Fe^{2+} ($3d^6$), Fe^{3+} ($3d^5$) and Co^{2+} ($3d^7$), a minor role being played by Co^{3+} ($3d^6$), Mn^{2+} ($3d^5$), as well as Cr^{2+} and Mn^{3+} ($3d^4$). The relevant ground states for an octahedral disposition of the ligands are LS $^1A_{1g}(t_{2g}^6)$ and HS $^5T_{2g}(t_{2g}^4 e_g^2)$ for iron(II), LS $^2T_{2g}(t_{2g}^5)$ and HS $^6A_{1g}(t_{2g}^3 e_g^2)$ for iron(III) and LS $^2E_g(t_{2g}^6 e_g)$ and HS $^4T_{1g}(t_{2g}^5 e_g^2)$ for cobalt(II). A spin-state change thus invariably implies a change in the population of the σ antibonding e_g^* orbitals which are directed toward the ligands. Along with this an opposite change in the occupancy of the t_{2g} orbitals occurs, affecting the electron backdonation between the metal ion and vacant π^* orbitals of the ligands. This is especially significant if the LS configuration involves a filled t_{2g}^6 subshell such as in iron(II) and cobalt(II) complexes. Both factors contribute to a change of the metal-ligand bond length which is well documented by the results of available X-ray structure studies [6]. The metal-ligand bond length variation is expected to be smaller for spin-state transitions of cobalt(II) complexes ($t_{2g}^6 e_g \leftrightarrow t_{2g}^5 e_g^2$) involving the transfer of a single electron ($\Delta S = 1$); whereas a larger change of bond length should be observed for iron(II) ($t_{2g}^6 \leftrightarrow t_{2g}^4 e_g^2$) and iron(III) ($t_{2g}^5 \leftrightarrow t_{2g}^3 e_g^2$) complexes where two electrons are involved in the transfer ($\Delta S = 2$). Indeed, an average metal-ligand bond length change between the LS and HS states $\overline{\Delta R} \simeq 0.17$ Å has been found for iron(II) complexes [6]. For complexes of iron(III) with dithiocarbamate ligands, the average bond length change results as $\overline{\Delta R} \simeq 0.15$ Å [7], whereas for [Fe^{III}–N_4O_2]-type complexes the mean value $\overline{\Delta R} \simeq 0.13$ Å is obtained [6]. Comparison of structural data for various square-pyramidal HS and LS cobalt(II) complexes [8] indicates $\overline{\Delta R}$ values of 0.10–0.14 Å for equatorial bonds while axial cobalt-ligand bonds vary by $+0.02$ to -0.07 Å, thus resulting in an average bond length change $\overline{\Delta R} \simeq 0.09$ Å. It must be remembered that the quoted values for the average bond length change should take account of the fact that, for a given metal ion, the bond length variation depends on the nature of the donor atoms. Additional factors influencing the extent of these changes are possible Jahn-Teller distortions (in the ^2E_g state of cobalt(II)) and the ability of the coordination sphere to reorganize. The latter will depend on the rigidity of the ligands, the geometry of the coordination unit and steric effects.

The above considerations should make it clear that the situation in compounds involved in a spin-state transition does not correspond to the point where, in a diagram of energy versus ligand field strength Dq (Tanabe-Sugano diagram [9]), the terms corresponding to the actual HS and LS states cross (crossover). Rather, the two states of the compound may be called *electronic isomers* since their geometrical and electronic structures are clearly separated on either side of the crossover point. This is well illustrated by the properties of the compound [$Fe(phen)_2(NCS)_2$] (phen = 1,10-phenanthroline). From the electronic spectrum of this [Fe^{II}–N_6]-type compound, the values $\Delta_{HS} \simeq 11\,900$ cm^{-1} and $\Delta_{LS} \simeq 16\,300$ cm^{-1} have been extracted [10]. On the other hand, the Fe–N bond lengths have been determined on the basis of the X-ray structure for polymorph II of the compound as [11] $\bar{R}_H = 2.156$ Å and $\bar{R}_L = 1.992$ Å. The average bond length change follows as $\overline{\Delta R} = 0.164$ Å. Similar values have been obtained for the closely related complex [$Fe(bpy)_2(NCS)_2$] (bpy = 2,2′-bipyridine) as [12] $\bar{R}_H \simeq 2.14$ Å and $\bar{R}_L \simeq 2.02$ Å, the average change being here $\overline{\Delta R} \simeq 0.12$ Å. In ligand field theory, Δ depends inversely on the fifth power of the metal-ligand distance R [13]:

$$\Delta = \frac{5}{3}\frac{Ze^2}{R^5}\langle r^4 \rangle \tag{1}$$

where Ze is the charge of the ligand and $\langle r^4 \rangle$ is the mean fourth power radius of the d electron. It is well known that this relation is applicable only within an electrostatic model and for perfect octahedral symmetry. If, however, for the sake of argument, the data derived for the above compound are introduced into Eq. (1), the relation is followed precisely [12]. This result confirms that the observed bond length differences lead to necessarily significantly different values of $\Delta = 10$ Dq for the two spin states, HS and LS.

So far, we have considered the phenomenon of spin-state transition employing both the properties of the macroscopic complexes as well as those of the isolated molecules. However, with respect to the nature of the energy variation at the transition, these two situations should be clearly distinguished. Thus, if the change of spin state is of purely molecular origin, the energy variation to be considered is that of the potential energy as a function of some of the internal coordinates $E = E(Q_i)$ or, in case of a single bond length change, $E = E(R)$. If, on the other hand, the change of spin state is associated with an assembly of interacting molecules, the energy variation of interest is that of the Gibbs free energy as a function of, e.g., the fraction of the HS species, $G = G(n_\mathrm{H})$, the transformation being characterized by:

$$\begin{array}{ccc} \mathrm{HS} & \rightleftharpoons & \mathrm{LS} \\ n_\mathrm{H} & & n_\mathrm{L} = 1 - n_\mathrm{H} \end{array} \tag{2}$$

Although the curves $E = E(Q_i)$ and $G = G(n_\mathrm{H})$ may be often of rather similar appearance, there are important differences in the details which should be observed.

Let us consider first an isolated molecule characterized by a potential energy with the two minima $E_\mathrm{H}(R)$ and $E_\mathrm{L}(R)$ corresponding to the HS and LS state, respectively. If the degeneracies of the two states are denoted as g_H and g_L, the probability to find the molecule in the state $j = \mathrm{H}, \mathrm{L}$ at the temperature T is given by:

$$w_j = \frac{g_j \exp[-E_j(R)/k_B T]}{g_\mathrm{H} \exp[-E_\mathrm{H}(R)/k_B T] + g_\mathrm{L} \exp[-E_\mathrm{L}(R)/k_B T]} \tag{3}$$

Here, the vibrational levels of the two electronic states have been taken to be identical, otherwise a summation over these levels would be required. For the HS state, in particular, we obtain:

$$w_\mathrm{H} = \frac{1}{1 + \dfrac{g_\mathrm{L}}{g_\mathrm{H}} \exp[\Delta E(R)/k_B T]} \tag{4}$$

where $\Delta E(R) = E_\mathrm{H}(R) - E_\mathrm{L}(R)$ and k_B is the Boltzmann constant.

If we consider an assembly of N molecules, each individual molecule may exist either in the HS state or in the LS state, the transformation for a fixed set of values of the external variables T, p, or x being characterized by Eq. (2). Here, N is the Avogadro number and the Gibbs free energy G per mole of the system may be expressed as:

$$G = n_\mathrm{H} G_\mathrm{H}^\circ + (1 - n_\mathrm{H}) G_\mathrm{L}^\circ + I(n_\mathrm{H}) - TS_\mathrm{mix} \tag{5}$$

In Eq. (5), G_H° and G_L° is the standard Gibbs free energy, in the absence of any interaction, of the HS and LS state, respectively, $I(n_\mathrm{H})$ is an interaction term and S_mix the mixing entropy. If a regular solution of molecules is assumed for the two states, S_mix is determined by [14, 15]

$$\begin{aligned} S_\mathrm{mix} &= k_B[N \ln N - n_\mathrm{H} N \ln N n_\mathrm{H} - (1 - n_\mathrm{H}) N \ln N(1 - n_\mathrm{H})] \\ &= -R[n_\mathrm{H} \ln n_\mathrm{H} + (1 - n_\mathrm{H}) \ln (1 - n_\mathrm{H})] \end{aligned} \tag{6}$$

Introducing the thermodynamical equilibrium condition:

$$\left(\frac{\partial G}{\partial n_{\mathrm{H}}}\right)_T = 0 \tag{7}$$

into Eq. (5) produces the relation:

$$\ln\left[\frac{n_{\mathrm{H}}}{1-n_{\mathrm{H}}}\right] = -\left[\Delta G^\circ + \frac{\partial I(n_{\mathrm{H}})}{\partial n_{\mathrm{H}}}\right]\bigg/RT \tag{8}$$

where $\Delta G^\circ = G^\circ_{\mathrm{H}} - G^\circ_{\mathrm{L}}$. It should be noted that for the limiting case of zero interaction between the molecules, Eq. (8) is equivalent to Eq. (4). Indeed, if $I(n_{\mathrm{H}}) = 0$, Eq. (8) gives:

$$n_{\mathrm{H}} = \frac{1}{1 + \exp[\Delta G^\circ/RT]} \tag{9}$$

and since $\Delta G^\circ = \Delta H^\circ - T\Delta S^\circ$, this may be written:

$$n_{\mathrm{H}} = 1\bigg/\left[1 + \exp\left(-\frac{\Delta S^\circ}{R}\right)\exp\left(\frac{\Delta H^\circ}{RT}\right)\right] \tag{10}$$

Here, $\Delta S^\circ = \Delta S_{\mathrm{el}} + \Delta S_{\mathrm{vib}}$, but since identical vibrational levels have been assumed for the two electronic levels, $\Delta S_{\mathrm{vib}} = 0$. In addition, ΔS_{el} is related to the electronic degeneracies of the two levels by $\Delta S_{\mathrm{el}} = R\ln(g_{\mathrm{L}}/g_{\mathrm{H}})$. Finally, if the transformation takes place at constant pressure, $\Delta H^\circ = N\Delta E$ where ΔE is defined as above. Consideration of the Gibbs free energy for an assembly of N molecules thus produces the same relation for the HS fraction n_{H} as the potential energy for the isolated molecule, if only the interaction between the molecules is neglected.

It is evident that a complete absence of any intermolecular interaction is rarely encountered. Therefore, the relative stability of the HS and LS isomers of a compound is determined in reality by the difference in Gibbs free energy $G = H - TS$ rather than the difference in potential energy. Conceptually it is clear that the HS isomer possesses greater entropy than the LS isomer, both from its higher spin degeneracy as well as from its larger vibrational partition function. It has been also outlined above that the two spin-state isomers are characterized by minima of the energy at significantly different internuclear distances, $\bar{R}_{\mathrm{H}}$ and $\bar{R}_{\mathrm{L}}$. The notion of real or avoided crossings of the potential energy surfaces $E(\bar{R}_i)$ of the two electronic isomers of the molecule (double well potential) is therefore useful for the description of spin-state transitions. Provided the active vibrational coordinate $\bar{R}$ is the totally symmetric metal-ligand stretching mode $\nu(^1A_{1g})$ which is nondegenerate, it is sufficient to consider a one-dimensional cross section through $E(\bar{R}_i)$. In practice, however, the cross section through the true Gibbs free energy surface $G(\bar{R}_i)$ (thermodynamic potential) is employed since this quantity is largely accessible on the basis of experimental data, cf. Fig. 1. The relevant parameters are: the barrier height between the minima, B; the zero-point energies, $(ZPE)_{\mathrm{H}}$ and $(ZPE)_{\mathrm{L}}$; the horizontal displacement of the

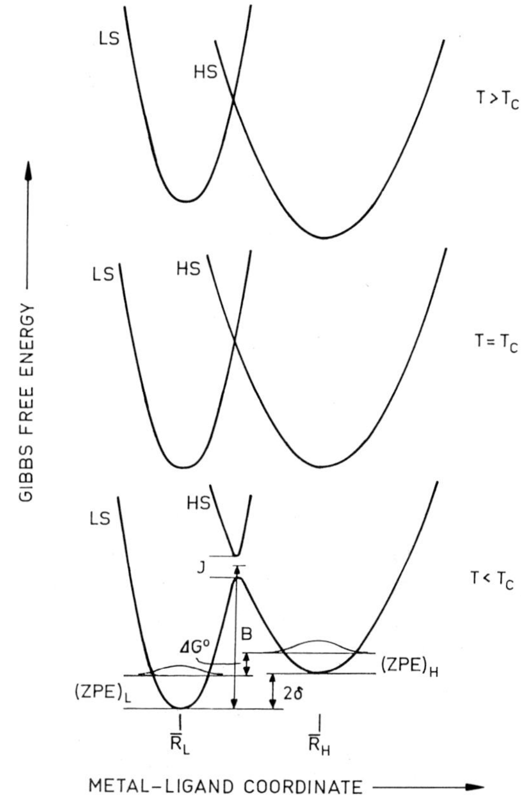

Fig. 1. Schematic one-dimensional cross section through the Gibbs free energy surface $G(\bar{R})$ of a spin-state transition system along the totally symmetric stretching coordinate. The situation for three characteristic temperatures is shown (B = barrier height, ZPE = zero-point energy, 2δ = asymmetry parameter, J = electronic coupling parameter, $\Delta G^\circ = G_H - G_L$)

minima, $\Delta R = \bar{R}_H - \bar{R}_L$; the asymmetry parameter, 2δ where $2\delta = \Delta G^\circ + (ZPE)_L - (ZPE)_H$ and $\Delta G^\circ = G_H - G_L$; the quadratic force constants determining the curvature at the minima, f_H and f_L; and the electronic coupling parameter at the (avoided) crossing, J. The construction of the thermodynamic potential curves will be demonstrated in Sect. 5 for a number of selected spin transition compounds in order to illustrate the situation at these transitions.

For a limited range of temperature, the quantities ΔH and ΔS may be considered as temperature independent. The difference of Gibbs free energy $\Delta G = \Delta H - T\Delta S = G_H - G_L$ and thus the relative position of the two minima will then show a significant temperature dependence as indicated in Fig. 1. At low temperature, the contribution of the term TS will be negligible, and the more stable isomer will be that of lower energy which is in general the LS state [16, 17]. If the difference in free energy becomes of the order of RT, $\Delta G = G_H - G_L \simeq RT$, the phenomenon of a spin-state transition may be observed.

For $\Delta G = \Delta H - T\Delta S = 0$ the conversion between the two spin states has achieved thermal equilibrium, and the fractions of both states have become equal, $n_H = n_L = \frac{1}{2}$. The transition temperature may be thus defined as $T_c = \Delta H/\Delta S$. At sufficiently high temperatures, the term TS in the Gibbs free energy will dominate and the stable isomer will be that of greater entropy which is in general the HS state.

The sensitivity of spin transition systems toward external perturbations as exemplified by the dependence of the electronic properties on temperature, pressure, and the concentration of solvent (in the solid state or in solution) is a consequence of the crossing of Gibbs free energy surfaces. Recently, the concept of "molecular bistability" has been introduced [18] in order to describe the property of certain molecular systems to assume one of two electronic states under the influence of a specific external perturbation. Additional examples of molecular bistability may be found in Jahn-Teller and mixed valence systems. An exciting perspective is the possible application of compounds showing this type of bistability in molecular electronic circuits and devices. The fields where these systems could become useful include switching, signal processing, sensor technology, and others. Of particular interest is the observation of hysteresis in these transitions which provides the possibility for passive information storage. Another aspect refers to the light-induced spin-state transitions discussed in Sect. 7 which may become useful for the construction of advanced materials in optical data storage.

Spin-state transitions have been studied by the application of numerous physical techniques such as the measurement of magnetic susceptibility, optical and vibrational spectroscopy, the ^{57}Fe-Mössbauer effect, EPR, NMR, and EXAFS spectroscopy, the measurement of heat capacity, and others. Most of these studies have been adequately reviewed. The somewhat older surveys [3, 19] cover the complete field of spin-state transitions. Several more recent review articles [20, 21, 22, 23, 24, 25] have been devoted exclusively to spin-state transitions in compounds of iron(II). Two reviews [26, 27] have considered *inter alia* the available theoretical models of spin-state transitions. Of particular interest is the determination of the X-ray crystal structures of spin transition compounds at two or more temperatures thus approaching the structures of the pure HS and LS electronic isomers. A recent survey [6] concentrates particularly on these studies.

This report has been written in order to demonstrate the nature of spin-state transitions and to review the studies of dynamical properties of spin transition compounds, both in solution and in the solid state. Spin-state transitions are usually rapid and thus relaxation methods for the microsecond and nanosecond range have been applied. The first application of relaxation techniques to the spin equilibrium of an iron(II) complex involved Raman laser temperature-jump measurements in 1973 [28]. The more accurate ultrasonic relaxation method was first applied in 1978 [29]. These studies dealt exclusively with the spin-state dynamics in solution and were recently reviewed by Beattie [30]. A recent addition to the study of spin-state transitions both in solution and the solid state has been achieved by the introduction of photoperturbation methods. Moreover, relaxation studies by the ^{57}Fe-Mössbauer effect are being used to an increasing extent to determine directly spin-state transition rates for solid iron(II) and iron(III) compounds. These studies are based on the analysis of line broadening according to the methods of Blume [31] and Blume and Tjon [32, 33]. We will therefore concentrate, in the present study, on these more recent developments. In addition, the theoretical analysis of the results by the various methods will be considered and an attempt will be made to compare the dynamics of spin equilibria in solution with the dynamics of spin-state transitions in solid complexes. Results of the dynamics studies in solution will be presented primarily in tabular form, details being available in the survey by Beattie [30].

The interconversion between different spin states is closely related to the intersystem crossing process in excited states of transition-metal complexes. Hence, much of the interest in the rates of spin-state transitions arises from their relevance to a better understanding of intersystem crossing phenomena. The spin-state change can alternatively be described as an intramolecular electron transfer reaction [34]. Therefore, rates of spin-state transitions may be employed to assess the effect of spin multiplicity changes on electron transfer rates. These aspects have been covered in some detail elsewhere [30].

2 The Variation of Molecular Volume and Some Theoretical Implications

The change of metal-ligand bond length which accompanies a spin-state transition invariably produces a change of the molecular volume of the complex, ΔV. This volume change may be determined on the basis of solid state or solution data. Thus, X-ray structure investigations of a spin transition compound at temperatures corresponding to the pure HS and LS state give the individual unit cell volumes of the two states, and from these a reasonable estimate of ΔV may be obtained as $\Delta V = (V_H - V_L)/Z$ where Z is the number of complex molecules per unit cell. It is necessary to correct for the contribution to the crystal volume of any associated cationic or anionic species as well as solvate molecules.

In a solution where a nonzero volume change between the electronic isomers, HS and LS, is encountered, the position of the spin equilibrium will depend on pressure. The volume change, usually denoted here ΔV°, may be obtained from the study of the pressure dependence of equilibrium properties such as the magnetic susceptibility or the electronic spectrum. In favorable cases, ΔV° values may be derived from the amplitude of sound absorption observed in ultrasonic relaxation measurements of a spin equilibrium as will be shown in the next section. A discussion of the changes of molecular volume which result for crystals and for solutions will be deferred until Sect. 9.

The change of molecular volume or the consequent volume change of the lattice have been a convenient starting point for several theoretical models of spin-state transitions. Let us consider in the following the Ising-type model as introduced by us [35] and subsequently modified by McGarvey [36]. According to this theory, the Gibbs free energy G per mole of a spin transition system is determined by Eq. (5). The interaction term is, in general, expected to assume the form:

$$I(n_{\mathrm{H}}) = I_{\mathrm{LL}}(1 - n_{\mathrm{H}})^2 + I_{\mathrm{LH}}n_{\mathrm{H}}(1 - n_{\mathrm{H}}) + I_{\mathrm{HH}}n_{\mathrm{H}}^2 \tag{11}$$

where I_{LL}, I_{LH}, and I_{HH} account for the interaction between two LS ions, between a LS and a HS ion, and between two HS ions, respectively. For convenience, Eq. (11) may be rewritten as:

$$I(n_{\mathrm{H}}) = J_0 + J_1 n_{\mathrm{H}} - J_2 n_{\mathrm{H}}^2 \tag{12}$$

where the negative sign of J_2 is chosen to conform to the original formulation [35] of the model. Introducing the equilibrium condition $(\partial G/\partial n_{\mathrm{H}})_{\mathrm{T}} = 0$ into Eq. (5) and Eq. (12) yields the HS fraction at equilibrium as:

$$n_{\mathrm{H}} = [1 + e^{(\Delta G^\circ + J_1 - 2J_2 n_{\mathrm{H}})/RT}]^{-1} \tag{13}$$

Since $\Delta G^\circ = \Delta H^\circ - T\Delta S^\circ$ where ΔH° and ΔS° are the enthalpy difference and the entropy difference, respectively, associated with the spin-state transition, Eq. (13) may be expressed as:

$$\ln\left(\frac{1 - n_{\mathrm{H}}}{n_{\mathrm{H}}}\right) = \frac{\Delta H^\circ + J_1 - 2J_2 n_{\mathrm{H}}}{RT} - \frac{\Delta S^\circ}{R} \tag{14}$$

This implicit equation is usually employed to derive values of n_{H} corresponding to the minima of G. Graphical methods for this purpose have been discussed by Drickamer and Frank [37]. The transition temperature T_c is defined for $n_{\mathrm{H}} = \frac{1}{2}$, and this corresponds according to Eq. (13) to:

$$\Delta G^\circ + J_1 - J_2 = \Delta H^\circ - T_c\Delta S^\circ + J_1 - J_2 = 0 \tag{15}$$

Thus, the transition temperature results as:

$$T_c = (\Delta H^\circ + J_1 - J_2)/\Delta S^\circ \tag{16}$$

The result Eq. (13) or Eq. (14) describes both gradual and abrupt transitions, an example for a specific set of parameter values being displayed in Fig. 2. Here, the HS fraction n_{H} is shown as a function of temperature for the values of J_2/RT_c attached to the individual curves. The curve characterized by $J_2 = 0$ corresponds to the case of a true spin-state equilibrium (zero interaction between the molecules), whereas increasingly higher values of J_2/RT_c produce a gradually more abrupt behavior at the transition. It can be shown that abrupt transitions result for $J_2/RT_c \geq 2.0$, whereas gradual transitions are obtained if $J_2/RT_c < 2.0$.

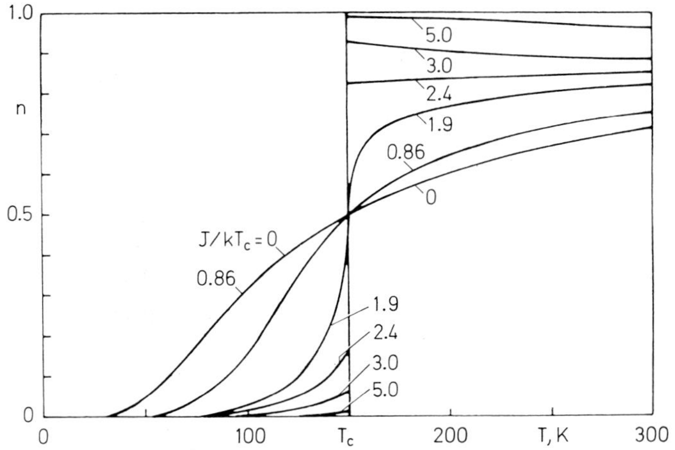

Fig. 2. Temperature dependence of the HS fraction n_H according to the Ising model. The employed parameter values are $T_c = 150$ K, $\Delta_1 = \Delta_2 = 500\ cm^{-1}$ and $\lambda = -100\ cm^{-1}$. Here, Δ_1 and Δ_2 are the orbital energy differences between the $^5B_2$ and $^5B_1$ levels and between the $^5B_3$ and $^5B_1$ levels, respectively, λ being the spin-orbit coupling constant. The model parameters Δ_1, Δ_2, and λ determine the value of ΔG. The levels result from the HS iron(II) ground state in orthorhombic symmetry according to $^5T_2 \rightarrow {}^5B_1 + {}^5B_2 + {}^5B_3$. The figures on the curves specify the values of J_2/RT_c. Notation in the figure uses n instead of n_H and J/kT_c instead of J_2/RT_c. According to Ref. [35]

Here, the orbital-splitting parameters $\Delta_1 = \Delta_2 = 500\ cm^{-1}$, the spin-orbit coupling constant $\lambda = -100\ cm^{-1}$ and $T_c = 150$ K have been assumed [35].

It has not been generally appreciated so far that several other models which have been suggested for spin-state transitions are mathematically equivalent to the Ising-type model discussed here. We list in Table 1 the expressions which result, within these models, for the equilibrium HS fraction n_H, and state the relations between the parameters used. Similar relations exist between the model expressions for the Gibbs free energy G. The similar mathematical form of the expressions resulting from the various models is clearly evident. The equivalence between the model of the author [35] and that of McGarvey [36] has been stated above. Taking the model of McGarvey as a reference, only the third term

Table 1. Comparison of model expressions

Author	Expression for equilibrium value of n_H	Parameter relation[a]
McGarvey et al. [36]	$n_H = [1 + e^{(\Delta G^\circ + J_1 - 2J_2 n_H)/RT}]^{-1}$	—
Zimmermann and König [35]	$n_H = [1 + e^{(\Delta - RT \ln \nu - 2Jn_H)/RT}]^{-1}$	$J_1 = 0$, $J_2 = J$
Drickamer [37, 38]	$n_H = [1 + e^{(\Delta G^\circ + \Gamma(1 - 2n_H))/RT}]^{-1}$	$J_1 = J_2 = \Gamma$
Spiering et al. [39, 40]	$n_H = [1 + e^{(\Delta G^\circ + \Delta(x) - 2n_H \Gamma(x))/RT}]^{-1}$	$J_1 = \Delta(x)$, $J_2 = \Gamma(x)$
Sorai and Seki [41]; Gütlich et al. [42]	$n_H = [1 + e^{n\Delta G^\circ/RT}]^{-1}$	$J_1 = J_2 = 0$

[a] Relative to the expression by McGarvey

of the interaction Eq. (12) is retained and thus $J_1 = 0$ and $J_2 = J$. Note that due to the assignment of energy zero at the fully LS state, the expression $\Delta - RT \ln \nu$ (where ν is the HS partition function and Δ the energy of the HS ground state) is equivalent to ΔG°. In the model developed by Drickamer [37, 38] the interaction term is chosen as $\Gamma(1 - 2n_H) = J_1 - 2J_2 n_H$ and thus the two interaction parameters become equal, $J_1 = J_2 = \Gamma$. In the model due to Spiering et al. [39, 40] it is $\Delta(x) - 2n_H\Gamma(x) = J_1 - 2J_2 n_H$, where x refers to the fraction of Fe atoms in a dilution experiment. The model is thus completely equivalent to that of McGarvey if only $J_1 = \Delta(x)$ and $J_2 = \Gamma(x)$. In the cluster model suggested by Sorai and Seki [41] and also applied by Gütlich et al. [42], zero interaction between the metal ions is assumed and therefore $J_1 = J_2 = 0$. The only model parameter is the number n of ions per cluster which modifies the free energy value ΔG°. Obviously, the similar mathematical form of the expressions is the result of application of the *Bragg and Williams approximation* [14, 15] familiar from the theory of alloys. Finally, the equivalence of expressions based on the molecular-field approximation [43, 44, 45] can be easily demonstrated [35].

In the Ising-type model, the change of molecular volume ΔV due to the LS↔HS transformation leads to a change of phonon frequencies of the lattice. The effect may be treated within the Debye approximation which requires that the interaction parameters J_1 and J_2 are replaced by J'_1 and J'_2 where:

$$J'_1 = J_1 + aNk_B\Delta\Theta$$
$$J'_2 = J_2 + bNk_B\Delta\Theta^2/\Theta \qquad (17)$$

Here, Θ is the Debye temperature for a particular value of n_H, $\Delta\Theta = (\Theta_L^2 - \Theta_H^2)/2\Theta_L \simeq \Theta_L - \Theta_H$ where Θ_L and Θ_H are the Debye temperatures for the LS and HS state, respectively, and N is the number of ions or radicals per molecule. Furthermore, $a(x)$ and $b(x)$ are functions of $x = \Theta_L/T$ which are defined as:

$$a(x) = \frac{9}{8} + \frac{3\pi^4}{5}x^{-4} - 9e^{-x}(x^{-1} + 3x^{-2} + 6x^{-3} + 6x^{-4})$$
$$- \frac{9}{8}e^{-2x}(4x^{-1} + 6x^{-2} + 6x^{-3} + 3x^{-4}) \qquad (18)$$

$$b(x) = \frac{5}{2}a - \frac{9}{4}(1 + 2e^{-x} + 2e^{-2x})$$

for $2.5 \leq x < \infty$ $(T \leq 0.4\Theta_L)$ and as:

$$a(x) = 3x^{-1} + \frac{3}{20}x - \frac{1}{560}x^3 + \frac{1}{30240}x^5$$

$$b(x) = 3x^{-1} + \frac{1}{560}x^3 - \frac{1}{15120}x^5 \qquad (19)$$

for $0 \leq x \leq 2.5$ $(T \geq 0.4\Theta_L)$. As a result of the account of lattice vibrations, the values of the model parameters are modified to some extent. Thus, for the example considered above ($\Delta_1 = \Delta_2 = 500\ \text{cm}^{-1}$, $\lambda = -100\ \text{cm}^{-1}$, $T_c = 150$ K), the Debye temperatures $\Theta_L = 150$ K and $\Theta_H = 130$ K have been chosen to agree with experiment. In this case, $J_1 = 0$ and $J_2' = J_2 + 40.6\ \text{cm}^{-1}$, thus resulting in a change of J_2 by $\sim 20\%$ [35].

The Ising-type model provides also a necessary though not sufficient condition for the observation of hysteresis. Thus, from the study of metastable minima of G for the case $J_1 = J_2 = J$, this condition results as [37]:

$$J \geq 2RT_c \tag{20}$$

In the following, we consider an example [46] characterized by the parameter values $\Delta = 519\ \text{cm}^{-1}$, $J_1 = 0$, $J_2 = 291\ \text{cm}^{-1}$ and $J_3 = -47\ \text{cm}^{-1}$ where the condition Eq. (20) is safely met. Here, J_3 represents a third order interaction in Eq. (12). Figure 3 shows the Gibbs free energy $G = G(n_H)$ for a number of temperatures between 105 and 130 K, the transition temperature having been chosen here as $T_c = 120$ K. For $T > T_c$, the minimum of G at $n_H \simeq 0.86$ is lower in energy than that close to $n_H = 0.10$. The former thus corresponds to a stable state and the compound is HS. For $T < T_c$, the minimum close to $n_H = 0$ is lower in energy and the stable state is LS. At $T = T_c$, the two minima are symmetrically disposed with regard to $n_H = 0.50$ and have the same Gibbs free

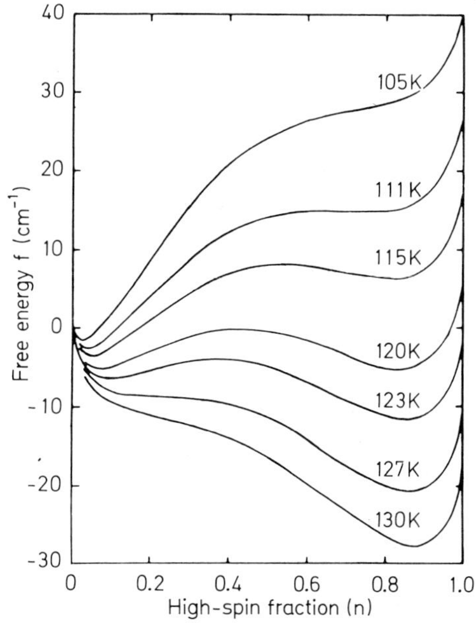

Fig. 3. Gibbs free energy G as a function of the HS fraction n_H for a series of temperatures in the region of $T_c = 120$ K. The employed parameter values are $\Delta = 519\ \text{cm}^{-1}$, $J_2 = 291\ \text{cm}^{-1}$ and $J_3 = -47\ \text{cm}^{-1}$. Here, Δ is the energy difference between the HS ($^5T_2$) and LS ($^1A_1$) ground state of iron(II), and J_3 corresponds to a third order interaction term $-J_3 n_H^3$ in Eq. (12). The figures on the curves specify the values of the temperature. Notation in the figure uses n instead of n_H and f instead of G. According to Ref. [46]

energy. In the region around T_c, hysteresis may eventually be observed. Thus, if T_c is approached by lowering of temperature, the system is initially in the HS state, even in close proximity of T_c. In order to achieve thermal equilibrium, the complete system has to accomplish the transition to LS when arriving at T_c. This transformation is unlikely to be performed instantaneously due to the large number of particles involved. The system may rather remain metastable in the HS state even below T_c until, after a sufficiently large overcooling, the transformation barrier is overcome. The same mechanism may occur, if T_c is approached by the increase of temperature. Thus a complete hysteresis loop may form. Tunneling through the barrier is a possibility of the system to suppress the hysteresis. Table 2 presents a compilation of experimental data which characterize the hysteresis at abrupt spin-state transitions. In the table, $T_c^{\uparrow}$ and $T_c^{\downarrow}$ denote the transition temperature for rising and lowering of

Table 2. Hysteresis effects associated with abrupt spin-state transitions

Compound[a]	$T_c^{\uparrow}$ K	$T_c^{\downarrow}$ K	ΔT_c K	$\bar{T}_c$ K	Method[b]	Ref.
$[Fe(bi)_3](ClO_4)^2$	114.8	108.3	6.5	111.6	X-ray	[48]
$[Fe(bt)_2(NCS)_2]$	181.9	172.3	9.3	177.1	X-ray	[49]
$[Fe(bt)_2(NCSe)_2]$	224.9	221.3	3.6	223.1	X-ray	[50]
$[Fe(bpy)_2(NCS)_2]$	212.6	212.2	0.4	212.4	DSC	[51]
$[Fe(phen)_2(NCS)_2]$, extracted	176.30	176.15	0.15	176.23	cal	[51]
$[Fe(4,7\text{-}(CH_3)_2\text{-phen})_2(NCS)_2]$	121.7	118.6	3.1	120.2	X-ray	[52]
· α-picoline, powdered	202	146	56	174	X-ray	[53]
$[Fe(phy)_2](ClO_4)_2$	256.1	248.0	8.1	252.1	DSC, X-ray	[54]
$[Fe(phy)_2](BF_4)_2$	286	277	9	281.5	DSC, X-ray	[55]
$[Fe(bpp)_2](BF_4)_2$	181.9	171.5	10.4	176.7	magnetism	[56]
$[Fe(paptH)_2](NO_3)_2$	263	229	34	246	magnetism	[57]
$[Fe(2\text{-pic})_3]Cl_2 \cdot H_2O$	295	204	90	249.5	magnetism	[58]
$[Fe\,L(CN)_2] \cdot H_2O$	222	207	15	214.7	ME, magnetism	[59]
[Fe(Hthpu)(thpu)]	235	225	10	230	magnetism	[60]
$[Fe(Hthpx)_2]Cl$	256	245	11	250.5	magnetism	[61]
[Fe(salen)NO]	180	170	10	175	ME	[62]
$[Co(H_2fsa_2en)(py)_2]$	127	115	12	121	DSC, X-ray	[63, 64]
$[Co(H_2fsa_2en)(4\text{-Bupy})_2]$	154	138	16	146	DSC	[65]
$[Co(H_2fsa_2en)(H_2O)_2]$	84.6	81.5	3.1	83	magnetism	[66]

[a] Ligand abbreviations: bi = 2,2′-bi-2-imidazoline; bt = 2,2′-bi-2-thiazoline; bpy = 2,2′-bipyridine; phen = 1,10-phenanthroline; phy = 1,10-phenanthroline-2-carbaldehyde phenylhydrazone; bpp = 2,6-bis(pyrazol-3-yl)pyridine; paptH = 2-(2-pyridylamino)-4-(2-pyridyl)thiazole; 2-pic = 2-picolylamine; L = macrocyclic ligand derived from condensation of 2,6-diacetylpyridine with 3,6-dioxaoctane-1,8-diamine; H_2thpu = pyruvic acid thiosemicarbazone; H_2thpx = pyridoxal thiosemicarbazone; salen = dianion of *N,N′*-ethylenebis(salicylideneimine); H_2fsa_2en = dianion of *N,N′*-ethylenebis(3-carboxysalicylaldimine).

[b] Method employed to establish first-order character of transition: cal = calorimetry; DSC = differential scanning calorimetry; X-ray = X-ray diffraction; ME = Mössbauer effect

temperature, respectively, ΔT_c the width of the hysteresis loop and $\bar{T}_c$ the average transition temperature.

An attempt has been made by Spiering et al. [39, 40] to relate the magnitude of the interaction parameter $\Gamma(x)$ as derived from experiment to the elastic interaction between HS and LS ions via an image pressure [47]. To this end, the metal atoms, inclusive of their immediate environments, in the HS and LS state are considered as incompressible spheres of radius r_H and r_L, respectively. The spheres are embedded in an homogeneous isotropic elastic medium, representing the crystal, which is characterized by specific values of the bulk modulus K and Poisson ratio σ where $0 \leq \sigma \leq 0.5$. The change of molecular volume ΔV as determined by X-ray diffraction may be related to the volume difference $\Delta v = v_H - v_L$ of the hard spheres by:

$$\Delta V = (V_H - V_L)/Z = \gamma_0(v_H - v_L) = \frac{4\pi}{3}\gamma_0[R_H^3 - (R_H - \delta R)^3] \qquad (21)$$

where $\gamma_0 = 3(1 - \sigma)/(1 + \sigma)$ is the so-called Eshelby constant. If the value $\delta R = 0.17$ Å is inserted into Eq. (21), radii between 2.1 and 2.4 Å are obtained. This demonstrates that the size of the hard spheres corresponds approximately to the metal-donor atom core of the complexes.

In order to derive an expression for the interaction parameter Γ on the basis of elasticity theory, the elastic energy of a single sphere of volume V_i is considered which is embedded in a spherical hole of volume V_0 in the elastic medium:

$$E_i = K(\gamma_0 - 1)(V_i - V_0)^2/2V_0 - K\gamma_0(\gamma_0 - 1)\,(V_i - V_0)^2/2V \qquad (22)$$

Here, the first term is the elastic energy stored in an infinite elastic medium, the second term is the change of the elastic energy subject to a hydrostatic pressure [47]:

$$P^I = K(\gamma_0 - 1)(V_i - V_0)/V \qquad (23)$$

(*image pressure*). The second term may thus be written as $-\frac{1}{2}P^I \Delta V_i$ where the volume change of the crystal ΔV_i is larger than the volume change $V_i - V_0$ of the hole by the factor γ_0:

$$\Delta V_i = \gamma_0(V_i - V_0) \qquad (24)$$

and V is the volume of the crystal. If there are already N spheres within the medium, Eq. (22) for the elastic energy per sphere may be generalized to:

$$E/N = K(\gamma_0 - 1)\sum_{i=1}^{N}(V_i - V_0)^2/2V_0 N - K\gamma_0(\gamma_0 - 1)\left[\sum_{i=1}^{N}(V_i - V_0)\right]^2\Big/2VN \qquad (25)$$

For the particular case where the HS and LS Fe atoms are diluted in a host metal M, there are three kinds of spheres with i = HS, LS, and M. Next we introduce the averages:

$$\sum_{i=1}^{N} (V_i - V_0)^\rho/N = x[n_\mathrm{H}(V_\mathrm{H} - V_0)^\rho + (1 - n_\mathrm{H})(V_\mathrm{L} - V_0)^\rho] + (1 - x)(V_\mathrm{M} - V_0)^\rho \quad (26)$$

where $\rho = 1$ or 2. The energy per iron sphere then results as:

$$e = E/Nx = \Delta E_s n_\mathrm{H} + 2\Gamma x(V_\mathrm{M} - V_\mathrm{L})/(V_\mathrm{H} - V_\mathrm{L})n_\mathrm{H} - \Gamma x n_\mathrm{H}^2 \quad (27)$$

Here, all terms independent of n_H have been omitted and the volume of the crystal belonging to one complex molecule has been denoted as $V_c = V/N$. In Eq. (27);

$$\Delta E_s = E_p[(V_\mathrm{H} + V_\mathrm{L})/V_0 - 2 - 2\gamma_0(V_\mathrm{M} - V_0)/V_c] \quad (28)$$

is the difference of the self-energies of HS and LS spheres in the elastic medium for vanishing iron content (mole fraction $x \to 0$) and:

$$E_p = K(\gamma_0 - 1)(V_\mathrm{H} - V_\mathrm{L})/2 \quad (29)$$

The second and third term in Eq. (27) account for the elastic interaction between the HS and LS spheres and:

$$\Gamma = K\gamma_0(\gamma_0 - 1)(V_\mathrm{H} - V_\mathrm{L})^2/2V_c \quad (30)$$

is the interaction constant. It may be shown, moreover, that the parameters Γ and Δ of the interaction term:

$$I(n_\mathrm{H}) = n_\mathrm{H}\Delta(x) - n_\mathrm{H}^2\Gamma(x) \quad (31)$$

in the free energy expression are related by:

$$\Delta(x) = 2q\Gamma(x)$$
$$q = (V_\mathrm{M} - V_\mathrm{L})/(V_\mathrm{H} - V_\mathrm{L}) \quad (32)$$

The model of Spiering et al. [39, 40] has been successfully applied to rationalize the effects of metal dilution [39, 40], pressure [67], and specific heat [68] on a series of spin transition compounds of the type $[Fe(2\text{-pic})_3]Cl_2 \cdot Sol$ where 2-pic = 2-aminomethylpyridine and Sol = CH_3OH or C_2H_5OH. It should be noted, however, that it is that part of the model based on Eq. (5) and Eq. (31) and thus in effect the Bragg and Williams approximation which is capable of reproducing the various experimental results. Obviously, for the dilution experiments a dependence of Δ and Γ, or any corresponding parameters, on the fraction x of Fe atoms has to be allowed for, cf. Eq. (31). Then almost any of the models listed in Table 1 would serve the purpose equally well. The studies yield empirical values for the interaction parameter Γ which may be compared with values derived on the basis of Eq. (30). Thus, from the experiment, $\Gamma_{obsd} = 103$ and $98\,cm^{-1}$ obtain for $[Fe(2\text{-pic})_3]Cl_2 \cdot CH_3OH$ if diluted in the corresponding Co or Zn compound [6], respectively, and $\Gamma_{obsd} = 130\,cm^{-1}$

results for $[Fe(2\text{-pic})_3]Cl_2 \cdot C_2H_5OH$ diluted in the corresponding Zn complex [40, 69]. On the other hand, the expression based on elasticity theory gives $\Gamma_{calcd} \simeq 18\ \mathrm{cm}^{-1}$ [39]. The calculated value of Γ is therefore too small by a factor of 5–10 compared with the experimentally based data. So far, the hard sphere model fails to reproduce the actual interaction of metal ions in a spin-state transition compound.

However, the calculation has taken account only of the change of the unit cell volume of the compound, $\Delta V_{HL} = V_{HS} - V_{LS}$ which occurs due to the spin-state transition, the individual HS and LS complexes being treated as incompressible spheres. On the basis of detailed X-ray structure investigations [6], the crystal experiences not only a change in size but also a change in shape in the course of the HS⇌LS transition. Every lattice vector $\mathbb{x}(T)$ may be expressed according to [70]:

$$\mathbb{x}(T) = (\mathbb{1} + \mathbb{\alpha} \cdot \Delta T + \mathbb{\varepsilon} \cdot \Delta n_H \cdot \mathbb{x}(T_0) \tag{33}$$

where $\Delta T = T - T_0$ and $\Delta n_H = n_H(T) - n_H(T_0)$. The tensor $\mathbb{\alpha}$ describes the thermal lattice expansion and tensor $\mathbb{\varepsilon}$ the deformation due to the spin-state transition. In the calculation performed above, only the trace of tensor $\mathbb{\varepsilon}$ has entered since $Tr(\mathbb{\varepsilon}) = \Delta V_{HL}/V_c$, whereas in general the full tensor $\mathbb{\varepsilon}$ would have to be taken into account.

In an attempt at a more complete calculation of the interaction energy, Willenbacher and Spiering [71] therefore considered the spin-changing ions as incompressible inclusions described by the elastic dipole tensors $\mathbb{P}^n$ ($n =$ HS, LS, and M) instead of isotropic spheres. Each type of defect is statistically distributed over the lattice sites, all defects being again embedded in an isotropic elastic medium. The interaction energy may be separated into a self-energy resulting from a stress-free boundary and a so-called direct interaction obtained by summation of the interaction of all pairs of defects. Thus, a separation into a long-range part (extending to the infinite) and a short-range part (proportional to R^{-3}) results. The first study [71] concentrated on the self-energy and improved the calculation considerably, although the size of the interaction constant Γ could not be explained by the image pressure. In a subsequent study [72] the direct interaction was computed. Collecting both contributions, consistency with the experimental results has been achieved.

Obviously, a more than qualitative agreement of the interaction constants Γ and energy shifts Δ cannot be expected, given a lattice approximated by a homogeneous isotropic elastic medium and the shape of the crystal taken to be a sphere. X-ray structure data for a large series of spin-state transition compounds [6] demonstrate the variability of the transitions, both as far as the magnitude of the volume change and the sort of deformation of the molecules is concerned. Necessarily, the observed interactions will be the result of the changes at the molecular level and will show a certain irregularity depending on the changes of the low-symmetry geometry of the molecules.

Of considerable interest is the pressure dependence of a spin-state transition which will be considered in Sect. 8 in some detail. It is assumed that the pressure dependence is determined by the volume change of the crystal per complex molecule, $\Delta V = (V_H - V_L)/Z$ and by the intermolecular interaction term $I(n_H)$ of Eq. (5). Within the above model, a useful expression between these quantities may be derived starting from the equilibrium relation for ΔG, i.e. Eq. (8), which is written here as:

$$\Delta G + \left(\frac{\partial I(n_H)}{\partial n_H}\right)_{p,T} + RT \ln K(T, p) = 0 \tag{34}$$

In Eq. (34), the equilibrium constant $K(T, p) = n_H/(1 - n_H)$ has been employed. The volume change ΔV is then introduced by the relation $\Delta G = \Delta \bar{E} + p\Delta V$ where $\Delta \bar{E}$ is the change of the mean energy of the system. If the expression Eq. (31) is now used for the quantity $I(n_H)$, we obtain:

$$\Delta \bar{E} + p\Delta V + \Delta - 2\Gamma n_H = -RT \ln K(T, p) \tag{35}$$

For the present purpose, the quantities Δ and Γ of Eq. (35) are considered as independent of temperature and pressure. If Eq. (35) is applied to measurements at two different pressures p and p_0 at a particular temperature T, the unknown change of mean energy $\Delta \bar{E}$ can be eliminated. The resulting equation may be written as:

$$\begin{aligned}(p - p_0)\Delta V + 2\Gamma \Delta n_H &= RT \ln[K(T, p_0)/K(T, p)] \\ &= \delta g\end{aligned} \tag{36}$$

where:

$$\Delta n_H = n_H(T, p_0) - n_H(T, p)$$

Thus, if the quantity δg is found to be a linear function of Δn_H, both the volume change ΔV and the interaction parameter Γ may be determined from the values of the equilibrium constant $K(T, p)$ for the pressures p and p_0.

Various other interactions have been considered as the driving force for spin-state transitions such as: the Jahn-Teller coupling between the d electrons and a local distortion [73], the coupling between the metal ion and an intramolecular distortion [74, 75, 76] or the coupling between the d electrons and the lattice strain [77, 78]. At present, based on the available experimental evidence, the contribution of these interactions cannot be definitely assessed. Moreover, all these models are mathematically rather ambitious and do not show the intuitively simple structure inherent in the effect of a variation of molecular volume considered here. Their discussion has to be deferred to a more specialized study.

3 Methods and Basic Relations for Relaxation Measurements of Spin-State Equilibria

Spin-state equilibria between low-spin (LS; S_1) and high-spin (HS; S_2) electronic isomers in solution may be characterized by:

$$^{(2S_1+1)}[\text{Me L}_n]^{m+} \underset{k_{HL}}{\overset{k_{LH}}{\rightleftharpoons}} {}^{(2S_2+1)}[\text{Me L}_n]^{m+} \tag{37}$$

$$K_{eq} = \frac{[\text{HS}]}{[\text{LS}]} = \frac{k_{LH}}{k_{HL}} \tag{38}$$

where $S_1 < S_2$ and $\Delta S = S_2 - S_1 = 2$ or 1. [HS] and [LS] denote the concentration of the HS and LS isomer, respectively, and K_{eq} is the equilibrium constant.

Spin-state equilibria possess a nonzero enthalpy of reaction, the reason being the larger entropy of the HS isomer as compared to the LS isomer. Since at equilibrium $\Delta G = 0$, ΔH is positive and thus the equilibria are temperature dependent and can be perturbed by a rapid change of temperature. Spin-state equilibria are also pressure dependent due to the volume change of the complex which accompanies the change of spin-state. Therefore, spin equilibria can be also perturbed by a rapid variation of pressure. The two methods that were employed to measure directly the forward and reverse rate constants k_{LH} and k_{HL} of the intersystem crossing process Eq. (37) utilize either a rapid change of temperature or a change of pressure.

The laser-stimulated Raman temperature-jump method uses the power of a laser to produce a temperature rise of a few degrees within the time of the laser pulse width. The 1.06 μ radiation of a Nd-glass laser is not absorbed by normal solvents. Therefore, liquid N_2 is used as the Raman absorber thus shifting the energy of the scattered laser beam by the N–N Raman stretching frequency to 1.41 μ. Hydroxylic solvents such as water and alcohols absorb at this frequency since it corresponds to the first overtone of the O–H stretch. The technique allows heating of hydroxylic solvents by 1–5° within 20 ns. The temperature jump produces changes in the population of the two spin states with consequent changes in the CT electronic absorption spectrum. The absorbance changes are followed spectrophotometrically. Further details both with respect to instrumentation [79] and experimental methods [28, 80] may be found in the literature.

The experimentally observed relaxation traces are displayed on an oscilloscope and photographed. The first-order relaxation time constants τ in ns are then determined from linear $\log(I_\infty - I)$ vs. time plots where I is the intensity of the light passing through the solution at time t and I_∞ is the limiting value of the intensity at long times. For the small intensity changes measured in these experiments, this is equivalent to plotting the log of the absorbance or concentration changes. Alternatively, the transient recording method may be employed. If $\tau \lesssim 30$ ns, the second method of moments integral deconvolution procedure has been used [81, 82]. Fundamental limitations on the resolution obtainable in the temperature-jump experiment arise from the finite duration of the heating pulse ($\sim$ 25 ns), the magnitude of the enthalpy change ΔH°, and the differences in the molar absorptivities of the HS and LS complexes $\Delta\varepsilon$.

The second method for the study of relaxation in spin-state equilibria makes use of the rapid change of pressure. Single-step pressure-jump relaxation requires an observation time of about 10^{-4} s which is too slow. However, the periodic pressure variation of a sound wave can be employed to study the dynamics of spin-state equilibria. Two different ultrasonic techniques have been used for the purpose, the resonator method and the pulse method. These two methods span different frequency ranges, the resonator method operating between below 1 and above 15 MHz (relaxation times 10^{-7} to 10^{-8} s), the pulse method spanning the range 15–150 MHz (relaxation times 10^{-8} to 10^{-9} s). Figure 4 shows the schematic disposition of an ultrasonic absorption apparatus. It consists of the ultrasonic cell, a pulsed oscillator, a wide-band rf amplifier, preamplifier, and a precision attenuator. For more details, we refer to the literature [29].

Perturbation of a spin-state equilibrium by ultrasound results in the absorption of sound, the absorption coefficient α (neper cm^{-1}) being determined as a function of frequency f (Hz) according to:

$$\alpha/f^2 = A(1 + \omega^2\tau^2)^{-1} + B \tag{39}$$

Here, B is the background absorption in the absence of chemical relaxation, ω the angular frequency of the inflection point of the absorption $\omega = 2\pi f$ (rad s^{-1}) and τ the relaxation time (s). The excess absorption due to chemical relaxation A is obtained from the low-frequency absorption where $\omega \ll \tau$ and is given by:

$$A = \frac{2\pi^2 \rho v}{RT}\left(\Delta V^\circ - \frac{\alpha_p}{\rho C_p}\Delta H^\circ\right)^2 \Gamma\tau \tag{40}$$

Here ρ is the solution density, v the sound velocity, α_p the coefficient of thermal expansion, C_p the specific heat, and Γ the concentration dependence of the equilibrium, $\Gamma^{-1} = [\mathrm{LS}]^{-1} + [\mathrm{HS}]^{-1}$. The measurement of ultrasonic relaxation thus enables the determination of both the relaxation time τ and the

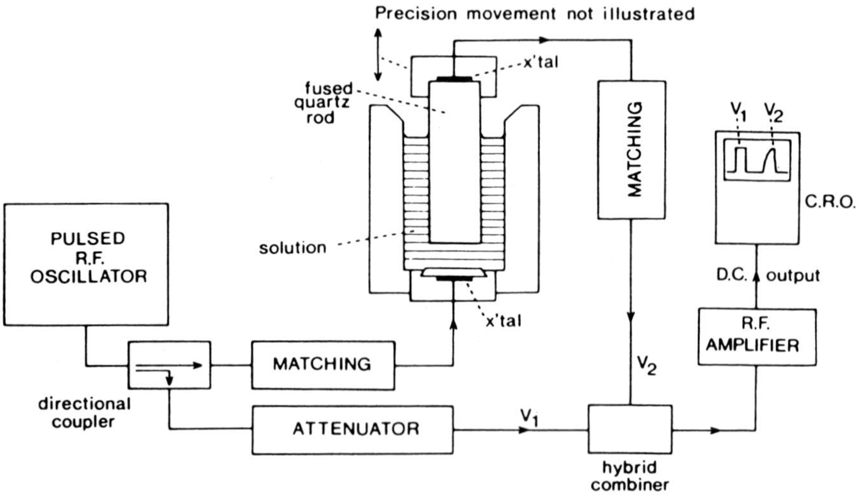

Fig. 4. Schematic diagram of experimental arrangement for pulsed ultrasonic absorption measurements. According to Ref. [29]

relaxation amplitude, A. The latter quantity may be used to evaluate the volume difference between the two electronic isomers, ΔV°. The values of the standard enthalpy change ΔH° and of Γ have to be obtained from a study of equilibrium properties.

It is thus evident that, in order to interpret the results obtained by one of the relaxation methods, a thorough investigation of the unperturbed equilibrium properties is required. In general, solution magnetic susceptibilities measured by the NMR method of Evans [83, 84] are used to this end, the equilibrium constant for the equilibrium in Eq. (37) being determined by:

$$K_{eq} = \frac{[HS]}{[LS]} = \frac{\mu_{eff}^2 - \mu_{LS}^2}{\mu_{HS}^2 - \mu_{LS}^2} \tag{41}$$

In Eq. (41), μ_{HS} and μ_{LS} are the limiting values for the magnetic moment of the HS and LS state, respectively. Relaxation processes are introduced by a slight disturbance of the equilibrium, the equilibrium concentration of the components being changed by the same amount x according to:

$$\begin{aligned} c_{LS} &= [LS] - x \\ c_{HS} &= [HS] + x \end{aligned} \tag{42}$$

The rate of formation of the HS component may then be written as:

$$\frac{dc_{HS}}{dt} = k_{LH}([LS] - x) - k_{HL}([HS] + x)$$

or (43)

$$\frac{dx}{dt} = -(k_{LH} + k_{HL})x$$

since $k_{LH}[LS] = k_{HL}[HS]$. Equation (43) gives on integration:

$$x = x_0 e^{-(k_{LH} + k_{HL})t} \tag{44}$$

The disturbance thus decays always according to first-order kinetics, the observed relaxation time τ being related to the rate constants k_{LH} and k_{HL} by:

$$\tau^{-1} = k_{LH} + k_{HL} \tag{45}$$

Consequently, the rate constants and lifetimes of the individual spin states may be calculated from the measured values of K_{eq} and τ according to:

$$\begin{aligned} k_{LH} &= [\tau(LS)]^{-1} = \frac{K_{eq}}{1 + K_{eq}}\tau^{-1} \\ k_{HL} &= [\tau(HS)]^{-1} = \frac{1}{1 + K_{eq}}\tau^{-1} \end{aligned} \tag{46}$$

The thermodynamic parameters ΔH° and ΔS° for the spin equilibrium are obtained from the temperature dependence of K_{eq} and thus from the variable-temperature magnetic susceptibility data using:

$$\ln K_{eq} = -\frac{\Delta G^\circ}{RT} = -\frac{\Delta H^\circ}{RT} + \frac{\Delta S^\circ}{R} \tag{47}$$

The temperature dependence of the rate constants follows the Arrhenius equation:

$$k = A\, e^{-\Delta E/RT} \tag{48}$$

where ΔE is the activation energy and A the frequency factor. The parameters of Eq. (48) can be evaluated by plotting the experimentally determined ln k as a function of $1/T$ (Arrhenius plot). According to the absolute reaction rate theory [85], the rate constant k is given as:

$$k = \frac{k_B T}{h} \kappa e^{-\Delta G^\ddagger/RT} = \frac{k_B T}{h} \kappa\, e^{-\Delta H^\ddagger/RT} e^{\Delta S^\ddagger/R} \tag{49}$$

where $\Delta G^\ddagger$ is the activation free energy, $\Delta H^\ddagger$ the activation enthalpy, and $\Delta S^\ddagger$ the activation entropy. In other words, $\Delta G^\ddagger$, $\Delta H^\ddagger$, and $\Delta S^\ddagger$ are the differences in free energy, enthalpy, and entropy, respectively, between the transition state and the reactants. The quantity $\Delta H^\ddagger$ may be obtained by plotting $\ln(k/T)$ versus $1/T$ (Eyring plot). Moreover, κ in Eq. (49) is the transmission coefficient which reflects the probability of intersystem crossing and $k_B T/h = 6 \times 10^{12}\ \mathrm{s}^{-1}$. The relationship between the parameters of Eq. (48) and Eq. (49) is:

$$A \simeq e \frac{k_B T}{h} \kappa e^{\Delta S^\ddagger/R} \qquad \Delta E \simeq \Delta H^\ddagger + RT \tag{50}$$

Equation (49) applies to both the forward and reverse rate constant, k_{LH} and k_{HL}. Consequently, the thermodynamic parameters for the intersystem crossing process are related according to:

$$\begin{aligned} \Delta G^\ddagger_{HL} &= \Delta G^\ddagger_{LH} - \Delta G^\circ \\ \Delta H^\ddagger_{HL} &= \Delta H^\ddagger_{LH} - \Delta H^\circ \\ \Delta S^\ddagger_{HL} &= \Delta S^\ddagger_{LH} - \Delta S^\circ \end{aligned} \tag{51}$$

It has been stated above that the difference of partial molar volumes of the LS and HS isomers ΔV° can be obtained from the relaxation amplitude A of ultrasonic absorption. An independent method for the determination of ΔV° is based on the pressure dependence of the equilibrium constant. The pressure derivative of K_{eq} being determined by:

$$\left(\frac{\partial \ln K_{eq}}{\partial p}\right)_T = -\frac{\Delta V^\circ}{RT} \tag{52}$$

ΔV° can be obtained from the slope of the straight line if log K_{eq} is plotted versus the pressure, p. If the plot is nonlinear, this may be due to a compressibility difference between the two spin isomers which results in a pressure dependence of ΔV°. In this case, the limiting value of ΔV° at 1 atm, ΔV_1°, can be determined from [86]:

$$-\frac{RT}{(p-1)}\ln\left(\frac{K_p}{K_1}\right) = \Delta V_1^\circ - \tfrac{1}{2}\Delta K_1^\circ(p-1) \tag{53}$$

where ΔK_1° is the change in compressibility ($cm^3\,mol^{-1}\,atm^{-1}$) and K_p and K_1 are the equilibrium constants at p and 1 atm, respectively.

The pressure dependence of the relaxation time τ follows the relation:

$$\left(\frac{\partial \ln \tau^{-1}}{\partial p}\right)_T = -\frac{\Delta V^\ddagger_{\tau^{-1}}}{RT} \tag{54}$$

and thus provides, if measured, the quantity $\Delta V^\ddagger_{\tau^{-1}}$ [87]. On the basis of the transition state theory, the pressure dependence of the rate constants for intersystem crossing may be written as:

$$\left(\frac{\partial \ln k_{LH}}{\partial p}\right)_T = -\frac{\Delta V^\ddagger_{LH}}{RT}$$
$$\left(\frac{\partial \ln k_{HL}}{\partial p}\right)_T = -\frac{\Delta V^\ddagger_{HL}}{RT} \tag{55}$$

where $\Delta V^\ddagger_{LH}$ and $\Delta V^\ddagger_{HL}$ are the activation volumes for the forward and reverse reaction. In particular, $\Delta V^\ddagger$ is the difference of partial molar volumes between the transition state and the reactant. These equations hold as long as $\Delta V^\ddagger$ remains independent of pressure, i.e., within a limited pressure range. The activation volumes are closely related to the quantities $\Delta V^\ddagger_{\tau^{-1}}$ and ΔV°. Thus, on the basis of Eq. (46), taking the logarithm and differentiating with respect to p gives:

$$\frac{\partial \ln k_{LH}}{\partial p} = \frac{\partial \ln \tau^{-1}}{\partial p} + \frac{K_{eq}}{1+K_{eq}}\frac{1}{K_{eq}^2}\frac{\partial K_{eq}}{\partial p}$$
$$\frac{\partial \ln k_{HL}}{\partial p} = \frac{\partial \ln \tau^{-1}}{\partial p} - \frac{K_{eq}}{1+K_{eq}}\frac{1}{K_{eq}^2}\frac{\partial K_{eq}}{\partial p} \tag{56}$$

Introducing Eq. (54) and Eq. (55) produces the relations:

$$\Delta V^\ddagger_{LH} = \Delta V^\ddagger_{\tau^{-1}} + \frac{1}{1+K_{eq}}\Delta V^\circ$$
$$\Delta V^\ddagger_{HL} = \Delta V^\ddagger_{\tau^{-1}} - \frac{K_{eq}}{1+K_{eq}}\Delta V^\circ \tag{57}$$

From the quantities $\Delta V^\ddagger_{\tau^{-1}}$, ΔV° and K_{eq}, therefore, the activation volumes $\Delta V^\ddagger_{LH}$ and $\Delta V^\ddagger_{HL}$ may be easily calculated. It should be noted that, similar to Eq. (51), these quantities are related by:

$$\Delta V^\ddagger_{HL} = \Delta V^\ddagger_{LH} - \Delta V^\circ \tag{58}$$

4. Relaxation of Spin-State Equilibria of Metal Complexes in Solution

Results of relaxation measurements on spin-state equilibria in solution are available for complexes of iron(II), iron(III), and cobalt(II). The results comprise values of relaxation time τ, rate constants for the forward and reverse reactions k_{LH} and k_{HL}, and activation parameters $\Delta H^{\ddagger}$ and $\Delta S^{\ddagger}$ for the two opposed reactions. In a number of cases, volumes of activation for the forward and reverse reaction, $\Delta V^{\ddagger}_{LH}$ and $\Delta V^{\ddagger}_{HL}$, were determined. In the following, the results for the complexes of each of the three metals will be separately considered and briefly summarized.

4.1 Complexes of Iron(II)

The most significant results obtained for complexes of iron(II) are collected in Table 3. The data derive from laser Raman temperature-jump measurements, ultrasonic relaxation, and the application of the photoperturbation technique. Where the results of two or three methods are available, a gratifying agreement is found. The rate constants span the narrow range between 4×10^5 and $2 \times 10^7\ s^{-1}$ which shows that the spin-state interconversion process for iron(II) complexes is less rapid than for complexes of iron(III) and cobalt(II).

Interesting conclusions may be derived from the consideration of quantitative aspects of the derived activation parameters [29]. For $Fe[HB(pz)_3]_2$ where $HB(pz)_3$ = hydrotris(pyrazolyl)borate, the activation enthalpy for the $^1A_1 \rightarrow {}^5T_2$ process of $\Delta H^{\ddagger}_{LH} = 5.64$ kcal mol^{-1} comprises 5.03 kcal mol^{-1} due to the endothermic enthalpy difference ΔH°, and thus the activation enthalpy for the reverse process is only $\Delta H^{\ddagger}_{HL} = 0.62$ kcal mol^{-1}. It follows that the rate constant k_{HL} is determined almost exclusively by an entropic barrier $-T\Delta S^{\ddagger} =$ 6.74 kcal mol^{-1}. According to Eq. (49), this term comprises the transmission coefficient κ and other entropy contributions, their magnitude being difficult to estimate separately. In conventional transition state theory, it is usually assumed that $\kappa = 1$. This assumption is most likely not applicable to the highly spin-forbidden process ($\Delta S = 2$) studied here. If it is assumed that the entire entropic barrier of 6.74 kcal mol^{-1} is due to κ, the value $\kappa = 10^{-4.9}$ is obtained. This value corresponds to equal structures of transition state and HS state. Such a situation is unlikely and thus a higher than the minimum value of κ is expected. Indeed, studies of the activation volumes of several iron(II) compounds (*vide infra*) indicate that the transition state lies about midway between the LS and HS states [89, 91]. The activation entropy will then contain a significant contribution due to changes in structure and vibrational properties of the complex which increases the value of κ. The $^5T_2 \rightarrow {}^1A_1$ process will be therefore less spin-forbidden and κ will be expected to be not smaller than of the

Table 3. Rate constants and activation parameters for the LS → HS and HS → LS unimolecular spin-state interconversions in iron(II) complexes

Compound[a]	Solvent	τ ns	k_{LH} s^{-1}	k_{HL} s^{-1}	$\Delta H^{\ddagger b}$ kcal mol^{-1}	$\Delta S^{\ddagger b}$ cal K^{-1} mol^{-1}	Ref.
$[Fe(6\text{-}Mepy)(py)_2tren](PF_6)_2$	CH_3OH-acetone/H_2O	120 ± 20	4×10^5	7.7×10^6			[88]
$[Fe(6\text{-}Mepy)_2(py)tren](PF_6)_2$	CH_3OH-acetone/H_2O	110 ± 30	4×10^6	5×10^6			[88]
$Fe[HB(pz)_3]_2$	CH_2Cl_2/CH_3OH	32 ± 10	1×10^7	2×10^7	5.64 ± 0.13	-9.0 ± 0.4	[28, 29]
					0.62 ± 0.13	-22.6 ± 0.4	
$[Fe(paptH)_2]Cl_2$	H_2O	41 ± 0.2	1.7×10^7	7.2×10^6	7.63 ± 0.25	0.09 ± 0.8	[29]
					3.72 ± 0.25	-14.7 ± 0.8	
$[Fe(phenmethoxa)_2](BF_4)_2$	acetone	110 ± 10	7.7×10^6		-7.8 ± 0.4		[89]
					2.1 ± 0.2		
$[Fe(pyimH)_3](BPh_4)_2$	CH_3OH/	48 ± 8	1.1×10^7	1.0×10^7	6.38 ± 0.28	-5.5 ± 1	[90, 91]
	20% CH_3CN				2.68 ± 0.14	-17.9 ± 1	
	CH_3CN				11.1 ± 0.7	9.8 ± 1	
					3.42 ± 0.24	-14.6 ± 1	
	acetone	45 ± 5			7.6 ± 0.7	-1.7 ± 1	
					3.84 ± 0.48	-13.4 ± 1	
$[Fe(pybimH)_3](BPh_4)_2$	CH_3OH/				6.95 ± 0.48	-1.4 ± 1	[89, 90, 91]
	20% CH_3CN				1.86 ± 0.12	-23.4 ± 1	
	CH_3CN				8.7 ± 0.7	5.0 ± 1	
					3.61 ± 0.24	-14.6 ± 1	
	acetone				7.95 ± 0.17	2.4 ± 1	
					3.25 ± 0.07	-16.2 ± 1	

Table 3. Continued

Compound[a]	Solvent	τ ns	k_{LH} s^{-1}	k_{HL} s^{-1}	$\Delta H^{\ddagger b}$ kcal mol^{-1}	$\Delta S^{\ddagger b}$ cal K^{-1} mol^{-1}	Ref.
$[Fe(ppa)_2](ClO_4)_2$	acetone				8.0 ± 1.0	-4.3 ± 1	[90, 92]
					3.2 ± 1.0	-16.7 ± 1	
$[Fe(tpmbn)](ClO_4)_2$	DMF				9.33 ± 0.48	2.4 ± 1	[90]
					2.17 ± 0.10	-15.5 ± 1	
$[Fe(tppn)](ClO_4)_2$	DMF				7.31 ± 0.36	-4.3 ± 1	[90]
					1.27 ± 0.48	-21.3 ± 1	
	CH_3CN				9.12 ± 0.38	4.0 ± 2	
					2.00 ± 0.10	-16.9 ± 2	
$[Fe(tpchxn)](ClO_4)_2$	DMF				13.28 ± 0.48	15.3 ± 1	[90]
					8.22 ± 0.28	0.48 ± 1	
	CH_3CN				6.21 ± 0.28	-6.4 ± 1	
					0.43 ± 0.10	-24.4 ± 1	
$[Fe(biz)_3](ClO_4)_2$	CH_3CN	27					[92]

[a] Ligand abbreviations: (6-Mepy)(py)$_2$tren and (6-Mepy)$_2$(py)tren = tris[4-[(6-*R*)-2-pyridyl]-3-aza-6-butenyl]amine, R = H or CH_3; HB(pz)$_3$ = hydrotris(pyrazolyl)borate; paptH = 2-(2-pyridylamino)-4-(2-pyridyl)thiazole; phenmethoxa = 3-[2-(1,10-phenanthrolyl)]-5-methyl-1,2,4-oxadiazole; pyimH = 2-(2′-pyridyl)imidazole; pybimH = 2-(2′-pyridyl)benzimidazole; ppa = N^2-(2-pyridylmethyl)picolineamidine; tpmbn = tetrakis(2-pyridylmethyl)-*meso*-2,3-butanediamine; tppn = tetrakis-(2-pyridylmethyl)-1-methyl-1,2-propanediamine; tpchxn = tetrakis(2-pyridylmethyl)-*trans*-1,2-cyclohexanediamine; biz = 2,2′-bi-1,4,5,6-tetrahydropyrimidine.

[b] First line lists values of $\Delta H^{\ddagger}_{LH}$ or $\Delta S^{\ddagger}_{LH}$, second line lists values of $\Delta H^{\ddagger}_{HL}$ or $\Delta S^{\ddagger}_{HL}$

order 10^{-2} [30]. The relation between the activation parameters for $Fe[HB(pz)_3]_2$ is displayed in Fig. 5. A similar analysis as above of the results for the iron(II) complex $[Fe(paptH)_2]Cl_2$ derived from the tridentate ligand paptH = 2-(2-pyridylamino)-4-(2-pyridyl)-thiazole produces a larger value of κ since appreciable activation enthalpies for the spin-state conversion in both directions are found [29]. The situation is illustrated in Fig. 6.

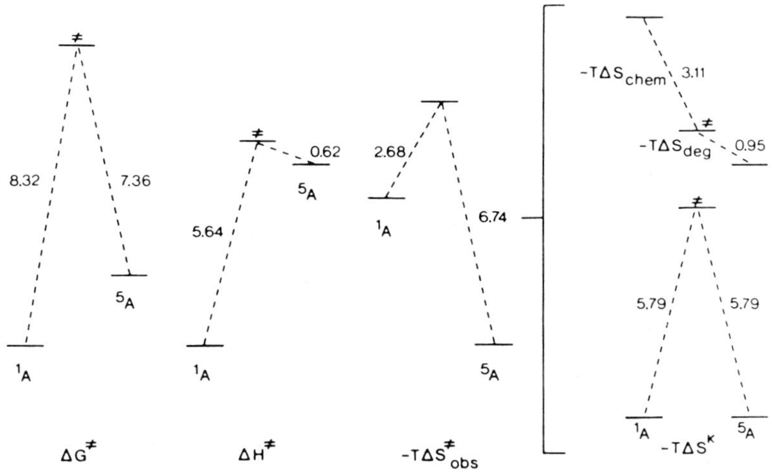

Fig. 5. Relation of activation parameters for $Fe[HB(pz)_3]_2$. The quantity $-T\Delta S^{\ddagger}_{obsd}$ is partitioned into $-T\Delta S^{\ddagger}_{deg}$ corresponding to spin degeneracy of $^5A_1$ state and $-T\Delta S^{\ddagger}_{chem}$ corresponding to change of structure and vibrational properties of the complex. According to Ref. [29]

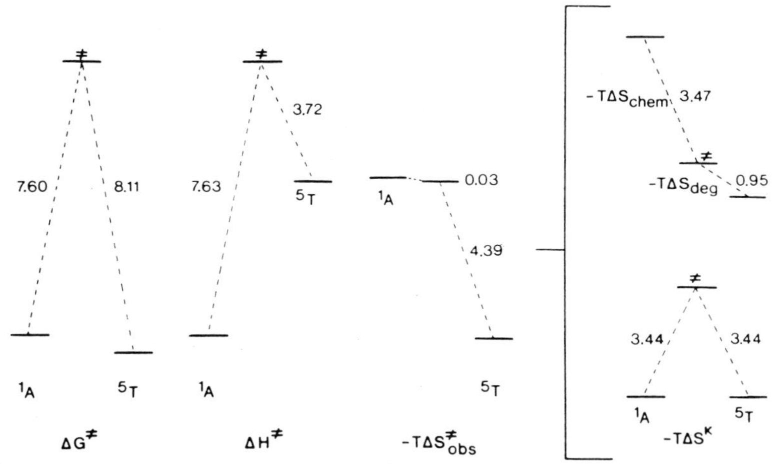

Fig. 6. Relation of activation parameters for $[Fe(paptH)_2]Cl_2$. The quantity $-T\Delta S^{\ddagger}_{obsd}$ is partitioned into $-T\Delta S^{\ddagger}_{deg}$ corresponding to spin degeneracy of $^5A_1$ state and $-T\Delta S^{\ddagger}_{chem}$ corresponding to change of structure and vibrational properties of the complex. According to Ref. [29]

In the $[Fe(pyimH)_2](BPh_4)_2$ and $[Fe(pybimH)_2](BPh_4)_2$ complexes of the bidentate ligands pyimH = 2-(2′-pyridyl)imidazole and pybimH = 2-(2′-pyridyl)benzimidazole a significant solvent dependence of molar volumes and activation parameters was observed [90, 91], cf. Table 4. On the other hand, the activation volume for the $^5T_2 \rightarrow {}^1A_1$ conversion $\Delta V^{\ddagger}_{HL}$ is solvent independent or shows only a slight variation for the pyimH and pybimH complex, respectively. The negative values of $\Delta V^{\ddagger}_{HL}$ indicate a significant contraction of the transition state relative to the HS isomer. The observed volume change for an activation step involves intrinsic factors such as the effect of the change of metal-ligand bond length and the effect of changes of solvation:

$$\Delta V^{\ddagger}_{obsd} = \Delta V^{\ddagger}_{intr} + \Delta V^{\ddagger}_{solv} \tag{59}$$

Virtually constant values of $\Delta V^{\ddagger}_{HL}$ in three different solvents suggest a rather small $\Delta V^{\ddagger}_{solv}$ which implies a similar extent of solvation for the transition state and the HS isomer. It follows that the major contribution to $\Delta V^{\ddagger}_{HL}$ will be from $\Delta V^{\ddagger}_{intr}$ which arises from Fe–N bond contraction in the $^5T_2 \rightarrow {}^1A_1$ activation step. The value of about $-5.4\ cm^3\ mol^{-1}$ for $\Delta V^{\ddagger}_{intr}$ indicates a transition state geometry intermediate between the $^1A_1$ and $^5T_2$ isomers. In support of this interpretation is the solvent dependence of $\Delta V^{\ddagger}_{LH}$ for the reverse process which reflects the opposing contributions of the intrinsic volume expansion and solvent contraction. The variation of $\Delta V^{\ddagger}_{HL}$ values for the $[Fe(pybimH)_2](BPh_4)_2$ complex is not surprising in view of greater steric demands and Fe–N

Table 4. Changes of molar volumes and volumes of activation for spin-state equilibria in iron(II) complexes[a]

Compound[b]	Solvent	ΔV° $cm^3\ mol^{-1}$	$\Delta V^{\ddagger}_{LH}$ $cm^3\ mol^{-1}$	$\Delta V^{\ddagger}_{HL}$ $cm^3\ mol^{-1}$
$[Fe(pyimH)_3](BPh_4)_2$	CH_3OH/20% CH_3CN	5.3 ± 0.2	0.0 ± 0.2	− 5.3 ± 0.3
	CH_3CN	14.3 ± 0.5	8.9 ± 0.4	− 5.4 ± 0.3
	acetone	10.3 ± 0.4	4.9 ± 0.3	− 5.4 ± 0.3
$[Fe(pybimH)_3](BPh_4)_2$	CH_3OH/20%CH_3CN	4.3 ± 0.4	0.2 ± 0.3	− 4.1 ± 0.3
	CH_3CN	12.4 ± 0.5	5.9 ± 0.4	− 6.4 ± 0.3
	acetone	9.6 ± 0.4	4.7 ± 0.4	− 4.9 ± 0.4
$[Fe(ppa)_2](ClO_4)_2$	acetone	8.7 ± 0.5	2.6 ± 0.5	− 6.1 ± 0.5
$[Fe(tpmbn)](ClO_4)_2$	DMF		10.5[c]	− 5.6[c]
$[Fe(tppn)](ClO_4)_2$	DMF	16.1 ± 2.0	7 ± 3	− 9 ± 3
	CH_3CN	10.7 ± 1.0	5.1 ± 0.7	− 5.6 ± 0.7
$[Fe(tpchxn)](ClO_4)_2$	DMF	15.5 ± 2.0	21 ± 3	5.7 ± 2.0
	CH_3CN	11.5 ± 1.0	5.4 ± 1.2	− 6.1 ± 1.2
$[Fe(phenmethoxa)_2](BPh_4)_2$	acetone	12.3 ± 1.0	3.9 ± 1.0	− 8.4 ± 2.0

[a] Data for $[Fe(phenmethoxa)_2](BPh_4)_2$ according to Ref. [89], all other data according to Ref. [90]
[b] Ligand abbreviations: pyimH = 2-(2′-pyridyl)imidazole; pybimH = 2-(2′-pyridyl)-benzimidazole; ppa = N^2-(2-pyridylmethyl)picolineamidine; tpmbn = tetrakis-(2-pyridylmethyl)-*meso*-2,3-butanediamine; tppn = tetrakis(2-pyridylmethyl)-1-methyl-1,2-propanediamine; tpchxn = tetrakis(2-pyridylmethyl)-*trans*-1,2-cyclo-hexanediamine.
[c] Preliminary data, calculated by using the ΔV° for the tppn complex in DMF

bond lengthening effects of the pybimH ligand. As far as [Fe(phenmethoxa)$_2$] $(BPh_4)_2$ is concerned [89], the negative value of $\Delta V^{\ddagger}_{HL}$ indicates that the volume of the activated complex is again substantially smaller than that of the respective HS($^5T_2$) isomer. Here, phenmethoxa denotes the tridentate ligand 3-[2-(1,10-phenanthrolyl)]-5-methyl-1,2,4-oxadiazole.

The detailed mechanism of the spin-state conversion, in particular whether a twisting coordinate or a motion along a radial coordinate is involved, is still under discussion. Theoretical considerations by Purcell [93], cf. Sect. 5, indicate that an energetically favorable pathway for the process may be provided by a pseudorotational mechanism via an intermediate trigonal-prismatic geometry which may even involve a spin triplet ground state. Inspection of Table 4 shows that the activation parameters for complex [Fe(tpchxn)]$(ClO_4)_2$ in DMF deviate significantly from the general trend. Thus $\Delta V^{\ddagger}_{HL}$ is the only positive value in the table, while $\Delta V^{\ddagger}_{LH}$ exhibits the largest positive value obtained with similar relations for $\Delta S^{\ddagger}_{HL}$ and $\Delta S^{\ddagger}_{LH}$. These results indicate that the HS $\rightarrow$ LS and LS $\rightarrow$ HS transformations both involve a significant volume expansion, and it has been suggested [90] that the energetically more favorable reaction path via a rhomboid twist may be blocked in this instance for sterical reasons. The inference is supported by the consideration of the related complexes [Fe(tpmbn)]$(ClO_4)_2$ and [Fe(tppn)]$(ClO_4)_2$ where the twisting motion is not impeded and where the activation parameters are in line with the values for other compounds. Consequently, the latter two complexes seem to represent the normal spin-state conversion mechanism which occurs along a coordinate combining angular and radial motions, whereas in [Fe(tpchxn)]$(ClO_4)_2$ the radial pathway is preferred since the full-twist motion is blocked.

4.2 Complexes of Iron(III)

The results of relaxation measurements on complexes of iron(III) as derived from ultrasonic and photoperturbation studies are listed in Table 5. Measurements have been also done by the laser Raman temperature-jump method [34, 80, 96, 97]. The results were found to be inconsistent with the more accurate ultrasonic experiments, however. In some cases, the expected absorbance changes in the electronic spectra were found to occur within the heating rise time of the laser pulse [30]. Therefore, the results of the laser Raman measurements have not been included in Table 5. Solvent dependence of activation parameters was observed for [Fe(Sal$_2$trien)]PF_6 using methanol and acetone [95]. The results suggest that the solvent plays a role in the interconversion dynamics, possibly due to spin-state related hydrogen bonding with coordinated N–H groups of the Sal$_2$trien ligand. The rate constants obtained vary between 6.1×10^7 and $3.2 \times 10^8\ s^{-1}$ but may not be representative. Apparently, the spin-state interconversion in iron(III) complexes is slightly more rapid than in complexes of iron(II).

Table 5. Rate constants and activation parameters for the LS → HS and HS → LS unimolecular spin-state interconversions in iron(III) complexes

Compound[a]	Solvent	τ ns	k_{LH} s^{-1}	k_{HL} s^{-1}	$\Delta H^{\ddagger b}$ kcal mol^{-1}	$\Delta S^{\ddagger b}$ cal K^{-1} mol^{-1}	Ref.
$[Fe(acac_2trien)]NO_3$	H_2O	2.11 ± 0.04	1.6×10^8	3.2×10^8	6.38 ± 0.17	0.34 ± 0.61	[94]
					2.40 ± 0.17	-11.59 ± 0.60	
$[Fe(Sal_2trien)]NO_3$	H_2O	5.33 ± 0.05	6.1×10^7	1.3×10^8	7.45 ± 0.29	2.1 ± 1.0	[94]
					2.34 ± 0.29	-13.6 ± 1.0	
$[Fe(Sal_2trien)]PF_6^c$	CH_3OH	5 ± 1			6.11 ± 0.21	-2.87 ± 1.0	[95]
					2.39 ± 0.21	-15.8 ± 1.0	
	acetone	3 ± 1			8.60 ± 0.17	6.68 ± 0.7	[95]
					4.00 ± 0.17	-9.79 ± 0.7	
$[Fe(5\text{-}OCH_3Sal_2trien)]PF_6$	CH_3OH/acetone	5 ± 1			6.21 ± 0.24	-1.67 ± 1	[95]
					1.91 ± 0.24	-18.6 ± 1	

[a] Ligand abbreviations: $acac_2trien$ = hexadentate Schiff base from triethylenetetramine (trien) and acetylacetone (acac); Sal_2trien = hexadentate Schiff base from triethylenetetramine (trien) and salicylaldehyde.
[b] First line lists values of $\Delta H^{\ddagger}_{LH}$ or $\Delta S^{\ddagger}_{LH}$, second line lists $\Delta H^{\ddagger}_{HL}$ or $\Delta S^{\ddagger}_{HL}$.
[c] Activation parameters are solvent dependent

The magnitude of the transmission coefficient κ for $[Fe(acac_2trien)]NO_3$ may be estimated in a manner similar to that for $Fe[HB(pz)_3]_2$. Here, $acac_2trien$ denotes the hexadentate Schiff-base ligand derived from triethylenetetramine (trien) and acetylacetone (acac). If the entire entropic barrier of 3.45 kcal mol^{-1} is assigned to κ, the minimum value of κ is obtained as $10^{-2.5}$. This is equivalent to assuming that the entropy of the transition state equals that of the HS state. In the actual complex, distortions of octahedral symmetry are expected to split the LS $^2T_2$ ground state according to $^2T_2 \rightarrow {}^2A_1 + {}^2E$. The splitting corresponds to introducing a degeneracy barrier into the spin-state interconversion process. The minimum probability of intersystem crossing, i.e. κ, then increases to $10^{-2.1}$ for a $^2A_1$ or $10^{-2.4}$ for a ^2E ground state. The detailed relations between the activation parameters are illustrated in Fig. 7. A similar situation is found for the complex $[Fe(Sal_2trien)]NO_3$ containing the hexadentate ligand Sal_2trien derived from triethylenetetramine (trien) and salicylaldehyde (Sal). Thus, if the entropic barrier of 4.05 kcal mol^{-1} is attributed completely to κ, the minimum value for κ results as $10^{-3.0}$. The introduction of a degeneracy barrier due to the splitting $^2T_2 \rightarrow {}^2A_1 + {}^2E$ increases the minimum intersystem crossing probability to $\kappa = 10^{-2.5}$ for a $^2A_1$ or $10^{-2.8}$ for a ^2E ground state. A graphical disposition of the activation parameters is shown in Fig. 8. These results indicate that the rate constants for intersystem crossing are determined by both electronic factors and the thermodynamic barrier associated with the structural reorganization.

Several observations indicate spin-state interconversion rates faster than could be measured with the applied technique. Thus an upper limit for the relaxation time of 30 ns has been set for $[Fe((CH_3)_2NC(S)S)_3]$ [34], whereas the estimate $\tau \leq 1$ ns has been obtained for the complex $[Fe(benzac_2trien)]PF_6$

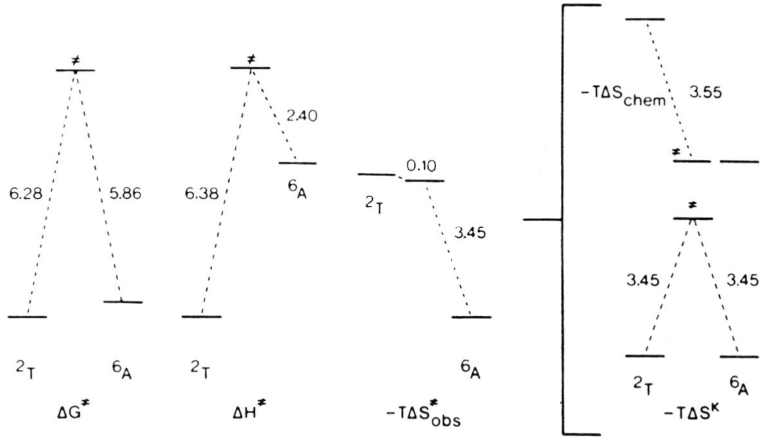

Fig. 7. Relation of activation parameters for $[Fe(acac_2trien)]NO_3$. The quantity $-T\Delta S^{\ddagger}_{obsd}$ is partitioned to give the minimum value of κ. According to Ref. [94]

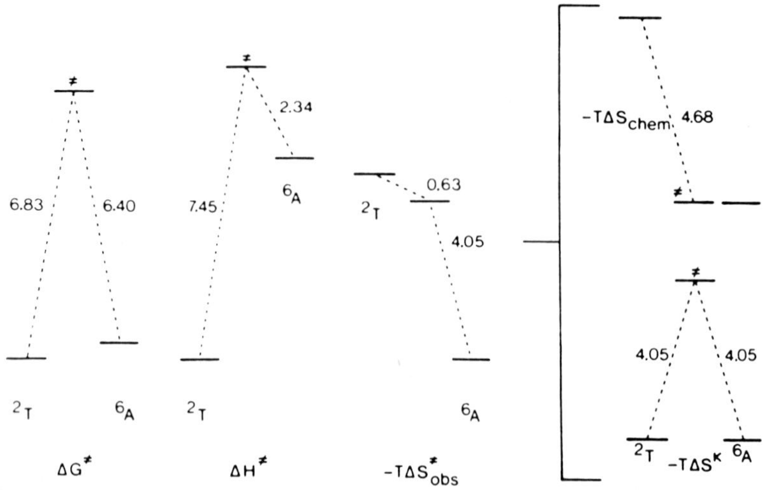

Fig. 8. Relation of activation parameters for [Fe(Sal_2trien)]NO_3. The quantity $-T\Delta S^{\ddagger}_{obsd}$ is partitioned to give the minimum value of κ. According to Ref. [94]

[98] where the hexadentate Schiff-base ligand derives from triethylenetetramine (trien) and benzoylacetone (benzac).

4.3 Complexes of Cobalt(II)

A very rapid relaxation is suggested for the spin-state equilibrium in [$Co(terpy)_2$]X_2 complexes where terpy = 2,2′,2″-terpyridine. Measurements by the laser Raman temperature-jump method on CH_3OH solutions of the complex yield a relaxation time $\tau \leq 15$ ns [34]. In ultrasonic measurements, no excess sound absorption has been observed, thus providing an upper limit of 0.2 ns for the relaxation time in water at 298 K [28]. A spin-state change faster than observed for iron(II) and iron(III) complexes is indeed expected. This is due to the fact that in cobalt(II) compounds only one electron is involved in the spin-state change since $\Delta S = 1$. Moreover, the ground states ^2E and $^4T_1$ can be directly mixed by spin-orbit coupling, the spin-state transition between the two states assuming adiabatic character and thus $\kappa = 1$.

An unusually slow relaxation has been observed for the 2,6-pyridinedicarboxaldimine cobalt(II) complex [Co(2,6-$(CH_3NH{=}CH)_2$py$)_2$]$(PF_6)_2$ in solution. Thus a relaxation time $\tau = 83$ ns has been reported [99], the rate constants being among the lowest found. It has been suggested that non-electronic factors such as partial ligand dissociation, steric effects or solvent interaction may be rate determining in this equilibrium.

5 Consequences with Respect to the Energetic and Molecular Properties of the Activated Complex

The change of spin state in a $3d$ transition metal complex in solution is accompanied by a sizeable reorganization of both the inner and outer coordination sphere around the metal ion. This reorganizational process is indeed expected due to a larger occupancy of the antibonding e_g^* orbitals which is accompanying the transition from the LS to the HS isomer. Increased population of the e_g^* levels is associated with an increase of the vibrational partition function as well as an increase in the molecular volume of the HS complex relative to that of the LS complex. The change of molecular volume ΔV has been verified for a number of spin transition compounds on the basis of multiple temperature single crystal X-ray diffraction [6]. The same information may be obtained from the pronounced temperature variation of the unit cell volume which is derived from the lattice constants and is usually more easily accessible [6, 70, 100]. Evidence for changes of the vibrational partition function has also been reported [101, 102]. The changes require that the nuclear configuration of the activated complex is situated somewhere between the equilibrium configurations of the two different spin states. Furthermore, these effects contribute to the intrinsic activation barrier for the transition. Since the interaction energy at the crossing point of the potential energies of the spin states is expected to be small, the rate constants for the spin conversion may involve an additional non-adiabatic factor.

It has been outlined above that, in the reorganization of the coordination sphere, the variation of the Me–L, i.e., metal-donor atom, distance is the most significant. The change often exceeds 0.10 Å and, for spin transition compounds of iron(II) e.g., a Me–L bond change by ~ 0.18 Å is frequently encountered [6]. In specific cases such as for $[Fe(dppen)_2Cl_2] \cdot 2(CH_3)_2CO$ where dppen = *cis*-1,2-bis(diphenylphosphino)ethylene, $Ph_2PCH{=}CHPPh_2$, changes as large as 0.28 Å have been reported [103]. It is evident that, due to greater occupancy of the formally antibonding e_g^* orbitals, the HS form invariably possesses longer Me–L bonds than the LS form. The occupancy of the nonbonding t_{2g} orbital set is of less significance in these structural considerations.

The difference in bond length between the HS and LS form of a metal complex may be calculated by expanding the potential energy difference of the two forms as a function of the totally symmetric metal-ligand stretching mode [104]. Assuming that the energy separation Δ between the d orbitals involved in the electron transfer depends inversely on some power n of the bond distance [13], such as $\Delta \simeq \frac{5}{3} Ze^2 \langle r^{n-1} \rangle / R^n$, the change in bond length ΔR is related to the force constant f of the symmetric stretch by:

$$\Delta R \simeq \frac{nm\Delta}{fR_0 N} \tag{60}$$

Here, m is the number of electrons involved in the change, R_0 the initial bond distance, and N the number of ligand atoms bonded to the metal. Covalent as well as electrostatic theories suggest that $n \simeq 5$. Note that for both Δ and f, the values appropriate for the initial state of bond length R_0 should be employed.

As briefly stated in the introduction, we may consider one-dimensional cross sections through the zero-order potential energy surfaces for the two spin states, cf. Fig. 9, in order to illustrate the spin interconversion process and the accompanying modification of molecular structure. The potential energy of the complex in the particular spin state is thus plotted as a function of the vibrational coordinate that is most active in the process, i.e., the metal-ligand bond distance, R. These potential curves may be taken to represent a suitable cross section of the metal 3N-6 dimensional potential energy hypersurface of the molecule. Each potential curve has a minimum corresponding to the stable

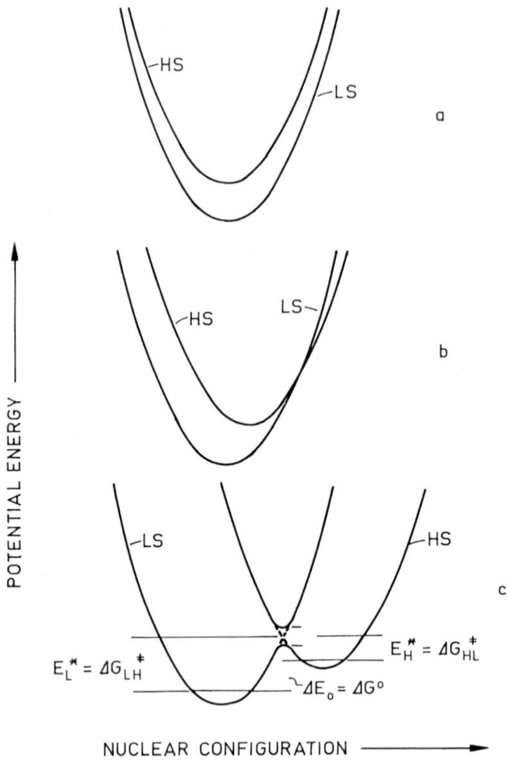

Fig. 9a–c. Relative positions of LS and HS potential energy surfaces for complexes showing spin-state equilibria associated with different amounts of geometric reorganization: **a** no intersection of potential surfaces; **b** intersection accompanied by moderate displacement of the minima of potential surfaces; **c** (avoided) intersection accompanied by sizeable displacement of the minima of potential surfaces. $\Delta E_0 = \Delta G^\circ$ is the difference of zero-point energies of LS and HS states, $E_L^* = \Delta G_{LH}^{\ddagger}$ and $E_H^* = \Delta G_{HL}^{\ddagger}$ are the energies of the intersection (transition state) relative to the zero-point energy of the LS and HS state, respectively

nuclear configuration of that state, i.e., the appropriate equilibrium Me–L distance, R_0. The potential curves will intersect in regions where the LS and HS states have the same nuclear configurations and energies, and the conversion occurs in this intersection region, the so-called activated complex.

Provided the potential curves are harmonic, relatively simple relationships can be set up. Thus the curves may be described by:

$$E_{LS} = \tfrac{1}{2} f_{LS} R^2$$

$$E_{HS} = \tfrac{1}{2} f_{HS} (R - \Delta R_0)^2 + E_0 \quad (61)$$

Here, f_{LS} and f_{HS} are the quadratic force constants which determine the curvature at the minima of the two curves and which are related to the corresponding vibrational frequency ν resulting from IR or Raman spectra by:

$$\nu = (1/2\pi)\sqrt{f/\mu} \quad (62)$$

where $\mu = m_1 m_2/(m_1 + m_2)$ is the reduced mass. In particular:

$$f_{LS}/f_{HS} = \nu_{LS}^2/\nu_{HS}^2 \quad (63)$$

The quantity R in Eq. (61) is the coordinate along the metal-ligand bond direction, $\Delta R_0 = R_H - R_L$ and $\Delta E_0 = 2\delta - (ZPE)_L + (ZPE)_H$ being the displacements of the configuration and zero-point energy, respectively, of the HS state minimum relative to the LS state minimum. In the present case, ΔR_0 is identical to the difference of metal-ligand bond distances of the two spin states in the solid complex and this value may be obtained from X-ray structural data. It has been outlined in the introduction that, for an assembly of molecules, it is the Gibbs free energy rather than the potential energy that should be considered in order to describe correctly spin-state equilibria. Thus we may identify ΔE_0 with ΔG°, the latter quantity being obtained on the basis of equilibrium studies. The energy at the intersection of the potential curves E_L^* relative to the zero-point energy of the LS state, i.e., $E_L^* = B - (ZPE)_L$, may be identified with $\Delta G_{LH}^{\ddagger}$, the energy of the activated complex determined from kinetic studies, and similarly the energy at the intersection E_H^* relative to the zero-point energy of the HS state, $E_H^* = B - (ZPE)_L - \Delta G_{HL}^\circ$, may be identified with $\Delta G_{HL}^{\ddagger}$. Notice that $\Delta G_{LH}^{\ddagger}$ and $\Delta G_{HL}^{\ddagger}$ are the barrier heights between the intersection point and the zero-point energy of the LS and HS state, respectively. The coordinate at the intersection R^* readily follows from Eq. (61) as:

$$R^* = \Delta R_0 \Bigg/ \left[1 + \frac{\nu_{LS}}{\nu_{HS}}\left(1 - \frac{\Delta E_0}{E^*}\right)^{1/2}\right] \quad (64)$$

Let us consider the possible relations of LS and HS potential energy surfaces as shown schematically in Fig. 9. As long as the zero-order or diabatic surfaces are considered, the electrons remain localized on the particular spin state, no electron transfer being possible. In order that a conversion between the LS and HS state takes place, electronic coupling of the states is required. This coupling effectively removes the degeneracy at the intersection of the zero-order surfaces and two new surfaces, the first-order or adiabatic states of the system, are formed, cf. Fig. 9c. The coupling of the zero-order states of energies H_{LL} and H_{HH} may be represented by the secular equation:

$$\begin{vmatrix} H_{LL} - E & H_{LH} \\ H_{HL} & H_{HH} - E \end{vmatrix} = 0 \tag{65}$$

where $H_{LH} = \langle \psi_L | H | \psi_H \rangle$ is the electronic coupling matrix element, H the Hamiltonian of the system, and the solutions $E_\pm$ are the energies of the resulting adiabatic states. At the point of intersection where $H_{LL} = H_{HH}$, these energies are:

$$E_+ = H_{LL} - |H_{LH}|$$

$$E_- = H_{HH} + |H_{LH}| \tag{66}$$

the separation of the adiabatic states at the intersection being equal to $2H_{LH} = J$. The magnitude of H_{LH} thus evidently determines the behavior of the system in the intersection region. If the splitting is small, there is a high probability that the system will reach the upper potential energy surface (*nonadiabatic process*). On the other hand, if the splitting at the intersection is large, the system will predominantly remain on the lower surface (*adiabatic process*). The probability of electron transfer in the intersection region may be expressed in terms of the transmission coefficient κ discussed in Sect. 4.

In the context of spin-state transitions, three situations characterized by increasing amounts of geometrical reorganization may be distinguished [34]. In Fig. 9a, the potential energy surfaces describing the LS and HS states do not intersect. Thus, a possible spin state conversion may be treated as a "normal" electronic transition. Figure 9c depicts the situation where the two potential surfaces do intersect and a splitting into two adiabatic surfaces is observed. Consequently, the spin state change can be described in terms of the motion of the system on a single (i.e. the lower) adiabatic potential energy surface. In Fig. 9b, an intermediate situation is displayed. It should be noted that the position of the crossing point is determined by the relative displacements of the minima on the two surfaces, i.e., by the amount of geometric reorganization. In particular, the two spin states in Fig. 9a show very similar bond distances and angles and therefore the minima occur at the same nuclear configuration. The spin states depicted in Fig. 9c have relatively different structures and thus the minima of their potential energy surfaces occur at different nuclear configurations.

In the following, we will employ the free energies of activation $\Delta G^\ddagger_{HL}$ and $\Delta G^\ddagger_{LH}$ resulting from relaxation kinetics, the free energies ΔG° obtained on the basis of equilibrium studies, and the molecular parameters ΔR and f or, instead of the latter, the underlying vibrational frequencies ν, to set up approximate (one-dimensional) potential energy surfaces for a few representative spin transition complexes. These pictorial descriptions may serve to illustrate in a very simplified way the energetic and configurational conditions which have to be met in order that a spin conversion can take place.

$Fe[HB(pz)_3]_2$. The energy barrier relative to the zero-point energies of the LS and HS state has been determined on the basis of relaxation measurements [29] as $\Delta G^{\ddagger}_{LH} = 8.32\ \text{kcal mol}^{-1} = 2910\ \text{cm}^{-1}$ and $\Delta G^{\ddagger}_{HL} = 7.36\ \text{kcal mol}^{-1} = 2574\ \text{cm}^{-1}$, respectively. The difference of zero-point energies of the two states results as $\Delta G^{\circ} = 0.96\ \text{kcal mol}^{-1} = 336\ \text{cm}^{-1}$ on the basis of equilibrium studies. The complex is at room temperature completely in the LS state. The average Fe–N bond length $\bar{R}_L = 1.973$ Å has been obtained from the single crystal X-ray structure study [105]. The Fe–N bond length in the HS state may be taken from the closely related HS complex $Fe[HB(3,5\text{-}(CH_3)_2pz)_3]_2$ as $\bar{R}_H = 2.172$ Å [105]. These values provide a reliable estimate for the difference of bond distances as $\Delta R = 0.199 \simeq 0.20$ Å. Two Fe–N stretching frequencies have been assigned to each of the complexes from FIR spectra employing the $^{54}Fe/^{57}Fe$ metal isotope technique [106]. For the LS $Fe[HB(pz)_3]_2$ complex, these frequencies are 434.5 and 399.5 cm^{-1} (average 417 cm^{-1}), whereas for the HS complex $Fe[HB(3,5\text{-}(CH_3)_2pz)_3]_2$, the frequencies 206.1 and 254.9 cm^{-1} (average 230.5 cm^{-1}) were obtained. The schematic potential energy diagram of Fig. 10 was constructed for ambient temperature on the basis of these parameter values. Consistent with the LS character of the solid compound, the minimum of the LS potential curve is at

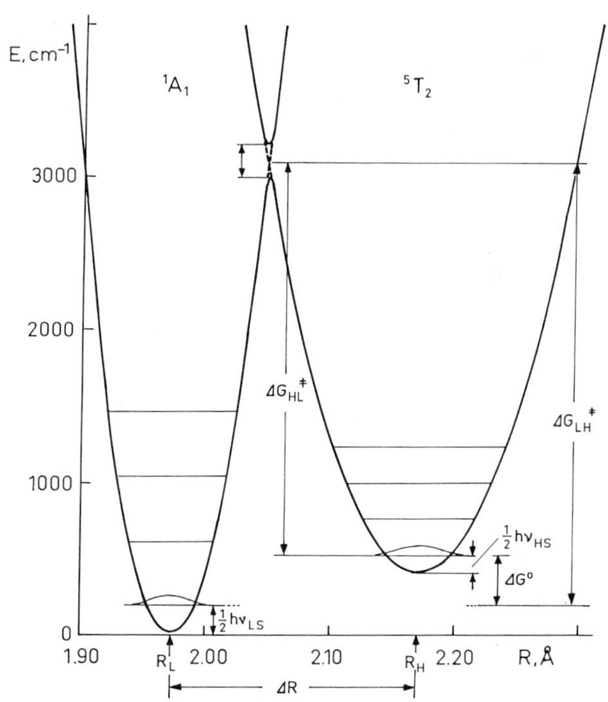

Fig. 10. Schematic potential energy diagram for the spin-state equilibrium $^5T_2 \rightleftharpoons {}^1A_1$ in the iron(II) complex $Fe[HB(pz)_3]_2$ at room temperature. The curves were calculated for $\Delta G^{\ddagger}_{LH} = 2910$, $\Delta G^{\circ} = 336$, $\tilde{\nu}_{LS} = 417$, $\tilde{\nu}_{HS} = 230.5\ \text{cm}^{-1}$, and $\Delta R = 0.20$ Å. The first vibrational levels are indicated

lower energy than that of the HS curve. As expected, the smaller width of the LS curve is an indication of the larger force constant and thus the stronger bonding in the LS $^1A_1$ state relative to the HS $^5T_2$ state. The ratio of the force constants follows readily from the known vibrational frequencies using Eq. (63) as $f_{LS}/f_{HS} = 3.27$. Individual force constants may be determined according to:

$$f_i = 4\pi^2 c^2 \mu \tilde{\nu}_i^2 \qquad (67)$$

where c is the velocity of light, μ the reduced mass of the vibrating ligand, and where the frequency $\tilde{\nu}_i$ is measured in cm^{-1}. Definition of the appropriate reduced mass for the tridentate organic N-donor ligands is difficult. We use in the following as representative for the ligand the mass of the atom directly bound to the metal, thus obtaining $\mu = 11.24$ amu for Fe^{II}–N bonds. Equation (67) then yields for the Fe–N stretching vibration the force constants:

$$f_{LS} = 1.15 \times 10^5 \text{ dyn cm}^{-1}$$

$$f_{HS} = 0.354 \times 10^5 \text{ dyn cm}^{-1}$$

[Fe(paptH)$_2$]Cl$_2\cdot$2H$_2$O. The energy barrier has been determined from relaxation measurements [29] as $\Delta G^{\ddagger}_{LH} = 7.60$ kcal mol^{-1} = 2660 cm^{-1} and $\Delta G^{\ddagger}_{HL} = 8.11$ kcal mol^{-1} = 2840 cm^{-1} relative to the zero-point energy of the LS and HS state, respectively. The difference of zero-point energies of the two states is negative, viz. $\Delta G^\circ = -0.51$ kcal mol^{-1} = -180 cm^{-1} in agreement with the HS character of the compound at ambient temperature. The average Fe–N bond length in the HS state has been assumed as equal to that in [Fe(paptH)$_2$](BF$_4$)$_2\cdot$3H$_2$O of $\bar{R} = 2.165$ Å. The room temperature X-ray structure of the latter compound has been recently reported [107]. The change of bond length for [Fe^{II}–N$_6$] type complexes on spin-state transition is often close to $\Delta R \simeq 0.20$ Å which value has been taken to apply for [Fe(paptH)$_2$]Cl$_2$ as well. Fe–N stretching frequencies have been assigned [108, 109] by the ^{54}Fe/^{57}Fe metal isotope technique for a number of spin transition complexes listed in Table 6. For the present purpose, we use the average values of $\tilde{\nu}_{LS}$ = 422 cm^{-1} and $\tilde{\nu}_{HS}$ = 231 cm^{-1}. The potential energy curves displayed in Fig. 11 have been obtained on the basis of the above parameter values. It is obvious that the resulting diagram is very similar to that of Fig. 10. Since the vibrational frequencies for the Fe–N stretch are almost the same, the force constants given above apply equally well to [Fe(paptH)$_2$]Cl$_2\cdot$2H$_2$O.

[Fe(acac$_2$trien)]NO$_3$. For the iron(III) complex of the hexadentate Schiff-base ligand acac$_2$trien, the barrier heights have been determined from ultrasonic relaxation [94] as $\Delta G^{\ddagger}_{LH} = 6.28$ kcal mol^{-1} = 2196 cm^{-1} and $\Delta G^{\ddagger}_{HL} = 5.85$ kcal mol^{-1} = 2046 cm^{-1}. The difference of zero-point energies has been obtained from equilibrium studies as $\Delta G^\circ = 0.43$ kcal mol^{-1} = 150 cm^{-1}. An X-ray structure determination has been performed for [Fe(acac$_2$trien)]PF$_6$ at room temperature where the [Fe^{III}–N$_4$O$_2$]-type complex is completely in the HS state [110]. The average Fe-ligand distances $\langle$Fe–N$\rangle_{HS}$ = 2.136 Å and $\langle$Fe–O$\rangle_{HS}$ = 1.930 Å are assumed to apply equally to

Table 6. Fe-Ligand stretching frequencies from FIR spectra of iron(II) and iron(III) spin transition complexes

Compound[a]	$\tilde{\nu}_{LS}$, cm^{-1}	$\tilde{\nu}_{HS}$, cm^{-1}	Ref.
$[Fe(phen)_2(NCS)_2]$	530; 375[b,c]	252.0; 220.0[b]	[108, 109]
$[Fe(phen)_2(NCSe)_2]$	529; 363[b,c]	228.0, 218[b]	[108, 109]
$[Fe(bpy)_2(NCS)_2]$	495; 384[b,c]	253.0; 235[b]	[108, 109]
average	422	231	
$Fe[HB(pz)_3]_2$	434.5; 399.5	257.5; 223	[111, 106]
$Fe[HB(3,5\text{-}(CH_3)_2pz)_3]_2$		206.1; 254.9	[106]
$[Fe(ptz)_6](BF_4)_2$	413	227	[112, 113]
$[Fe((CH_3)_2NC(S)S)_3]$	338.0	235.0	[114]
$[Fe((C_2H_5)_2NC(S)S)_3]$	336.7	236.4	[114]

[a] Ligand abbreviations: phen = 1,10-phenanthroline; bpy = 2,2′-bipyridine; $HB(pz)_3$ = hydrotrispyrazolylborate; $HB(3,5\text{-}(CH_3)_2pz)_3$ = 3,5-dimethyl-hydrotrispyrazolylborate; ptz = 1-propyltetrazole.
[b] First frequency listed corresponds to Fe–N(NCS) or Fe–N(NCSe) stretching vibration; second frequency has been assigned to Fe–N(phen) or Fe–N(bpy) stretching vibration.
[c] Values listed are in each case the average of two close lying frequencies, viz. 530 cm^{-1} for $[Fe(phen)_2(NCS)_2]$ is the average of 532.6 and 528.5 cm^{-1}, and similarly 375 cm^{-1} is the average of 379.0 and 371.0 cm^{-1}

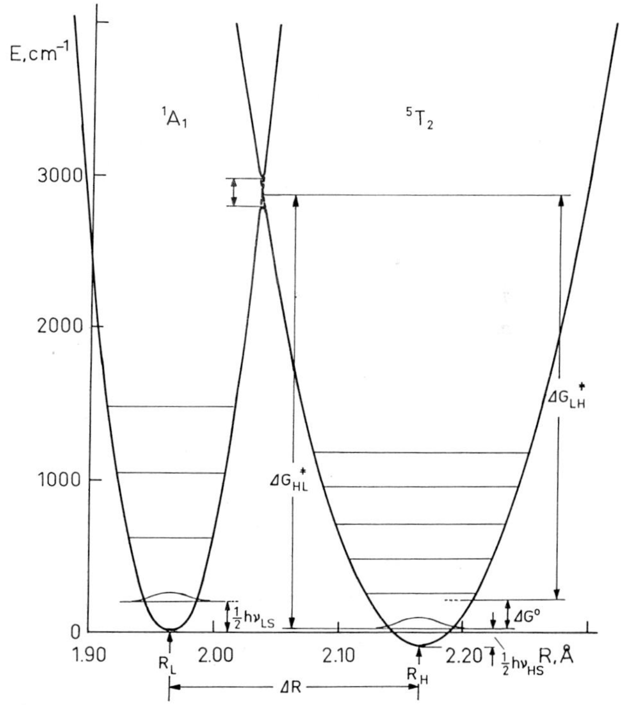

Fig. 11. Schematic potential energy diagram for the spin-state equilibrium $^5T_2 \rightleftharpoons {}^1A_1$ in the iron(II) complex $[Fe(paptH)_2]Cl_2 \cdot 2H_2O$ at room temperature. The curves were calculated for $\Delta G^{\ddagger}_{LH} = 2660$, $\Delta G^{\circ} = -180$, $\tilde{\nu}_{LS} = 422$, $\tilde{\nu}_{HS} = 231$ cm^{-1}, and $\Delta R \simeq 0.20$ Å. The first vibrational levels are indicated

the HS isomer of the present complex. The X-ray structure of the related complex [Fe(Sal$_2$trien)]Cl·2H$_2$O may be considered as example for the LS state, the average bond distances having been obtained as $\langle$Fe–N$\rangle_{LS}$ = 1.968 Å and $\langle$Fe–O$\rangle_{LS}$ = 1.884 Å [109]. This gives the differences of bond lengths as $\Delta R_{Fe-N} = 0.168$ Å and $\Delta R_{Fe-O} = 0.046$ Å. If the spin conversion is considered on the basis of simplified HS and LS potential curves, the displacement of the curve minima is determined by the mean value $\overline{\Delta R} = 0.127$ Å $\simeq 0.13$ Å. The Fe^{III}–N stretching frequencies are expected to be higher than corresponding frequencies for iron(II). In complete absence of such data, there is little other choice than to employ the average of the frequencies of Table 6. In this case, the potential curves shown in Fig. 12 are obtained. Consistent with the LS behavior of the compound, the minimum of the LS potential curve is at lower energy than that of the HS curve. For the iron(III) complex [Fe(Sal$_2$trien)]NO$_3$·H$_2$O, rather similar values for the energy barrier and the difference of zero-point energies have been determined [94]. All other data being equal, the potential energy diagram will closely resemble that of Fig. 12.

In a discussion of the mechanism of spin-state interconversion on the basis of the simplified potential energy curves presented above, considerable attention

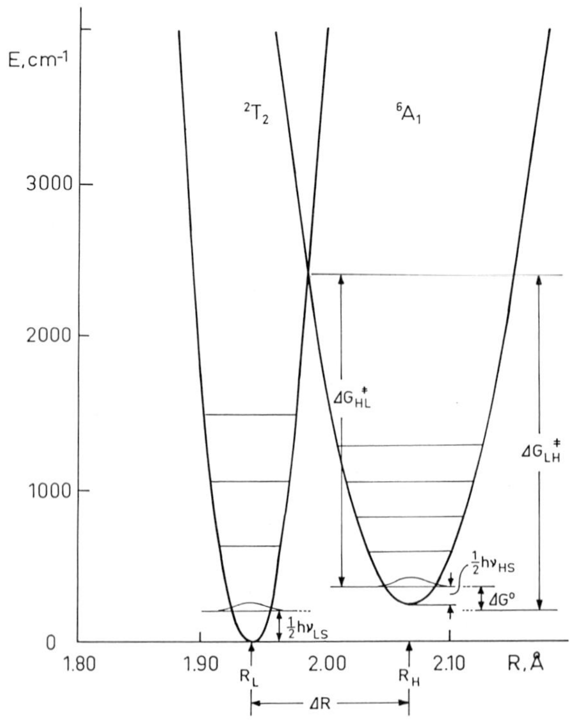

Fig. 12. Schematic potential energy diagram for the spin-state equilibrium $^6A_1 \rightleftharpoons {}^2T_2$ in the iron(III) complex [Fe(acac$_2$trien)]NO$_3$ at room temperature. The curves were calculated for $\Delta G^{\ddagger}_{LH} = 2196$, $\Delta G^{\circ} = 150$, $\tilde{\nu}_{LS} = 422$, $\tilde{\nu}_{HS} = 231$ cm^{-1}, and $\Delta R = 0.13$ Å. The first vibrational levels are indicated

should be given to κ, the transmission coefficient. Thus, in terms of absolute reaction rate theory [85], the rate constant for the spin-state conversion is determined by Eq. (49). The quantity κ in this equation reflects the probability of the system remaining on the lower potential energy surface, i.e., the probability for adiabatic behavior. An estimate for the transmission coefficient may be obtained from the semiclassical Landau-Zener formula [115]:

$$\kappa = P_{ad} = 1 - \exp\left[-\frac{4\pi^2}{hv|\Delta s|}\varepsilon^2\right]$$
$$\simeq \frac{4\pi^2}{hv|\Delta s|}\varepsilon^2 \tag{68}$$

In Eq. (68), v is the velocity with which the atom or submolecular fragment moves through the intersection region and is given by:

$$v = \left[\frac{2(E - V_c)}{\mu}\right]^{1/2} \tag{69}$$

Here, E is the total energy of the molecule along the reaction coordinate, V_c the potential energy at the intersection, and μ the reduced mass of the moving fragment. Moreover, Δs in Eq. (68) is the difference in slope of the potential surfaces at the intersection and ε is the interaction energy between the two states. A value $\kappa \ll 1$ implies that the system is highly nonadiabatic with the reactant infrequently crossing onto the potential energy surface of the product, i.e., instead of remaining on the lower potential curve of Fig. 9c, the reactant follows the dotted line onto the upper energy surface. If the energy separation between the two surfaces increases, the spin interconversion process becomes more adiabatic, κ becomes larger and the system remains on the lower surface more frequently. With the assumption of $v = 3 \times 10^4$ cm s^{-1}, $|\Delta s| = 100 \times 10^8$ kcal mol^{-1} cm^{-1} [116] and $\varepsilon \simeq 10$ cm^{-1}, the transmission coefficient is estimated as $\kappa \simeq 10^{-3}$ which is of the same order of magnitude as the values derived from the kinetics measurements.

The assessment of κ is of some importance since it relates to the question as to how much if any of the free energy of activation barrier is due to the spin-forbidden character of the transition. From the experimental point of view, Eq. (49) shows that the transmission coefficient κ and the activation entropy $\Delta S^‡$ appear in the temperature-independent part of the rate constant and thus cannot be separated without additional assumptions. Possible approaches to the partition of $-T\Delta S^‡_{obsd}$ have been discussed in Sect. 4 for spin transition complexes of iron(II) and iron(III). If the assumption is made that the entropy of activation is completely due to κ, minimum values between $10^{-2.5}$ and 10^{-5} are obtained for iron(II) and values between 10^{-2} and 10^{-3} for iron(III). There is an increase of entropy for the transition LS → HS and thus the above assumption implies that the transition state resembles the HS state. On the other hand, volumes of activation indicate that the transition state should be about midway between the LS and HS state. This appears indeed more reasonable and has the consequence that the observed activation entropy will contain contributions from increased solvation and decreased vibrational partition function for the process of HS state compression to the transition state. The minimum value of κ thus increases and should not be lower than about 10^{-2}. The conclusion from the analysis of the transmission coefficient κ is therefore that these spin-state equilibria are only weakly spin forbidden. For the $\Delta S = 2$ equilibria, in particular, the activation free energy is dominated by the inner-sphere reorganization energy which derives from the appreciable changes in metal-ligand bond length associated with the transfer of the two antibonding e_g^* electrons.

It is thus evident that the experimental results considered in sect. 4 above are fully consistent with the interpretation based on absolute reaction rate theory. Alternatively, consistency is equally well established with the quantum mechanical treatment of Buhks et al. [117] which will be considered in Sect. 6. This treatment considers the spin-state conversion in terms of a radiationless non-adiabatic multiphonon process. Both approaches imply that the predominant geometric changes associated with the spin-state conversion involve a radial compression of the metal-ligand bonds (for the HS $\rightarrow$ LS transformation).

An alternative mechanism that has been suggested [93, 118] for the intersystem crossing process is based on a twist movement of the octahedral ligand arrangement. Two modes designated by M'_3 and M''_3 and illustrated in Fig. 13

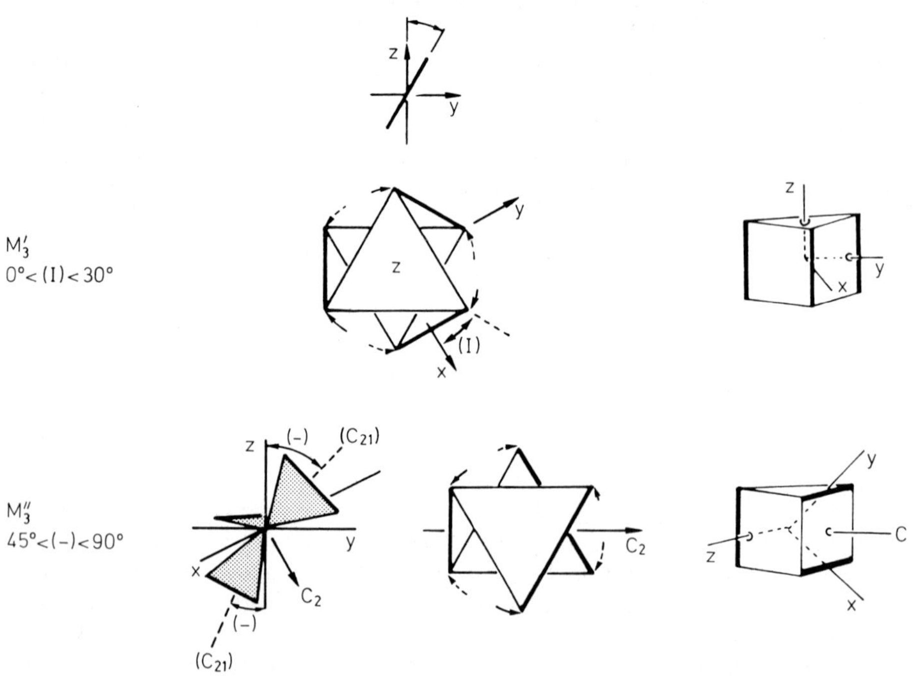

Fig. 13. Transition-state structures for pseudorotation by the trigonal twist (M'_3) and rhombic twist (M''_3) mechanism. According to Ref. [93]

have in fact been studied [93]. The M'_3 mode characterized by the angle ϕ (*trigonal twist*) corresponds to a pathway leading to an antiprismatic arrangement of ligands [119], the symmetry being D_3 at all points along the reaction coordinate. The M''_3 mode characterized in terms of the coordinate θ of the intraligand C_2 axis (*rhombic twist*) [120] breaks down the initial D_3 symmetry to C_2 along the reaction coordinate. The transition state structure is D_{3h} and C_{2v} for the M'_3 and M''_3 twist, respectively. Most of the spin transition complexes would not be capable of following fully a M'_3 or M''_3 twist movement, the reason being ligand rigidity or interligand repulsion. It has been pointed out [93], however, that for the very small LS–HS energy separation found in these complexes, even the execution of a small fraction of a pseudo-rotation might effect spin-state crossing.

A differentiation between the two mechanisms seems to be possible if the activation volumes determined from the pressure dependence of spin relaxation rates are considered. Results are available for a significant number of spin transition complexes, cf. Table 4. In general, the activation volumes $V^{\ddagger}_{LH}$ for these complexes are positive, whereas the activation volumes $V^{\ddagger}_{HL}$ for the reverse process are negative, thus indicating that the volume of each activated complex is substantially smaller than the volume of the corresponding HS isomer. It follows that the transition state geometry for these iron(II) complexes should be situated somewhere between the geometries of the $^1A_1$ and $^5T_2$ isomer. On the other hand, the mechanism based on a twist of the ligand field would be expected to show positive activation volumes for both the LS → HS and HS → LS transformation. This expectation is a result of the displacement of solvent molecules from the cavities between the chelate rings [121, 122]. Thus based on the available evidence, a pure twisting mechanism is inconsistent with results for activation volumes. However, see also discussion in Sect. 4.

6 Description in Terms of a Radiationless Nonadiabatic Multiphonon Process

An alternative theoretical approach to spin conversion processes in solutions of transition metal complexes is the description in terms of a radiationless nonadiabatic multiphonon process. The spin conversion is assumed to take place between two distinct zero-order spin states characterized by slightly different nuclear equilibrium configurations. The two configurations are separated by an energy barrier which is large relative to the thermal energy $k_B T$, the configurational difference being manifested by an elongated metal-ligand bond in the HS state, as compared to the LS state. Within this conceptual framework, a quantum mechanical calculation of spin conversion rates may be accomplished [117].

6.1 Outline of the Method

The theoretical method employed is based on and largely similar to the theory of electron-transfer reactions in solution [123, 124, 125]. Thus the intramolecular spin conversion may be described as a transition between an initial manifold of states $\psi_i(\vec{\mathbf{r}}, q_c)\chi_{\kappa_i}(q_c)\phi_{n_i}(q_s)$ with energies $E_i + E^c_{\kappa_i} + E^s_{n_i}$ and a final manifold of states $\psi_f(\vec{\mathbf{r}}, q_c)\chi_{\kappa_f}(q_c - \Delta_c)\phi_{n_f}(q_s - \Delta_s)$ with energies $E_f + E^c_{\kappa_f} + E^s_{n_f}$. Here, ψ_i and ψ_f denote the electronic wave functions in the initial and final states, respectively, $\bar{\mathbf{r}}$ the electronic coordinates, and q_c and q_s the nuclear coordinates with respect to internal modes and solvent modes, respectively. Further, χ_κ and ϕ_n are the harmonic oscillator nuclear states of the internal modes and of the solvent. Due to configurational changes and different interactions with the solvent, the oscillators describing the harmonic nuclear potential surfaces are displaced in the final state by Δ_c and Δ_s relative to the initial state, respectively. The product wavefunctions $\psi_i(\vec{\mathbf{r}}, q_c)\chi_{\kappa_i}(q_c)$ and $\psi_f(\vec{\mathbf{r}}, q_c)\chi_{\kappa_f}(q_c - \Delta_c)$ thus represent the pure spin molecular Born-Oppenheimer states. Finally, E_i and E_f are the electronic energies in the initial and final states, respectively, whereas $(E^c_{\kappa_i} + E^s_{n_i})$ and $(E^c_{\kappa_f} + E^s_{n_f})$ denote the energies of the vibrational levels in both states. Here, E^c corresponds to the separate contribution from the internal modes and E^s to the contribution from the solvent.

It should be noted that, due to the effect of spin-orbit interaction $\mathbf{H}_{so}$, the correct initial and final states are not exactly the pure spin states. The admixture with higher electronic states ψ_m may be ignored only if there exists a direct coupling between the initial and final pure spin states. Otherwise, the wave function for the initial state is obtained to first order of perturbation theory as:

$$\psi_i' = \psi_i + \sum \frac{\langle \psi_i | \mathbf{H}_{\text{so}} | \psi_m \rangle}{(E_i - E_m)_{q_i}} \psi_m \tag{70}$$

where the denominator is evaluated at the minimum energy configuration of the initial state. A similar expression determines the wave function for the final state.

The microscopic rate constant is derived from the quantum mechanical transition probability by considering the system to be initially present in one of the vibronic levels on the initial potential surface. The initial level is coupled by spin-orbit interaction to the manifold of vibronic levels belonging to the final potential surface. The microscopic rate constant is then obtained, following the Fermi-Golden rule, as:

$$W = \frac{2\pi}{\hbar} g_f |V|^2 \sum_{\kappa_f n_f} (\chi_{\kappa_i} | \chi_{\kappa_f})^2 (\phi_{n_i} | \phi_{n_f})^2 \times \delta(\Delta E + E^c_{\kappa_f} + E^s_{n_f} - E^c_{\kappa_i} - E^s_{n_i}) \tag{71}$$

Here, V is the electronic coupling matrix element:

$$V = \langle \psi_i' | \mathbf{H}_{so} | \psi_f' \rangle$$
$$= \langle \psi_i | \mathbf{H}_{so} | \psi_f \rangle + \sum_m \langle \psi_i | \mathbf{H}_{so} | \psi_m \rangle \langle \psi_m | \mathbf{H}_{so} | \psi_f \rangle$$
$$\times [(E_i - E_m)_{q_i}^{-1} + (E_f - E_m)_{q_f}^{-1}] \tag{72}$$

g_f the degeneracy of thc final electronic state, $\Delta E = E_f - E_i$ the electronic energy gap and $\delta(x)$ the Dirac delta function:

$$\delta(x) = \frac{1}{2\pi} \int_{-\infty}^{+\infty} e^{ikx}\, dk = \begin{cases} 0 & \text{if } x \neq 0 \\ \infty & \text{if } x = 0 \end{cases} \tag{73}$$

Assuming that the spin conversion is a nonadiabatic process, the macroscopic rate constant may be expressed, following Levich [125], in terms of the thermally averaged transition probability, the averaging being extended over the initial vibronic levels, as:

$$k = (Z_i^s Z_i^c)^{-1} \sum_{\kappa_i n_i} \exp[-(E_{n_i}^s + E_{\kappa_i}^c)/k_B T]\, W \tag{74}$$

In Eq. (74), $Z_i^s = \Sigma_{n_i} \exp(-E_{n_i}^s/k_B T)$ and $Z_i^c = \Sigma_{\kappa_i} \exp(-E_{\kappa_i}^c/k_B T)$ are the partition functions for the nuclear motion of the solvent and for the internal motion of the complex, respectively. The rate constant may be conveniently expressed as:

$$k = (2\pi/\hbar) g_f |V|^2 G \tag{75}$$

where V is the electronic coupling term Eq. (72) and where the contribution of the inner vibrations and the solvent modes is collected in the thermally averaged nuclear Franck-Condon vibrational overlap factor G defined as:

$$G = (Z_i^s Z_i^c)^{-1} \sum_{\kappa_i n_i \kappa_f n_f} \exp[-(E_{n_i}^s + E_{\kappa_i}^c)/k_B T]$$
$$\times |(\chi_{\kappa_i}|\chi_{\kappa_f})(\phi_{n_i}|\phi_{n_f})|^2\, \delta(\Delta E + E_{n_f}^s + E_{\kappa_f}^c - E_{n_i}^s - E_{\kappa_i}^c) \tag{76}$$

In order to simplify the expression for G, one has to employ a sufficiently simple model for the vibrational modes of the system. In the present case, the solvent contribution to the rate constant is expressed by a single parameter E_s, the solvent reorganization energy. In addition, frequency changes between the initial and final states are neglected and it is assumed that only a single internal mode with frequency ω and with the displacement Δr is contributing to G. Thus the expression for G reduces to [124]:

$$G(\Delta E, \omega, S, E_s, T) = \frac{1}{\hbar\omega}\left(\frac{x}{2\pi q}\right)^{1/2} \sum_{m=-\infty}^{+\infty} I_m\left(\frac{S}{\sin hx}\right)$$
$$\times \exp[-S \cot hx - mx - (p + q - m)^2 x/2q] \tag{77}$$

In Eq. (77), $x = \hbar\omega/2k_BT$ is the reduced internal frequency, $q = E_s/\hbar\omega$ the reduced solvent reorganization energy, $p = \Delta E/\hbar\omega$ the reduced electronic energy gap and $I_m(z)$ the modified Bessel function of order m. The quantity S is a coupling parameter which defines the contribution of the change in the internal normal mode:

$$S = m\omega(\Delta r)^2/2\hbar \tag{78}$$

In the conversion case, the solvent reorganization energy is very small and thus the one-mode expression [124] for the vibrational overlap factor G is generally adequate such that:

$$G = \frac{1}{\hbar\omega}\exp[-S\coth hx - px]\, I_p\left(\frac{S}{\sin hx}\right) \tag{79}$$

Figure 14 shows that almost the same numerical values for G are obtained from Eq. (79) as from Eq. (77) with $q = 0.3$.

A certain ambiguity arises in the proper choice of the thermodynamic parameter p, since entropy changes due to solvent orientation are neglected. The available experimental data (cf. Sect. 4) indicate, however, that the free energy of reaction for systems showing a spin change is close to zero. The numerical analysis has been therefore performed for the specific case $p = 0$, for which value the rate constant in Fig. 15 has been computed as a function of S and $\hbar\omega/k_BT$.

Following Eq. (75), the rate constant for spin conversion may be expressed as a product of the electronic matrix element V and the nuclear Franck-Condon

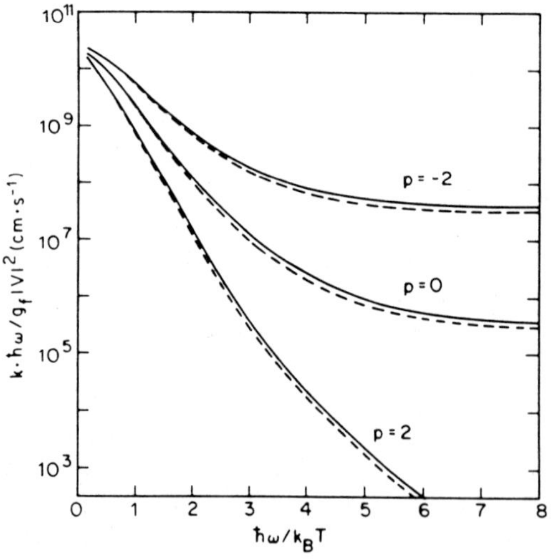

Fig. 14. Dependence of the quantity $(\hbar\omega/g_f|V|^2)k$ where k is the rate constant for spin conversion on $\hbar\omega/k_BT$ as calculated from the single-mode approximation Eq. (79) (*solid lines*) and the full expression Eq. (77) for the value $q = 0.3$ of the solvent reorganization parameter (*dashed lines*). Data are given for $p = -2.0, 0, +2.0$ and for the value of the coupling parameter S = 15. According to Ref. [117]

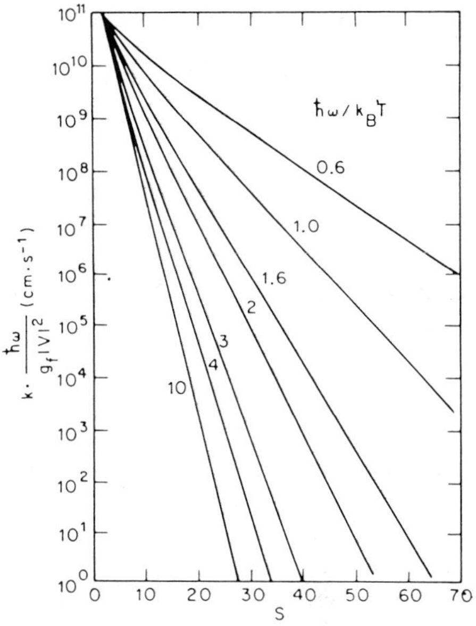

Fig. 15. Dependence of the quantity $(\hbar\omega/g_f |V|^2)k$, where k is the rate constant for spin conversion on the coupling parameter S. Data are given for various values of $\hbar\omega/k_B T$. According to Ref. [117]

factor G. These quantities separate the electronic and nuclear contributions to the process and will now be considered separately in detail.

6.2 Calculation of Electronic Matrix Elements

The calculations are based on the assumption of octahedral symmetry of the complex. Essentially two types of spin conversion may be distinguished: (i) in the configurations d^4 and d^7, a single electron is transferred between the t_{2g} and e_g orbitals thus producing a first term in Eq. (72) which is different from zero; (ii) in the configurations d^5 and d^6, two electrons are involved in the transition. Here, it turns out that the first term in Eq. (72) vanishes, whereas the second term is different from zero.

d^6 Ions Fe(II), Co(III). The matrix element of spin-orbit interaction between the LS ground term $^1A_{1g}(t_{2g}^6)$ and the HS ground term $^5T_{2g}(t_{2g}^4 e_g^2)$ is zero. The only term which has nonvanishing matrix elements with both $^1A_{1g}$ and $^5T_{2g}$ terms is the $^3T_{1g}$ arising from the electron configuration $t_{2g}^5 e_g$. If the spin-orbit coupling constant is denoted with ζ, the matrix elements are [126]:

$$\langle {}^1A_{1g} | \mathbf{H}_{so} | {}^3T_{1g} \rangle = -\sqrt{6}\,\zeta$$

$$\langle {}^5T_{2g} | \mathbf{H}_{so} | {}^3T_{1g} \rangle = \sqrt{3}\,\zeta \qquad (80)$$

Employing Eq. (3) one then obtains:

$$V = -3\sqrt{2}\zeta^2\left(\frac{1}{\Delta E_1} + \frac{1}{\Delta E_2}\right) \tag{81}$$

In Eq. (81), ΔE_1 is the energy difference between the $^3T_{1g}$ and $^1A_{1g}$ states at the equilibrium configuration of $^1A_{1g}$, while ΔE_2 is the energy difference between $^3T_{1g}$ and $^5T_{2g}$ at the equilibrium configuration of $^5T_{2g}$. Since we have assumed, in the present case, $p = 0$, the energies of $^1A_{1g}$ and $^5T_{2g}$ are approximately the same and thus $\Delta E_1 \simeq \Delta E_2$. An estimate of this energy difference may be derived on the basis of the d^6 energy diagram of Tanabe and Sugano [9] calculated with the assumption of $C/B = 4.81$. From the diagram, $\Delta E = 7.6B$ is obtained at the crossover of ground terms. The values of the Racah parameters B and C are decreased in the complex as compared to the free ion values [127, 128] as is the spin-orbit coupling parameter ζ. The decrease is due to covalency effects and is related to coefficient a in the molecular orbital:

$$\psi_{\mathrm{MO}} = a\psi_{\mathrm{metal}} + b\psi_{\mathrm{ligand}} \tag{82}$$

Since B and C are proportional to a^4, while ζ is proportional to a^2 [13], the covalency effects in Eq. (81) are expected to cancel each other. The use of the free ion values for B and ζ in estimating the numerical value of the matrix element V is thus justified. The free ion values of $B = 1058\ \mathrm{cm}^{-1}$ [126] and $\zeta = 400\ \mathrm{cm}^{-1}$ [129] then give $|V| = 170\ \mathrm{cm}^{-1}$ for Fe(II), whereas $B = 1065\ \mathrm{cm}^{-1}$ [9] and $\zeta = 580\ \mathrm{cm}^{-1}$ [129] produce $|V| = 380\ \mathrm{cm}^{-1}$ for Co(III).

$\mathbf{d^5}$ Ions Fe(III), Mn(II). The only terms which have nonvanishing matrix elements with both ground terms $^2T_{2g}$ (t_{2g}^5) and $^6A_{1g}$ $(t_{2g}^3e_g^2)$ are the spin-orbit components $^4T_{1g}$ (Γ_7) and $^4T_{1g\,5/2}(\Gamma_8)$ of the intermediate configuration term $^4T_{1g}\,(t_{2g}^4e_g)$ [126]:

$$\begin{aligned}
\langle ^6A_{1g}|\mathbf{H}_{\mathrm{so}}|^4T_{1g}(\Gamma_7)\rangle &= -\sqrt{2}\,\zeta \\
\langle ^2T_{2g}|\mathbf{H}_{\mathrm{so}}|^4T_{1g}(\Gamma_7)\rangle &= \sqrt{3}\,\zeta \\
\langle ^6A_{1g}|\mathbf{H}_{\mathrm{so}}|^4T_{1g\,5/2}(\Gamma_8)\rangle &= -\sqrt{2}\,\zeta \\
\langle ^2T_{2g}|\mathbf{H}_{\mathrm{so}}|^4T_{1g\,5/2}(\Gamma_8)\rangle &= -\sqrt{6/5}\,\zeta
\end{aligned} \tag{83}$$

If the energy difference within the spin-orbit multiplet of the $^4T_{1g}$ is neglected and ΔE used to designate the energy relative to the other two terms, Eq. (72) gives:

$$V = 1.8\zeta^2/\Delta E \tag{84}$$

From the Tanabe-Sugano diagram calculated for $C/B = 4.48$ [9], the average energy difference at the crossover follows as $\Delta E = 8.4B$. Employing the free ion values $B = 1015\ \mathrm{cm}^{-1}$ [9] and $\zeta = 460\ \mathrm{cm}^{-1}$ [129] for Fe(III), one obtains for the matrix element $|V| = 44\ \mathrm{cm}^{-1}$.

d^7 Ion Co(II). The HS ground term $^4T_{1g}(t_{2g}^5e_g^2)$ is split by spin-orbit interaction into the levels $_{3/2}\Gamma_8$ and $_{5/2}\Gamma_8$ with energies $-\zeta/3$ and $\zeta/2$, respectively [126]. The matrix elements of $\mathbf{H}_{so}$ with the $^2E_g(t_{2g}^6e_g)$ LS ground term are $\zeta/\sqrt{5}$ and $3\zeta/\sqrt{5}$. However, the ^2E_g ground state which is not split by spin-orbit coupling is expected to undergo a large distortion due to the Jahn-Teller effect. Thus the state is split into two and the value of the electronic coupling is difficult to estimate except to say that it has to be of the order of ζ. The free ion value for Co(II) being 515 cm^{-1} [129], we thus may take $|V| \simeq 500$ cm^{-1}.

6.3 *Calculation of Rate Constants*

The rate constants for the processes considered in this section are determined by Eq. (75) and Eq. (79). However, all the complexes for which measurements of rates are available deviate significantly from strict octahedral symmetry. In order to determine the coupling parameter S, the contributions of all modes which are characterized by a displacement $\Delta r \neq 0$ have to be summed up:

$$S = \sum_{\kappa} m_\kappa \omega_\kappa (\Delta r_\kappa)^2/2\hbar \tag{85}$$

It will be assumed that the dominant contribution to S is from the metal-ligand bonds of symmetry A_{1g}. The numerical value of S is then evaluated by using the frequencies of these bands, employing the average value where the vibrational frequencies ω_κ for the HS and LS states are known separately. This is in keeping with the approximations introduced in the derivation of Eq. (77) above. In addition, the mass m_κ associated with these modes is approximated by the mass of the atom which is directly bound to the metal. The change of metal-ligand bond length Δr_κ is estimated, where not directly available, from structural data on similar complexes. The resulting value of S is utilized, in conjunction with the value of $x = \hbar\omega/k_B T$, to obtain the quantity $(\hbar\omega/|V|^2 g_f)k$ on the basis of the relationships illustrated in Fig. 15. The rate constant k then follows by employing the results of the previous section for the electronic matrix element V and the electronic degeneracy g_f of the final state.

Iron(III) Complexes. The measured spin conversion rates are $k_{LH} = 1.6 \times 10^8$ s^{-1} and $k_{HL} = 3.2 \times 10^8$ s^{-1} for [Fe(acac$_2$trien)]NO$_3$ and $k_{LH} = 6.1 \times 10^7$ s^{-1} and $k_{HL} = 1.3 \times 10^8$ s^{-1} for [Fe(Sal$_2$trien)]NO$_3 \cdot 1.5$H$_2$O both in aqueous solution at 25°C [94]. In order to obtain estimates for the values of the rate constants on the basis of the multiphonon theory of radiationless transitions, values of the parameters m_κ, ω_κ and Δr_κ are required. The measured complexes are of the type [Fe^{III}–N$_4$O$_2$]. Estimates for the changes of bond lengths upon spin conversion may be obtained from the comparison of HS [Fe(acac$_2$trien)] complexes with LS [Fe(Sal$_2$trien)] complexes [110] as 0.17 Å and 0.04 Å for the Fe^{III}–N and Fe^{III}–O bonds, respectively. The Fe-donor atom stretching frequency is taken as $\hbar\omega = 500$ cm^{-1} for both bonds [130]. Inserting the values of m_κ, $\hbar\omega$, and Δr_κ into Eq. (78) gives $S = 13$, and using Fig. 15 for $\hbar\omega/k_B T \sim 2.5$ one obtains the result as $(\hbar\omega/|V|^2 g_f)k \sim 10^8$ cm s^{-1}. In complexes of Fe(III), the degeneracy is $g_f = 6$ for both HS and LS states, and taking $|V| = 44$ cm^{-1} from the results of the previous section, one arrives at $\hbar\omega/|V|^2 g_f \simeq 0.04$ cm and finally the rate constant follows as $k \simeq 2 \times 10^9$ s^{-1}.

For [Fe(dtc)$_3$] no excess ultrasonic absorption has been observed [29] thus indicating a rate constant greater than 10^9 s^{-1}. The metal-ligand stretching frequency has been assigned as $\hbar\omega_{\text{HS}} = 236.4$ cm^{-1} for the HS state and $\hbar\omega_{\text{LS}} = 336.7$ cm^{-1} for the LS state [114] with an average value of 286.5 cm^{-1}. Employing the change in Fe–S bond length of 0.116 Å [6, 131], one obtains $S = 11$, $\hbar\omega/k_B T \sim 1.4$ and thus the rate constant is estimated as $k \simeq 5 \times 10^{10}$ s^{-1}.

Iron(II) Complexes. Spin conversion rates have been measured as $k_{\text{LH}} = 5 \times 10^6$ s^{-1} and $k_{\text{HL}} = 2.5 \times 10^7$ s^{-1} for Fe[HB(pz)$_3$]$_2$ in THF and $k_{\text{LH}} = 1.7 \times 10^7$ s^{-1} and $k_{\text{HL}} = 7.2 \times 10^6$ s^{-1} for [Fe(paptH)$_2$]Cl$_2$ in water [29]. The measured complexes are of the type [Fe^{II}–N_6] and thus $m = 14$. The Fe–N stretching frequencies are available for the HS Fe[HB(3,5-(CH$_3$)$_2$pz)$_3$]$_2$ complex as $\hbar\omega = 206.1$ and 254.9 cm^{-1}, whereas for the LS Fe[HB(pz)$_3$]$_2$ complex $\hbar\omega = 434.5$ and 399.5 cm^{-1} [106]. For the present purpose, we take the average value $\hbar\omega = 324$ cm^{-1}. An estimate for Δr may be obtained from the difference in Fe–N bond lengths of these two complexes, i.e. 2.172 Å for the HS Fe[HB(3,5-(CH$_3$)$_2$pz)$_3$]$_2$ and 1.973 Å for the LS Fe[HB(pz)$_3$]$_2$, as $\Delta r \simeq 0.20$ Å [105]. This value is also in reasonable agreement with recent results of multiple temperature X-ray structure studies [6]. The values of m, $\hbar\omega$, and Δr when inserted into Eq. (78) produce $S = 16$, and employing Fig. 15 for $\hbar\omega/k_B T \sim 1.6$, one obtains $(\hbar\omega/|V|^2 g_f)k \sim 1 \times 10^8$ cm s^{-1}. Taking now $|V| = 170$ cm^{-1} from the results of the previous section, it follows that $k/g_f = 1 \times 10^{10}$. The complexes for which spin conversion rates are available show considerable distortions from octahedral symmetry. Thus the $^5T_{2g}$ ground term will be split into at least two states, leaving an orbital singlet with a spin degeneracy $g_f = 5$ lowest. We then arrive at the estimate of $k \sim 4 \times 10^{10}$ s^{-1} which is somewhat higher than the experimental values quoted above. The assumption of $\Delta r = 0.26$ Å on which the original estimate of $k \sim 1 \times 10^9$ s^{-1} was based [117] is not supported by the more recent results of X-ray structure studies.

Cobalt(II) Complexes. Spin conversion rates greater than 10^9 s^{-1} have been estimated for [Co(terpy)$_2$](PF$_6$)$_2$ [29]. Although $\Delta r \simeq 0.04$ Å is available for the [Co(terpy)$_2$] complex [132] there is a large uncertainty concerning the magnitude of electronic coupling. Therefore, no attempt has been made to calculate the rate constant.

In the following, a brief account of the approximations inherent in the quantum mechanical description of rate constants will be given. As far as applicability of the nonadiabatic formalism is concerned, two criteria have been devised [117] which are well obeyed. Equations (71) and (76) are quite general, although drastic assumptions regarding nuclear motion are involved in the derivation of Eqs. (77) and (79) from the above. Thus, the number of harmonic oscillators is strictly limited and configurational changes are described by displacements of the potential surface minima, while frequency changes of the initial and final states are neglected. The calculation of the electronic coupling term V is based on an octahedral ligand field, although much lower molecular symmetries are established or expected for the compounds in question. The numerical calculation of the nuclear term G is fraught with difficulties as not all input data required are known with certainty. In the calculation of the coupling term S, only the metal-ligand vibrational modes have been considered. It is clear that several other modes of the chelate ring may contribute to S. Moreover, the atomic masses m of the ligand atoms directly bound to the metal have been used. Since these atoms are located in a rigid aromatic ring, a higher effective mass should be expected. In this case, the value of S will become larger than that used in the calculations and, consequently, the rate constant smaller. It is evident that these modifications of the procedure would considerably improve the agreement between calculated and measured values of the rate constants.

7 Relaxation of Thermally Quenched or Optically Excited High-Spin States Resulting from the Spin-State Transition in Solid Metal Complexes

7.1 Survey of Available Results

Recently, the HS → LS relaxation kinetics for a number of spin-state transformations in solid metal complexes has been studied [57, 59, 133]. In these instances, rapid cooling of the system from a higher, usually ambient, temperature to liquid nitrogen or lower temperature results in an effectively trapped HS species. At some higher temperature, thermal relaxation of the trapped HS state to the stable LS state sets in, and this process may be followed by conventional measurements such as the time-dependence of the Mössbauer effect or that of the magnetic susceptibility. In general, the temperature range available for these measurements is rather small, typically between 15 and 20 K, the relaxation outside this region being either too slow (no time dependence observed) or too rapid to allow reasonable measurement with the method chosen.

Alternatively, optical excitation of the LS state in a solid metal complex which is involved in a thermally driven spin-state transition may result in the formation of a trapped HS state, the study of the kinetics of the HS → LS relaxation being then possible [134]. This process is initiated, e.g., in an iron(II) complex, by irradiation into the spin-allowed $^1A_1 \rightarrow {}^1T_1$ absorption band at temperatures well below the thermal transition temperature T_c. The proposed mechanism for the subsequent conversion to the HS state involves intersystem crossing from the excited $^1T_1$ state to the intermediate $^3T_1$ state, followed by decay to the $^5T_2$ state where a nonequilibrium population is accumulated. The effect has been termed "light-induced excited spin-state trapping" (LIESST). The mechanism and other competitive processes are indicated in Fig. 16. Apparently, the radiative and nonradiative processes in the excited states are more rapid than the relaxation time of the spin-state equilibrium. The quantum efficiency of LIESST is considerable, estimates being $\Phi_L = k_{eff}/k_1 \simeq 0.50$ and $\Phi_R \simeq 0.20$ [135, 135a]. Here k_{eff} is the effective rate constant describing the exponential buildup of the HS state or the concomitant depletion of the LS state and k_1 the rate of excitation. At sufficiently low temperature, the nonequilibrium population of the HS state can be trapped for extended periods of time. The process is reversible in that the $^5T_2$ state may be converted back to the $^1A_1$ state by irradiation into the $^5T_2 \rightarrow {}^5E$ absorption band. Thus Φ_R is the efficiency for the reverse LIESST. Eventually, the relaxation of the trapped $^5T_2$ state to the stable $^1A_1$ state may be studied at elevated temperatures by measuring the time dependence of the optical density at a fixed wavelength. Again, only a limited range of temperatures is available for these studies.

Under favorable conditions, rates of intersystem crossing between HS and LS states for solid complexes may be obtained directly, if a suitable spectroscopic technique is applied. Conditions and details of this method will be discussed in Sect. 8.

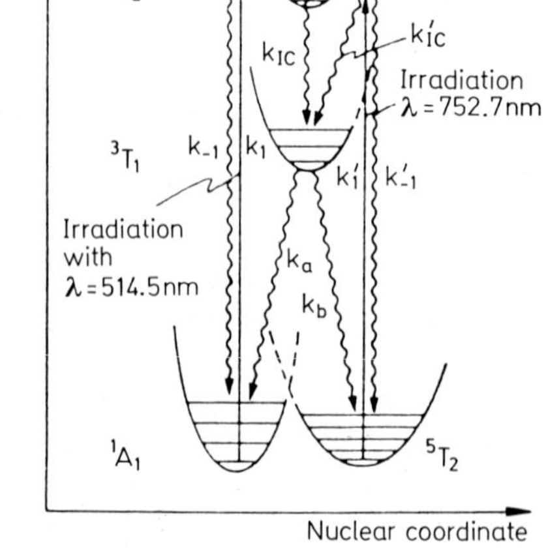

Fig. 16. Schematic potential energy diagram of the low-lying ligand field states for d^6 spin-state transition compounds. *Full lines* indicate the mechanism of forward and reverse LIESST effect. According to Ref. [135]

7.2 *Studies on Individual Systems*

In what follows, the studies on specific spin transition complexes by the methods discussed in Sect. 7.1 above, will be considered in some detail.

$[Fe(paptH)_2](NO_3)_2$. The iron(II) complex of the tridentate ligand paptH = 2-(2-pyridylamino)-4-(2-pyridyl)thiazole shows, in the solid state, a $^5T_2 \leftrightarrow {}^1A_1$ spin-state transition centred at about 246 K [57, 136]. The transition is associated with a pronounced hysteresis of $\Delta T_c = 34$ K ($T_c^{\uparrow} = 263$ K, $T_c^{\downarrow} = 229$ K) thus indicating that it is thermodynamically first-order despite its overall gradual appearance. Rapid cooling of the compound to 77 K results in the $^5T_2$ state remaining trapped due to the relatively slow rate of the HS$\leftrightarrow$LS transformation. At a somewhat elevated temperature, the metastable $^5T_2$ state relaxes back to the stable $^1A_1$ state. The kinetics of the HS ($^5T_2$) $\rightarrow$ LS($^1A_1$) transformation has been followed at 160, 170, and 180 K by measurements of the Mössbauer effect. The relaxation follows a first order rate law, the resulting rate constants and lifetimes being listed in Table 7. From the Arrhenius plot, the activation energy has been determined as $\Delta E = 7.5 \pm 1.8$ kcal mol^{-1}, the frequency factor as $A = 1.1 \times 10^6$ s^{-1} and the activation entropy at 160 K as $\Delta S^{\ddagger} = -21.6$ cal K^{-1} mol^{-1}.

$[Fe(bpp)_2](BF_4)_2$. The spin-state transition in the iron(II) complex derived from the tridentate ligand bpp = 2,6-bis(3-pyrazolyl)pyridine is of the discontinuous type [133] the observed hysteresis with transition temperatures of $T_c^{\uparrow} = 183$ K and $T_c^{\downarrow} = 173$ K indicating that the transition is first-order. If the compound is rapidly cooled to 77 K, the $^5T_2$ state remains trapped at the low temperature. At a somewhat higher temperature, the metastable $^5T_2$ state relaxes back to the stable $^1A_1$ state. The kinetics of the HS($^5T_2$) $\rightarrow$ LS($^1A_1$) relaxation has been followed by measurements of the magnetic susceptibility as a function of time at the temperatures of 99, 104, 109, and 114 K. The transformation follows first-order kinetics, the rate constants and lifetimes being listed in Table 7. From the temperature dependence of the rate constant, the activation energy has been determined as $\Delta E = 4.66$ kcal mol^{-1} and the frequency factor as $A = 2.3 \times 10^6$ s^{-1}.

$[FeL(CN)_2] \cdot H_2O$. The iron(II) complex of the macrocyclic Schiff-base ligand L derived from the condensation of 2,6-diacetylpyridine with 3,6-dioxaoctane-1,8-diamine shows as a solid a complicated behavior with temperature [59]. In the higher temperature region, a reversible transition between the pure HS $^5T_2$ state and a mixture of approximately equal proportions of HS and LS molecules is observed. The transition is associated with a hysteresis of $\Delta T_c = 15$ K width ($T_c^{\uparrow} = 222$ K, $T_c^{\downarrow} = 207$ K) and is thus thermodynamically first-order. Rapid cooling to 77 K results in the HS state remaining trapped, relaxation to the LS $^1A_1$ state setting in at temperatures above $\sim$110 K. The kinetics of the HS($^5T_2$) $\rightarrow$ LS($^1A_1$) relaxation has been studied between 115.7 and 130.0 K on the basis of Mössbauer effect measurements. The

Table 7. Relaxation times τ and corresponding rate constants k_{HL} for the HS → LS conversion in solid iron(II) complexes

Compound	T K	τ s	k_{HL} s^{-1}
$[Fe(paptH)_2](NO_3)_2$	160	14460	6.92×10^{-5}
	170	5590	1.79×10^{-4}
	180	1160	8.59×10^{-4}
	300[a]		$\sim 1 - 10$
$[Fe(bpp)_2](BF_4)_2$	99	9615	1.04×10^{-4}
	104	2300	4.35×10^{-4}
	109	885	1.13×10^{-3}
	114	418	2.39×10^{-3}
	300[a]		$\sim 10^3$
$[FeL(CN)_2]\cdot H_2O$	115.7	14771	6.77×10^{-5}
	120.6	4310	2.32×10^{-4}
	123.4	2392	4.18×10^{-4}
	126.2	1208	8.28×10^{-4}
	130.0	433	2.31×10^{-3}
	300[a]	–	$\sim 10^4$

[a] Extrapolated from low temperature values

transformation follows a rate law of second order, values of the rate constants being listed in Table 7. The activation energy has been found as $\Delta E = 7.3 \pm 0.4$ kcal mol^{-1}, the frequency factor as $A = 3.7 \times 10^9\ s^{-1}$ and the activation entropy at 115.7 K as $\Delta S^{\ddagger} = -4.8$ cal K^{-1} mol^{-1}.

The relaxation rate constants for the HS → LS conversion in the solid metal complexes $[Fe(paptH)_2](NO_3)_2$, $[Fe(bpp)_2](BF_4)_2$ and $[Fe\,L(CN)_2]\cdot H_2O$ show a remarkable similarity of the data, viz. Table 7. Thus all the measured k_{HL} values span the range 10^{-3} to $10^{-5}\ s^{-1}$. This is of course due to the temperature range which was chosen such that the kinetic measurements could be conveniently performed with the appropriate technique. At lower temperatures, the relaxation rate is in general so slow that a time dependence is not observed; whereas at higher temperatures the rate is too rapid to allow a measurement with acceptable accuracy. An extrapolation of the values of k_{HL} to ambient temperature promises little reliability due to the large range of extrapolation. The only conclusion to which some confidence may be attached is that the HS → LS relaxation would proceed in the solid complexes at 300 K with a rate constant k_{HL} between 1 and $10^4\ s^{-1}$. This is considerably less than the rates observed for the HS → LS conversion in solutions of metal complexes where rate constants in the range 10^6 to $10^8\ s^{-1}$ were obtained.

$[Fe(ptz)_6](BF_4)_2$ and $[Zn_{1-x}Fe_x(ptz)_6](BF_4)_2$. The solid iron(II) complex of the unidentate ligand ptz = 1-propyltetrazole shows a reasonably sharp spin-state transition at about 130 K [112]. Optical excitation into the spin-allowed $^1A_1 \rightarrow {}^1T_1$ absorption band produces via LIESST the HS $^5T_2$ state which remains trapped at temperatures below 50 K. Relaxation of the metastable $^5T_2$ state to the LS $^1A_1$ state has been studied by following the changes of optical excitation with time at 53, 57, 60, and 61 K [137]. The sigmoidal characteristics of the curves indicates a strong deviation from first-order kinetics. A similar study of the diluted complex $[Zn_{1-x}Fe_x(ptz)_6]\,(BF_4)_2$ shows that for $x \leq 0.1$ the relaxation of the light-induced $^5T_2$ state to the LS $^1A_1$ state may be described by a first-order rate law [134]. From the temperature dependence of the rate constants over the range 63 to 72 K, the activation energy $\Delta E = 2.32 \pm 0.09\ \mathrm{kcal\,mol^{-1}}$ and the frequency factor $A \simeq 10^5\ \mathrm{s^{-1}}$ have been obtained. In the general case, the activation energy is assumed to be a function of the LS fraction n_L and the mole fraction x of Fe(II) in the diluted complex:

$$\Delta E(n_L, x) = E_0 + n_L(\Delta E^\circ(x)) \tag{86}$$

A linear regression of $\Delta E(n_L, x)$ for $n_L < 0.8$ was performed for each value of x, the slope giving $\Delta E^\circ(x)$ which is found to be linear in x. From the linear dependence of $\Delta E^\circ(x)$ on x, the original relaxation curves can be retraced. Inserting Eq. (86) into the modified Arrhenius equation:

$$k = k(n_L, x) = A\exp[-\Delta E(n_L, x)/k_B T] \tag{87}$$

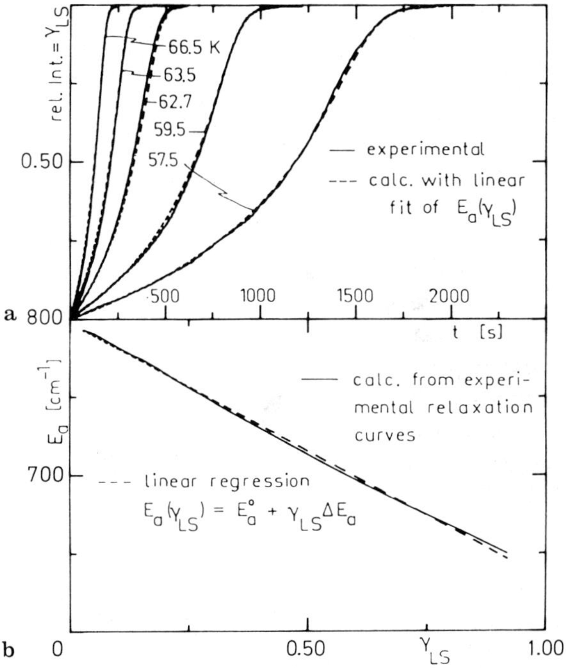

Fig. 17. High-spin → low-spin relaxation of $[Fe(ptz)_6](BF_4)_2$: **a** normalized relaxation curves at temperatures between 57.5 and 66.5 K; **b** activation energy ΔE as a function of n_L. The *solid line* represents a simultaneous fit of the five experimental curves displayed under **a**. The *broken line* is the result of a linear regression determined by $\Delta E^\circ = -164\ \mathrm{cm^{-1}} = -0.47\ \mathrm{kcal\,mol^{-1}}$ as slope and $\Delta E = 797\ \mathrm{cm^{-1}} = 2.28\ \mathrm{kcal\,mol^{-1}}$ as intercept. Notation in the figure uses γ_L, E_a, ΔE_a instead of n_L, ΔE, ΔE°, respectively. According to Ref. [134]

provides values of $k(n_L, x)$ which can be used to calculate curves of $n_L(t)$ by the numerical integration of the rate law. The relaxation curves for the pure $[Fe(ptz)_6](BF_4)_2$ displayed in Fig. 17 have been calculated using $\Delta E = 2.28\ \text{kcal mol}^{-1}$ and $\Delta E^\circ(x = 1) = -0.47\ \text{kcal mol}^{-1}$. The relaxation curves of Fig. 18 have been obtained for a series of mixed crystals, the reduced activation energy $\Delta E(n_L, x)/\Delta E(n_L \to 0, x)$ being shown as derived from experiment and as resulting from a linear regression. The dependence of ΔE on n_L and x postulated in Eq. (86) is a consequence of the change in volume $\Delta V = V_H - V_L$ accompanying the spin-state transition. The volume change exerts an internal pressure, called image pressure (cf. Eq. (23)) [47], to which, in the mixed crystal, a contribution due to the imbalance of V_H and V_{Zn} may be added thus giving:

$$\Delta p \simeq \Delta V n_L x + (V_H - V_{Zn})(1 - x) \tag{88}$$

The pressure Δp is not expected to affect the probability of intersystem crossing and thus the frequency factor A to any large extent. It is rather the activation energy ΔE that becomes a function of Δp and, since the second term in Eq. (88) is rather small, a function of n_L and x,

$$\Delta E = \Delta E(\Delta p) \simeq \Delta E(n_L, x) \tag{89}$$

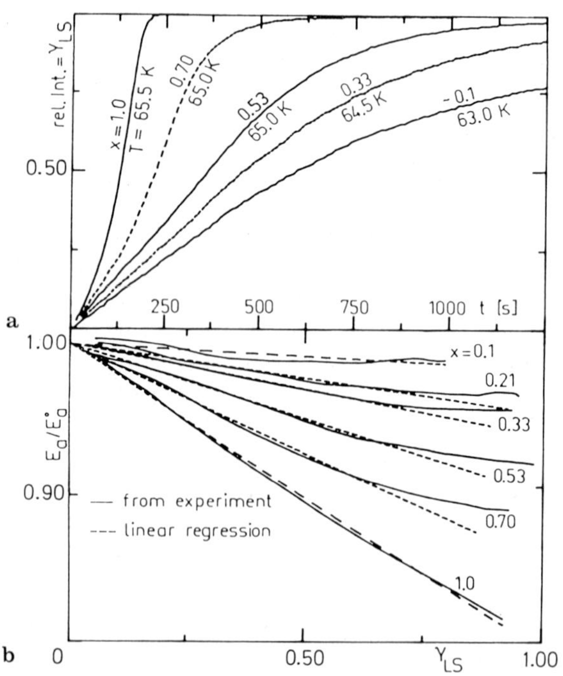

Fig. 18a, b. High-spin → low-spin relaxation of $[Zn_{1-x}Fe_x(ptz)_6](BF_4)_2$: **a** normalized relaxation curves for concentrations of Fe(II) between $x = 0.1$ and 1.0 at ≈ 65.0 K; **b** normalized activation energies $\Delta E(n_L, x)/\Delta E(n_L \to 0, x)$ for various values of x. Solid lines are derived from the experiment, broken lines are the result of a linear regression. Notation in the figure uses γ_L, E_0 instead of n_L, ΔE, respectively. According to Ref. [134]

It is concluded that the cooperative effect observed is of long-range nature and therefore of elastic rather than of electronic origin. Recently, the additional suggestion has been made [138] that, due to intermolecular interactions in the crystal environment of $[Fe(ptz)_6](BF_4)_2$, domains of iron(II) complexes interconvert together. The observed kinetics would then correspond to a first- or higher-order phase transition rather than to the kinetics which are characteristic for the conversion of isolated molecules.

8 Rates of Spin-State Transition for Solid Metal Complexes Based on the Line Shape Analysis of Mössbauer Spectra

8.1 Lifetime Limits and the Rate of Spin-State Conversion

The description of states participating in a spin-state transition as electronic isomers with discrete nuclear configurations, in particular different metal-ligand distances, requires that separate electronic and vibrational spectra of the two spin states exist. Indeed, a superposition of the individual vibrational spectra of the two states is in general observed, the relative contribution of the states being a function of temperature [41, 139, 140, 141, 142]. This observation sets a lower limit for the spin-state lifetime longer than the nuclear vibrational period, i.e., $\tau_L = 1/k_{LH} > 10^{-13}$ s. For molecules which are not EPR silent, this limiting value may be better defined. In the case where the EPR spectra of both spin-state isomers are observed, the lifetime is determined by $\tau_L = 1/k_{LH} > 10^{-10}$ s.

The simultaneous observation of the two EPR spectra has been reported in particular for several tris(dithiocarbamato)iron(III) complexes $[Fe(R_2NC(S)S)_3]$ where R = cyclohexyl [143], hydroxyethyl [144], and *n*-butyl [145, 146]. In addition, a considerable number of iron(III) complexes of the type $[Fe^{III}–N_4O_2]$ has been found which show EPR spectra of both the HS and LS isomers. These comprise $[Fe(X\text{-SalEen})_2]Y_2$ where X-SalEen is the Schiff-base ligand obtained by condensation of X-substituted salicylaldehyde and *N*-ethylethylenediamine [147] and similar compounds [100, 148, 149, 150, 151]. For the cobalt(II) complex $[Co(terpy)_2](ClO_4)_2$, it is not completely clear whether the two observed EPR spectra are due to HS and LS states related by a spin-state transformation [152].

On the other hand, NMR spectra appear in general as the average of the spectra of the two spin states [36, 153]. This observation determines an upper limit for the spin-state lifetime shorter than the nuclear spin relaxation time $\tau_L = 1/k_{LH} < 10^{-4}$ s. In general, therefore, either the superposition or the average of the particular spectroscopic properties of the two spin states is observed, subject to the relative magnitude of lifetime of the excited spectroscopic state and the rate of spin-state conversion. The rate k_{LH} is thus estimated to span the range:

$$10^4\ \mathrm{s}^{-1} < k_{LH} < 10^{10}\ \mathrm{s}^{-1}$$

A unique situation is encountered if ^{57}Fe-Mössbauer spectroscopy is applied for the study of spin-state transitions in iron complexes. The half-life of the excited state of the ^{57}Fe nucleus involved in the Mössbauer experiment is $t_{1/2} = 0.977 \times 10^{-7}$ s which is related to the decay constant k by $t_{1/2} = \ln 2/k$. The lifetime $\tau = 1/k$ is therefore $\tau_n^{Fe} = 1.410 \times 10^{-7}$ s which value is just at the centre of the range estimated for the spin-state lifetime $\tau_L = 1/k_{LH}$. Thus both the situations discussed above are expected to appear under suitable conditions in the Mössbauer spectra. The quantity of importance is here the nuclear Larmor precession frequency ω_n. If the spin-state lifetime $\tau_L = 1/k_{LH}$ is long relative to the nuclear precession time $1/\omega_n$, i.e. $\tau_L \gg 1/\omega_n$, individual and sharp resonance lines for the two spin states are observed. On the other hand, if the spin-state lifetime is short and thus $\tau_L \ll 1/\omega_n$, averaged spectra with intermediate values of quadrupole splitting ΔE_Q and isomer shift δ^{IS} are found. For the intermediate case where $\tau_L \simeq 1/\omega_n$, broadened and asymmetric resonance lines are obtained. These may be the subject of a lineshape analysis that will eventually produce values of rate constants for the dynamic spin-state interconversion process. The rate constants extracted from the spectra will be necessarily of the order of 10^5–10^7 s^{-1}.

Two different approaches have been followed to calculate the lineshapes within a relaxation model. According to a phenomenological approach based on the modified Bloch equations [154, 155], the intensity distribution of the theoretical Mössbauer spectrum may be written as [156]:

$$I(\omega) = \sum_{1,2} K'[(1 + \tau\Gamma)P_i + Q_i R_i/(P_i^2 + R_i^2)] \tag{90}$$

where the subscript i refers to an individual Mössbauer transition and:

$$P_i = \tau\{\Gamma^2 - [\tfrac{1}{2}(\omega_i^H + \omega_i^L) - \omega]^2 + \tfrac{1}{4}(\omega_i^H - \omega_i^L)^2\} + \Gamma$$
$$Q_i = \tau[\tfrac{1}{2}(\omega_i^H + \omega_i^L) - \omega - \tfrac{1}{2}(n_H - n_L)(\omega_i^H - \omega_i^L)]$$
$$R_i = [\tfrac{1}{2}(\omega_i^H + \omega_i^L) - \omega](1 + 2\tau\Gamma) + \tfrac{1}{2}(n_H - n_L)(\omega_i^H - \omega_i^L)] \tag{91}$$

Here, n_H and n_L are the populations of the HS and LS state, respectively, τ the relaxation time, $\tau = \tau_L\tau_H/(\tau_L + \tau_H)$ where τ_L and τ_H are the lifetimes of the individual spin states, Γ the linewidth, and K' the relative intensity of the i-th transition.

In the stochastic theory of lineshape developed by Blume [31], the spectral lines are calculated under the influence of a time-dependent Hamiltonian. The method has been successfully applied to a variety of dynamic effects in Mössbauer spectra. We consider here an adaptation due to Blume and Tjon [32, 33] for a Hamiltonian fluctuating between two states with axially symmetric electric field gradients (efg's), the orientation of which is parallel or perpendicular to each other. The present formulation is applicable for states with the same isomer shifts and the same linewidths such as is often found for iron(III), the relevant algebra being considered in the next section in some detail.

8.2 *The Relaxation Theory of Blume and Tjon*

The stochastic theory of lineshape has been developed by Anderson and Weiss [157], by Kubo [158], and by Kubo and Tomita [159] in order to treat the narrowing of spectral lines by exchange or motion, a generalized formulation having been subsequently presented by Blume [31]. We consider below an application of the theory of Blume to the specific problem of relaxation between LS and HS states in Mössbauer spectra of powder materials which is based on the formulation by Blume and Tjon [32, 33]. Accordingly, the probability of emission of a photon of wave vector $\mathbb{k}$ and frequency ω is given as [160]:

$$W_{fg}(\mathbb{k}) = \frac{|\langle g|\mathbf{H}^{(+)}|f\rangle|^2}{(\omega + E_g - E_f)^2 + \frac{1}{4}\Gamma^2} \tag{92}$$

Where $\mathbf{H}^{(+)}$ is the Hamiltonian for the interaction of the nucleus with the photon, $|f\rangle$ is the initial (excited) state of the nucleus, $|g\rangle$ the final (ground) state and Γ the inverse of the natural lifetime of state $|f\rangle$. We introduce an integral form of the denominator of Eq. (92) by:

$$\begin{aligned} 1/[(\omega + E_g - E_f)^2 + \tfrac{1}{4}\Gamma^2] &= \\ &= \frac{2}{\Gamma} Re \int_0^\infty e^{i(\omega + E_g - E_f)t - \frac{1}{2}\Gamma t}\, dt \end{aligned} \tag{93}$$

This relation can be easily verified by employing the Euler formula $e^{\pm i\theta} = \cos\theta \pm i\sin\theta$ for the pure imaginary part of the exponential function and by observing the definite integrals:

$$\int_0^\infty e^{-ax}\cos bx\, dx = \frac{a}{a^2 + b^2}$$

$$\int_0^\infty e^{-ax}\sin bx\, dx = \frac{b}{a^2 + b^2} \tag{94}$$

In order to introduce a time-dependence of the Hamiltonian in Eq. (92), use is made of the exponential functions e^{iE_gt} and e^{-iE_ft} of Eq. (93) such that we obtain:

$$\begin{aligned} W_{fg}(\mathbb{k}) &= \frac{2}{\Gamma} Re \int_0^\infty e^{(i\omega t - \frac{1}{2}\Gamma t)} \\ &\quad \times \langle g|\mathbf{H}^{(+)}|f\rangle^* \langle g|e^{iE_gt}\mathbf{H}^{(+)}e^{-iE_ft}|f\rangle\, dt \\ &= \frac{2}{\Gamma} Re \int_0^\infty e^{(i\omega t - \frac{1}{2}\Gamma t)} \\ &\quad \times \langle f|\mathbf{H}^{(-)}|g\rangle \langle g|\mathbf{U}^+(t)\mathbf{H}^{(+)}\mathbf{U}(t)|f\rangle\, dt \end{aligned} \tag{95}$$

Here $\mathbf{H}^{(-)} = \mathbf{H}^{(+)\dagger}$ and:

$$\mathbf{U}(t) = e^{-i\mathbf{H}t} \tag{96}$$

is the time evolution operator, $\mathbf{U}^{\dagger}(t)$ its adjoint and $\mathbf{H}$ the total Hamiltonian of the system. Evidently, $e^{\pm i\mathbf{H}t}$ is defined by the exponential series $\Sigma \pm i\mathbf{H}t/n!$ and thus:

$$e^{\pm i\mathbf{H}t}|g\rangle = e^{\pm iE_g t}|g\rangle \tag{97}$$

and similarly for $|f\rangle$. We introduce the abbreviation $\mathbf{H}^{(+)}(t) = \mathbf{U}^{\dagger}(t)\mathbf{H}^{(+)}\mathbf{U}(t)$ and express the average by:

$$\langle T\rangle = \sum_f p_f \langle f|T|f\rangle \tag{98}$$

where p_f is the probability that the initial state $|f\rangle$ occurs. The experimentally observed emission probability is then obtained by averaging Eq. (95) over all possible initial states $|f\rangle$ and summarizing over all final states $|g\rangle$ as:

$$W(\mathbb{k}) = \sum_{fg} p_f W_{fg} = \frac{2}{\Gamma} Re \int_0^{\infty} e^{(i\omega t - \frac{1}{2}\Gamma t)} \langle \mathbf{H}^{(-)}\mathbf{H}^{(+)}(t)\rangle \, dt \tag{99}$$

If the nucleus in the emitter is under the influence of time-dependent forces, it is convenient to modify Eq. (99) by introducing, instead of Eq. (96), the time-ordered operator:

$$\mathbf{U}(t) = \exp\left[-i\int_0^t \mathbf{H}(t')\,dt'\right] \tag{100}$$

where $\mathbf{H}(t')$ is now an explicitly time-dependent Hamiltonian. In order to describe a system undergoing fluctuations, the Hamiltonian of the emitter is assumed to be a random function of time. The probability of emission of a photon $\mathbb{k}$ is then the stochastic average of Eq. (99):

$$W(\mathbb{k}) = \frac{2}{\Gamma} Re \int_0^{\infty} e^{(i\omega t - \frac{1}{2}\Gamma t)} [\langle \mathbf{H}^{(-)}\mathbf{H}^{(+)}(t)\rangle]_{\text{av}} \, dt \tag{101}$$

Here, $[\;]_{\text{av}}$ denotes the average over the stochastic degrees of freedom in the Hamiltonian, $[\langle \mathbf{H}^{(-)}\mathbf{H}^{(+)}(t)\rangle]_{\text{av}}$ being known as the correlation function.

The quadrupole interaction operator in a Mössbauer experiment may be expressed as:

$$\mathbf{H}_Q = \frac{eQV_{zz}}{4I(2I-1)}\left[(3\mathbf{I}_z^2 - \mathbf{I}^2) + \frac{\eta}{2}(\mathbf{I}_+^2 + \mathbf{I}_-^2)\right] \tag{102}$$

Here, $\mathbf{I}^2$, $\mathbf{I}_z$, and $\mathbf{I}_{\pm}$ are angular momentum operators, Q is the quadrupole moment of the nucleus, V_{zz} the z component, and η the asymmetry parameter of the electric field gradient (efg) tensor. We wish to construct the Hamiltonian for a nucleus if the efg jumps at random between HS and LS states. For this purpose, a random function of time $f(t)$ is introduced which can assume only the two possible values ± 1. For convenience of presentation we assume equal isomer shifts and zero η values for the HS and LS state. In the following, two special cases are considered with respect to the orientation of the efg's of the HS and LS state:

(1) *Stochastic model if the efg's for the* HS *and* LS *state are parallel.* The Hamiltonian for a nucleus with an excited state of spin $I_1 = \frac{3}{2}$ and a ground state of spin $I_0 = \frac{1}{2}$ such as ^{57}Fe may be written as [161]:

$$\mathbf{H} = \mathbf{H}_0 + \tfrac{1}{12}[(1 + f(t))\Delta E_Q^{\mathrm{L}} + (1 - f(t))\Delta E_Q^{\mathrm{H}}](3\mathbf{I}_z^2 - \mathbf{I}^2) \tag{103}$$

where $\mathbf{H}_0$ is the unperturbed Hamiltonian and ΔE_Q^{L} and ΔE_Q^{H} is the quadrupole splitting for the LS and HS state, respectively. Putting $q_1 = \frac{1}{12}(\Delta E_Q^{\mathrm{L}} + \Delta E_Q^{\mathrm{H}})$ and $q_2 = \frac{1}{12}(\Delta E_Q^{\mathrm{L}} - \Delta E_Q^{\mathrm{H}})$, Eq. (103) may be rewritten as:

$$\mathbf{H} = \mathbf{H}_0 + [q_1 + q_2 f(t)](3\mathbf{I}_z^2 - \mathbf{I}^2) \tag{104}$$

The correlation function then obtains as:

$$[\langle \mathbf{H}^{(-)}\mathbf{H}^{(+)}(t)\rangle]_{\mathrm{av}} = \frac{1}{2I_1 + 1} \sum_{m_0 m_1 m_0' m_1'} [\langle I_1 m_1 | \mathbf{H}^{(-)} | I_0 m_0\rangle$$

$$\times \langle I_0 m_0 | \mathbf{U}^{\dagger}(t) | I_0 m_0' \rangle \langle I_0 m_0' | \mathbf{H}^{(+)} | I_1 m_1' \rangle \langle I_1 m_1' | \mathbf{U}(t) | I_1 m_1 \rangle]_{\mathrm{av}} \tag{105}$$

In Eq. (105) it has been assumed that the various initial m_1 sublevels are equally probable, i.e. $p_f = 1/(2I_1 + 1)$. Moreover, $|I_1 m_1\rangle$ and $|I_0 m_0\rangle$ denote the excited and the ground state of the nucleus, respectively, and $\mathbf{U}(t)$ is determined according to Eq. (100). Since a $I_0 = \frac{1}{2}$ state has no effect in the I_0 matrix element, we obtain:

$$\langle I_0 m_0 | \mathbf{U}^{\dagger}(t) | I_0 m_0' \rangle = \exp[iE_0 t]\delta_{m_0 m_0'}$$

$$\langle I_1 m_1' | \mathbf{U}(t) | I_1 m_1 \rangle = \exp\left[-i\left\{ E_1 t + q_1(3I_z^2 - I^2)t + \right.\right.$$

$$\left.\left. + q_2(3I_z^2 - I^2)\int_0^t f(t')\,dt' \right\}\right]\delta_{m_1 m_1'} \tag{106}$$

where δ is the Kronecker delta. If the probability of emission for a powder sample is considered, only the diagonal element of Eq. (106) survives and thus Eq. (105) reduces to:

$$[\langle \mathbf{H}^{(-)}\mathbf{H}^{(+)}(t)\rangle]_{\mathrm{av}} = \frac{1}{4}\sum_{m_0 m_1} |\langle I_0 m_0 | \mathbf{H}^{(+)} | I_1 m_1 \rangle|^2$$

$$\times \exp\left[-i(E_1 - E_0)t - iq_1\left(3m_1^2 - \frac{15}{4}\right)t \right]$$

$$\times \left[\exp\left\{ -iq_2\left(3m_1^2 - \frac{15}{4}\right)\int_0^t f(t')\,dt' \right\} \right]_{\mathrm{av}} \tag{107}$$

where for ^{57}Fe, $E_1 - E_0 = \omega_0 = 14.4$ keV is the frequency of the unsplit line. The remaining problem is to determine the stochastic average of the last term in

Eq. (107). The solution has been given by Abragam [162] as:

$$\left[\exp\left\{-i\alpha\int_0^t f(t')\,dt'\right\}\right]_{av} =$$

$$= \sum_{ij} p_i(j|\exp[(-i\alpha\mathbb{F} + \mathbb{W})t]|i) \tag{108}$$

where $\alpha = q_2(3m_1^2 - \frac{15}{4})$ and p_i is the population of state i. In addition, $\mathbb{W}$ is the matrix of transition probabilities per unit time and $\mathbb{F}$ is the matrix where the diagonal elements are the permissive values of $f(t)$. In the present case where $f(t) = \pm 1$, the matrices $\mathbb{W}$ and $\mathbb{F}$ are 2×2 and thus the equation may be easily evaluated. Introducing Eq. (107) and Eq. (108) into Eq. (101) we obtain:

$$W(\mathbb{k}) = \frac{1}{2\Gamma} Re \sum_{m_0 m_1} |\langle I_0 m_0|\mathbf{H}^{(+)}|I_1 m_1\rangle|^2$$

$$\times \int_0^\infty \exp\left[-i\left(\omega_0 - \omega + q_1\left(3m_1^2 - \frac{15}{4}\right)\right)t - \frac{1}{2}\Gamma t\right]$$

$$\times \sum_{ij} p_i(j|\exp[(-i\alpha\mathbb{F} + \mathbb{W})t]|i)\,dt \tag{109}$$

This expression may be cast into a simpler form by defining the quantity:

$$G_{m_1}(p) = \int_0^\infty \exp\left[-i\left(\omega_0 - \omega + q_1\left(3m_1^2 - \frac{15}{4}\right)\right)t - \frac{1}{2}\Gamma t\right]$$

$$\times \sum_{ij} p_i(j|\exp[(-i\alpha\mathbb{F} + \mathbb{W})t]|i)\,dt \tag{110}$$

Employing the abbreviation $p = -i\{\omega - \omega_0 - q_1(3m_1^2 - \frac{15}{4})\} + \frac{1}{2}\Gamma$, integration of Eq. (110) between the given limits produces:

$$G_{m_1}(p) = \sum_{ij} p_i(j|p\mathbb{1} - \mathbb{W} + i\alpha\mathbb{F})^{-1}|i) \tag{111}$$

The probability of emission Eq. (109) may then be written as:

$$W(\mathbb{k}) = \frac{1}{2\Gamma} Re \sum_{m_0 m_1} |\langle I_0 m_0|\mathbf{H}^{(+)}|I_1 m_1\rangle|^2 G_{m_1}(p) \tag{112}$$

It may be shown [31] that, if the fluctuations are described by a stationary Markoff process, the diagonal elements of matrix $\mathbb{W}$ are determined by the off-diagonal elements W_{ij} according to:

$$W_{ii} = -\sum_i W_{ij} \quad (j \neq i) \tag{113}$$

Here, the W_{ij} represent the probability per unit time that the system performs a transition from state i to state j. In the present case of relaxation between the two states denoted as LS and HS, we may associate the matrix elements W_{12} and W_{21} with the rate constants k_{LH} and k_{HL} for the conversion process LS → HS and HS → LS, respectively. The diagonal elements of the relaxation matrix $\mathbb{W}$

are then given by:

$$W_{11} = -k_{LH} \quad \text{and} \quad W_{22} = -k_{HL} \tag{114}$$

and matrix $\mathbb{W}$ assumes the simple form:

$$\mathbb{W} = \begin{pmatrix} -k_{LH} & k_{LH} \\ k_{HL} & -k_{HL} \end{pmatrix} \tag{115}$$

The rate constants, and thus the matrix elements of matrix $\mathbb{W}$, conform to the requirement of detailed balance:

$$n_H k_{HL} = n_L k_{LH} = (1 - n_H) k_{LH} \tag{116}$$

where $n_L = 1 - n_H$ and n_H is the HS fraction. The complete matrix $(p\mathbb{1} - \mathbb{W} + i\alpha\mathbb{F})$ in Eq. (111) results thus as:

$$\begin{pmatrix} p + i\alpha + k_{LH} & -k_{LH} \\ -k_{HL} & p - i\alpha + k_{HL} \end{pmatrix} \tag{117}$$

and matrix inversion finally produces the expression:

$$G_{m_1}(p) = [n_H(p + i\alpha + k_{HL} + k_{LH}) + n_L(p - i\alpha + k_{HL} + k_{LH})]/ \\ [(p - i\alpha + k_{HL})(p + i\alpha + k_{LH}) - k_{HL}k_{LH}] \tag{118}$$

The real part of $G_{m_1}(p)$ determines the line shape for variable relaxation times between the LS and HS states, a representative example being displayed in Fig. 19.

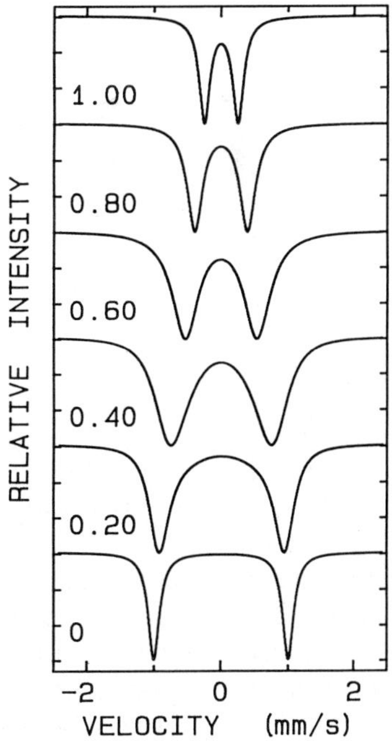

Fig. 19. Calculated lineshapes for various values of n_H for the spin-state transition between LS and HS states. The parameter values employed in the calculation are $\Delta E_Q^L = 2.00\ \mathrm{mm\,s^{-1}}$, $\Delta E_Q^H = 0.50\ \mathrm{mm\,s^{-1}}$, $\delta_L^{IS} = \delta_H^{IS} = 0$, $\Gamma = 0.20\ \mathrm{mm\,s^{-1}}$ and $\tau_L = 2 \times 10^7$ s. The calculation has been performed for the efg's of HS and LS state parallel and of the same sign

(2) *Stochastic model if the efg of the* HS *state is perpendicular to that of the* LS *state.* In this case, the Hamiltonian may be written as [161]:

$$\mathbf{H} = \mathbf{H}_0 + \tfrac{1}{12}(1 + f(t))\Delta E_Q^{\mathrm{L}}(3I_z^2 - I^2) + \\ + \tfrac{1}{12}(1 - f(t))\Delta E_Q^{\mathrm{H}}(3I_x^2 - I^2) \quad (119)$$

the quantities having the same meaning as above. The calculation proceeds in analogy to case (1) and is not given here in detail. The expression for the probability of emission is identical to Eq. (112), the quantity $G_{m_1}(p)$ being determined by:

$$G_{m_1}(p) = \sum_{ij} p_i(j|[\tilde{A}(p)^{-1} + 12q_3^2\tilde{B}(p)]^{-1}|i) \quad (120)$$

where $p = -i(\omega - \omega_0) + \frac{1}{2}\Gamma$, $\tilde{A}(p) = (p\mathbb{1} - \mathbb{W} - i\beta\mathbb{F}_1)^{-1}$, $\tilde{B}(p) = \mathbb{F}(p\mathbb{1} - \mathbb{W} + i\beta\mathbb{F}_1)^{-1}\ \mathbb{F}$ and $\mathbb{F}$, $\mathbb{F}_1$, and $\mathbb{W}$ are given as:

$$\mathbb{F} = \begin{pmatrix} 0 & 0 \\ 0 & 2 \end{pmatrix} \qquad \mathbb{F}_1 = \begin{pmatrix} q_2 + q_1 & 0 \\ 0 & q_2 - q_1 \end{pmatrix}$$

$$\mathbb{W} = \begin{pmatrix} -k_{\mathrm{LH}} & k_{\mathrm{LH}} \\ k_{\mathrm{HL}} & -k_{\mathrm{HL}} \end{pmatrix} \quad (121)$$

Here, $q_1 = \frac{1}{12}(\Delta E_Q^{\mathrm{L}} + \frac{1}{2}\Delta E_Q^{\mathrm{H}})$, $q_2 = \frac{1}{12}(\Delta E_Q^{\mathrm{L}} - \frac{1}{2}\Delta E_Q^{\mathrm{H}})$, $q_3 = \frac{1}{16}\Delta E_Q^{\mathrm{H}}$ and $\beta = -(3m_1^2 - \frac{15}{4})$. Performing the calculations indicated in Eq. (120) one obtains the final expression:

$$\begin{aligned} G_{m_1}(p) = {} & \frac{1}{c[d + 48bq_3^2/(ab - k_{\mathrm{HL}}k_{\mathrm{LH}})] - k_{\mathrm{HL}}k_{\mathrm{LH}}} \\ & \times \left[n_{\mathrm{L}}\left(d + \frac{48bq_3^2}{ab - k_{\mathrm{HL}}k_{\mathrm{LH}}} + k_{\mathrm{LH}}\right) + n_{\mathrm{H}}(k_{\mathrm{HL}} + c)\right] \\ & + \frac{1}{b[a + 48cq_3^2/(cd - k_{\mathrm{HL}}k_{\mathrm{LH}})] - k_{\mathrm{HL}}k_{\mathrm{LH}}} \\ & \times \left[n_{\mathrm{L}}\left(a + \frac{48cq_3^2}{cd - k_{\mathrm{HL}}k_{\mathrm{LH}}} + k_{\mathrm{LH}}\right) + n_{\mathrm{H}}(k_{\mathrm{HL}} + b)\right] \end{aligned} \quad (122)$$

where $a = p + 3i(q_2 - q_1) + k_{\mathrm{HL}}$, $b = p + 3i(q_2 + q_1) + k_{\mathrm{LH}}$, $c = p - 3i(q_2 + q_1) + k_{\mathrm{LH}}$, $d = p - 3i(q_2 - q_1) + k_{\mathrm{HL}}$, the other quantities having been defined above. The real part of $G_{m_1}(p)$ again gives the lineshape for variable relaxation times between the HS and LS states.

Let us consider in the following, instead of intermediate relaxation, the limit of fast relaxation where a nucleus experiences the average of the efg between the HS and LS states. Figure 20 shows the result of the calculation of quadrupole splitting as a function of the HS fraction, n_{H}. If the signs of the efg for the LS and the HS state are the same, the quadrupole splitting varies linearly with n_{H}. If, on the other hand, the signs of the efg for the LS and HS state are opposite, the quadrupole splitting exhibits a minimum for a particular value of n_{H}. It is

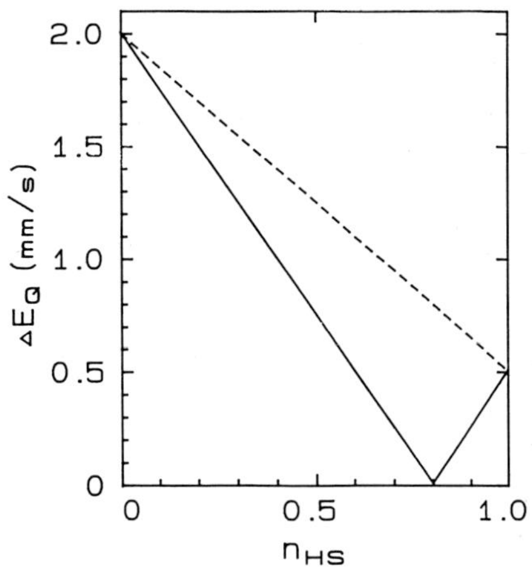

Fig. 20. The variation of quadrupole splitting as a function of the HS fraction n_H (in percent) in the limit of fast spin interconversion. The parameter values used are $\tau_L = 1 \times 10^{-8}$ s, $\Delta E_Q^L = 2.00$ mm s^{-1}, $\Delta E_Q^H = 0.50$ mm s^{-1}, $\delta_L^{IS} = \delta_H^{IS} = 0$. The *dashed line* applies to $\mathrm{sgn}(V_{zz}^H) = \mathrm{sgn}(V_{zz}^L)$, the *full line* to $\mathrm{sgn}(V_{zz}^H) \neq \mathrm{sgn}(V_{zz}^L)$

evident that, from the temperature dependence of the quadrupole splitting, the relative sign of the efg of the two spin states may be conveniently determined.

8.3 A Simple Example of Line Shape Analysis: Spin-State Relaxation in the Nitrosyl Complex Fe(J-mph)NO

The nitrosyl iron complex Fe(*J*-mph)NO shows a spin-state transition between the $S = \frac{3}{2}$ and $S = \frac{1}{2}$ ground states. The structure of the complex is characterized by:

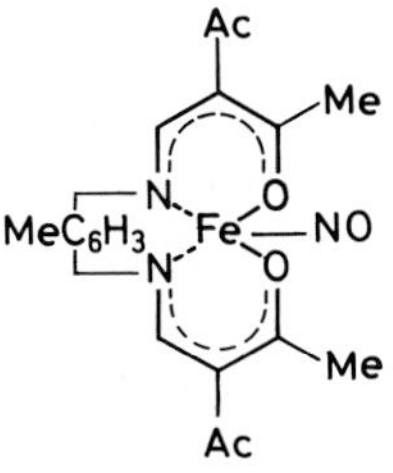

J-mph denoting the four-dentate N_2O_2 Jäger-type ligand 4-methyl-1,2-phenylenebis(diacetylacetaldimine) [163]. Based on magnetic susceptibility measurements between 80 and 320 K, the transition is centred at $T_c \simeq 212$ K and is incomplete at both high and low temperatures. In order to reproduce the Mössbauer spectra between 84 and 319 K, three models of increasing sophistication were employed [164]:

(i) Since only two lines are observed over the complete range of temperature, the spectra were fitted by two independent Lorentzians, cf. Fig. 21.

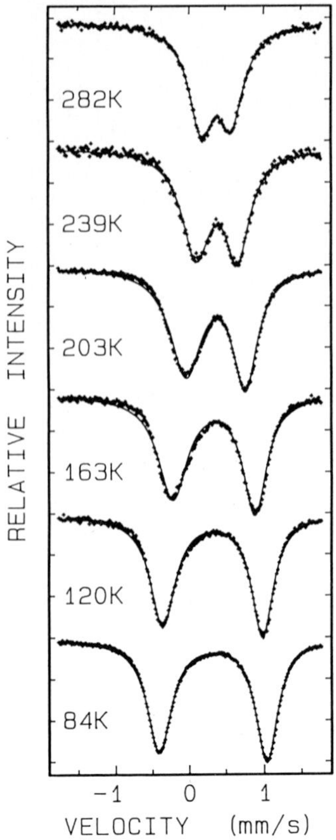

Fig. 21. Mössbauer spectra of Fe(J-mph)NO between 84 and 282 K. *Solid lines* are the result of a fit by two independent Lorentzians

This procedure is applicable if the relaxation between the spin states is fast ($\tau \leq 1 \times 10^{-8}$ s) and thus the quadrupole doublets of the two spin states collapse into one. It is found that, in the intermediate temperature range, the widths of the two lines are significantly enlarged. This shows that the assumption of fast relaxation strictly does not apply. In spite of this, the areas of the lines can be well reproduced within the Debye model employing the same Debye temperature for both spin states, $\Theta_D \simeq 123$ K.

(ii) At low temperatures, the spectra can be fitted by two symmetrical quadrupole doublets with equal line width. This procedure is applicable if the relaxation between the spin states is slow ($\tau \geq 1 \times 10^{-5}$ s) and thus both quadrupole doublets are observed simultaneously. Again the line widths are found to increase with increasing temperature indicating that the assumption of slow relaxation is not correct. Based on equal Debye-Waller factors for the two spin states, the fraction n_I of the $S = \frac{3}{2}$ state has been extracted from the obtained fits. The resulting values are displayed in Fig. 22 as a function of μ_{eff}^2. The plot is essentially linear and its extrapolation to $n_I = 1.0$ and $n_I = 0$ produces the magnetic moments $\mu_{IS} = 3.86\ \mu_B$ and $\mu_{LS} = 1.51\ \mu_B$, respectively,

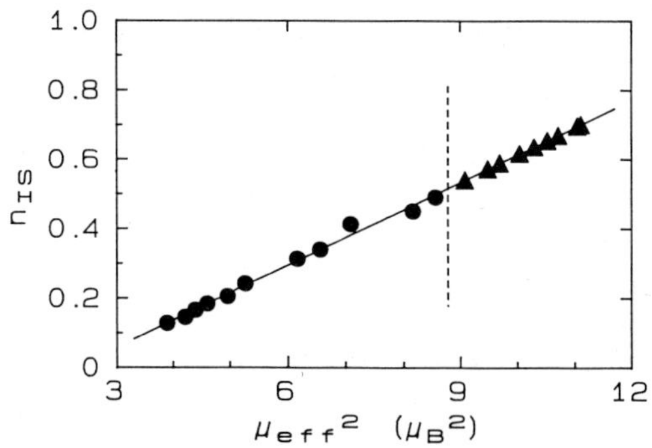

Fig. 22. The fraction n_I of the $S = 3/2$ state as a function of μ_{eff}^2. The values at the lower temperatures (●) result from a fit of the Mössbauer spectra of Fe(J-mph)NO by two symmetrical quadrupole doublets. The values at the higher temperatures (▲) are extrapolated. According to Ref. [164]

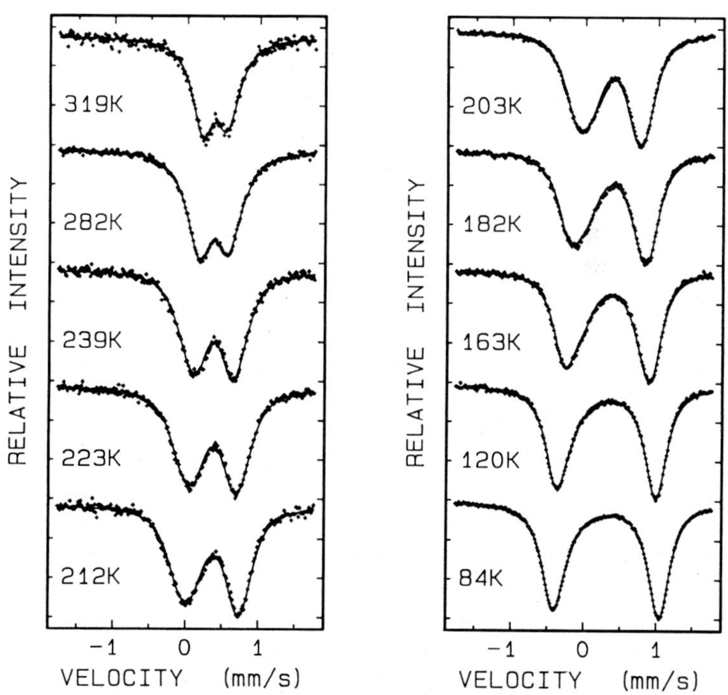

Fig. 23. Mössbauer spectra of Fe(J-mph)NO between 84 and 319 K. *Solid lines* result from a fit by a two-state relaxation model based on the stochastic theory of lineshapes. According to Ref. [164]

values which are very close to the spin-only moments of the corresponding spin states.

(iii) Since the numerical fits according to (i) and (ii) are not completely satisfactory, the system provides an example for intermediate electronic relaxation (1×10^{-5} s $\geq \tau \geq 1 \times 10^{-8}$ s). The spectra were therefore reproduced by a two-state relaxation model based on the stochastic theory of lineshape [161, 165]. This model requires the same principal axis system for the efg of the two spin states as well as the same asymmetry parameter η. In addition, it has been assumed that the Debye-Waller factors and the line widths for both doublets are equal. Employing the values of n_I obtained according to (ii), the resulting spectra are shown as the solid lines in Fig. 23. The fit is satisfactory over the complete temperature range between 84 and 319 K. The resulting values of the rate constant k_{IL} for the transition from the IS($S = \frac{3}{2}$) state to the LS($S = \frac{1}{2}$) state are listed in Table 8. The rate constants k_{IL} and k_{LI} are related by the requirement of

Table 8. Values of the rate constant k_{IL} of Fe(J-mph)NO derived on the basis of a two-state relaxation model[a]

T K	n_{IS}	$10^{-6}\ k_{IL}$ s^{-1}
319	0.700	6.4
282	0.669	6.1
239	0.588	5.7
223	0.539	5.5
212	0.498	5.4
203	0.466	4.8
182	0.382	4.6
163	0.308	4.2
120	0.184	3.2
84	0.128	2.0

[a] According to Ref. [164]

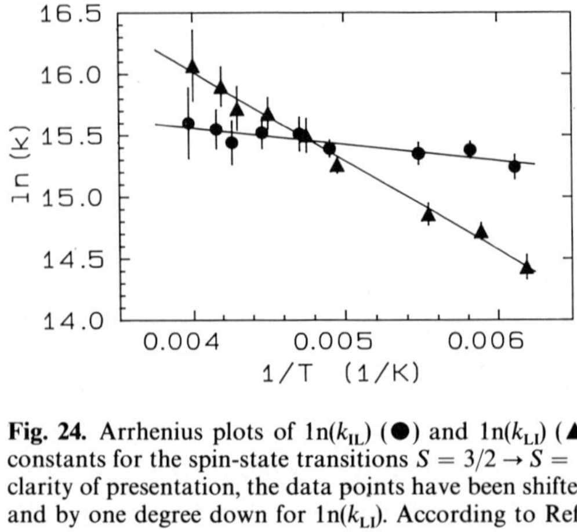

Fig. 24. Arrhenius plots of $\ln(k_{IL})$ (●) and $\ln(k_{LI})$ (▲) versus $1/T$ where k_{IL} and k_{LI} are the rate constants for the spin-state transitions $S = 3/2 \rightarrow S = 1/2$, and $S = 1/2 \rightarrow S = 3/2$, respectively. For clarity of presentation, the data points have been shifted in temperature by one degree up for $\ln(k_{IL})$ and by one degree down for $\ln(k_{LI})$. According to Ref. [164]

detailed balance, $k_{\mathrm{IL}} n_{\mathrm{I}} = k_{\mathrm{LI}}(1 - n_I)$. Figure 24 shows that the dependence of $\ln(k_{\mathrm{IL}})$ and $\ln(k_{\mathrm{LI}})$ on $1/T$ is linear within the temperature range 163 to 251 K. From the linear relation, the activation energies of the process have been derived as $\Delta E_{\mathrm{IL}} = 1.10 \pm 0.20$ kJ mol^{-1} and $\Delta E_{\mathrm{LI}} = 6.09 \pm 0.22$ kJ mol^{-1}. These values are rather small if compared with activation energies for HS↔LS transitions in complexes of iron(II) (see Sect. 8.4 below). A possible explanation may be based on the fact that Fe-ligand bonds in nitrosyl iron complexes decrease by about 0.1 Å on passing from the IS to the LS state, whereas the corresponding change for a HS↔LS transition in iron(II) complexes is about twice as large [6].

8.4 Individual Spin-State Interconversion Rate Studies

In this section, the available studies of spin-state relaxation in iron(II) and iron(III) complexes are reviewed in as far as the results are based on the analysis of Mössbauer effect lineshapes. As a rule, this analysis produces the rate constants for the spin-state interconversion process in the solid complexes. Several different approaches were followed. Thus, both the modified Bloch equations Eq. (90) and Eq. (91) as well as the stochastic theory of lineshape were employed, studies by the latter method being distinguished according to whether the analysis has been started from the more general formulation by Blume [31] or whether an adaptation of the stochastic model due to Blume and Tjon [32, 33] has been used. The application of the Blume and Tjon method to the treatment of spin-state transitions as formulated by Maeda et al. [161, 166] employs the restriction to axially symmetric efg's and to substances with essentially identical isomer shifts. This is a situation often exemplified by complexes of iron(III). The restriction to identical isomer shifts may be easily relaxed [167] if some minor algebraic manipulations are applied. Within this modification, the model may be applied to most compounds of iron(II). The restriction concerning the orientation of the efg's is relaxed in cases where one of the spin states is characterized by a single unsplit line, and thus iron(II) complexes with this particular property were preferably studied. Additional modifications of the theoretical treatment as well as the introduction of supplementary physical methods, such as the application of external pressure, add to the individual character of the studies. In the following, the results of the relevant investigations on complexes of iron(II) and iron(III) are discussed in the order of increasing complexity.

$[Fe(acpa)_2]BPh_4 \cdot H_2O$. The iron(III) complex of the tridentate ligand acpaH = *N*-(1-acetyl-2-propylidene)-2-pyridylmethylamine shows, between 80 and 300 K, an effective magnetic moment that is gradually decreasing with decreasing temperature. The Mössbauer spectra between 78 and 189 K are typical for LS iron(III), whereas above 200 K a single asymmetric doublet is found which shows substantial broadening of one line [156]. This behavior is characteristic for a relaxation mechanism where the spin interconversion rate is rapid or comparable to the quadrupole precession time. The Mössbauer spectra above 189 K were simulated using a formalism based on the modified Bloch equations, viz. Fig. 25. The numerical fit is reasonable at low temperatures, i.e. for $T < 268$ K, whereas at higher temperatures increasing population of the $^6A_1$ state contributes to the linewidth via spin-spin and spin-lattice relaxation. The relaxation times are temperature dependent and vary between $\tau = 6.9 \times 10^{-7}$ and $\tau = 1.6 \times 10^{-5}$ s. The HS populations n_H and lifetimes of the LS state $\tau(^2T_2)$ resulting from the calculations are listed in Table 9. The rate constants for the intersystem crossing process:

$$^2T_2(Fe^{3+}; LS) \underset{k_{HL}}{\overset{k_{LH}}{\rightleftharpoons}} {}^6A_1(Fe^{3+}; HS)$$

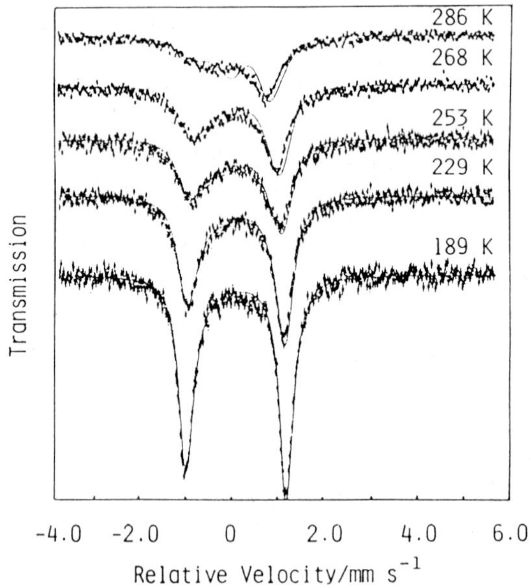

Fig. 25. ^{57}Fe-Mössbauer spectra of $[Fe(acpa)_2]BPh_4 \cdot H_2O$ between 189 and 286 K. *Full curves* have been calculated on the basis of the modified Bloch equations using the parameter values of Table 9. According to Ref. [156]

Table 9. High-spin populations, $^2T_2$ lifetimes and rate constants k_{LH} for the intersystem crossing in $[Fe(acpa)_2]BPh_4 \cdot H_2O$ according to a numerical simulation of the Mössbauer spectra[a]

T K	n_H	$\tau(^2T_2)$ s	k_{LH} s^{-1}
286	0.60	3.4×10^{-7}	2.9×10^6
268	0.42	4.0×10^{-7}	2.5×10^6
253	0.38	5.0×10^{-7}	2.0×10^6
229	0.24	7.4×10^{-7}	1.4×10^6
189	0.10	1.4×10^{-5}	7.1×10^4

[a] According to Ref. [156]

are defined as $k_{LH} = [\tau(^2T_2)]^{-1}$ and $k_{HL} = [\tau(^6A_1)]^{-1}$. The activation energy has been obtained from the Arrhenius equation as $\Delta E = 240$ kcal mol^{-1}. It is concluded that the rapid spin interconversion in the solid state proceeds essentially by an intramolecular mechanism, although the effect of a phonon interaction cannot be excluded.

The Mössbauer spectra of the complex [Fe(acpa)$_2$]PF$_6$ shown in Fig. 26 have also been interpreted on the basis of a relaxation mechanism [168]. For the calculations, the formalism using the modified Bloch equations again was employed. The resulting correlation times $\tau = \tau_L\tau_H/(\tau_L + \tau_H)$ are temperature dependent and span the range between 1.9×10^{-6} s at 110 K and 0.34×10^{-6} s at 285 K. Again the correlation times are reasonable only at low temperatures, whereas around 200 K increase of the population of the $^6A_1$ state contributes to line broadening via spin-spin or spin-lattice relaxation.

[Fe(mim)$_2$(salacen)]PF$_6$. The iron(III) complex of the tetradentate ligand salacenH$_2$ = ethylene(*N*-acetylacetonylideneimine) (*N*′-α-methylsalicylideneimine) shows, between 78 and 326 K, a gradual $^6A_1 \leftrightarrow {}^2T_2$ spin-state transition in the solid state (mim = *N*-methylimidazole) [166]. The Mössbauer spectrum at 78 K shows a single doublet characteristic for LS iron(III), whereas the spectrum at 250 K can be fitted to two doublets of Lorentzian lineshape, one for the LS, the other for the HS isomer. At 275 K, again a single doublet, although of

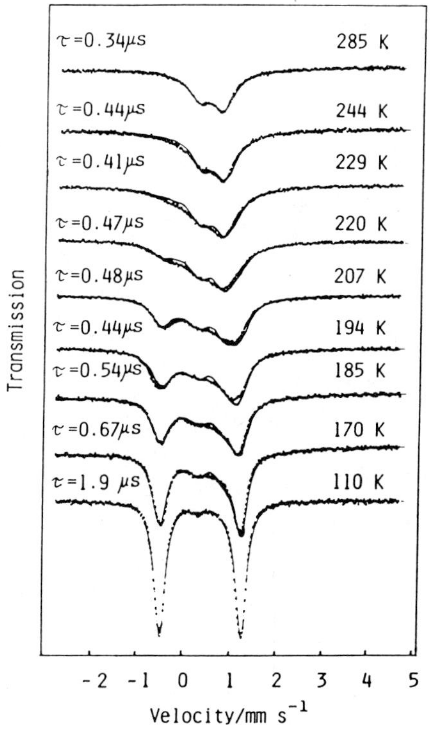

Fig. 26. ^{57}Fe-Mössbauer spectra of [Fe(acpa)$_2$]PF$_6$ between 110 and 285 K. *Full curves* are calculated on the basis of the modified Bloch equations using the correlation times marked on the spectra. According to Ref. [168]

somewhat broader lines, is adequate to fit the experimental spectrum, the spectra at even higher temperatures being affected by line broadening and inward coalescence. The stochastic theory of Blume [32, 33] was applied in order to simulate the time-averaged Mössbauer spectra assuming equal sign of the efg's for the LS and the HS state. This assumption is supported by the fact that the experimental quadrupole splitting ΔE_Q decreases linearly with increasing temperature (cf. Fig. 20). In the calculations, $\Delta E_Q(^2T_2) = 2.36\ \mathrm{mm\,s^{-1}}$, $\Delta E_Q(^6A_1) = 1.00\ \mathrm{mm\,s^{-1}}$ and a constant line width $\Gamma = 0.28\ \mathrm{mm\,s^{-1}}$ were assumed. Values of the isomer shift δ^{IS} were taken from the experimental spectra, whereas values of the HS fraction n_H were estimated from magnetic susceptibility data. The spectra calculated on the basis of Eq. (118) are displayed, for a number of temperatures, in Fig. 27. The resulting values of the relaxation time τ defined as $\tau = (k_{LH} + k_{HL})^{-1} = \tau_L\tau_H/(\tau_L + \tau_H)$ are listed in the caption. The numerical fit is reasonably accurate at low temperatures, whereas increase of the $^6A_1$ population causes an increase of line width due to both spin-spin and spin-lattice relaxation. It is evident that the relaxation rate $k = \tau^{-1}$ is temperature dependent and varies between $3.1 \times 10^6\ \mathrm{s^{-1}}$ at 221 K and $6.7 \times 10^6\ \mathrm{s^{-1}}$ at 298 K.

[Fe(2-pic)$_3$](PF$_6$)$_2$. The iron(II) complex of the bidentate ligand 2-pic = 2-amino-methylpyridine shows a gradual $^5T_2 \leftrightarrow {}^1A_1$ spin-state transition in the solid state which extends over the temperature range 50 to above 300 K [167]. The Mössbauer spectrum at 80 K displays a unique singlet characteristic for LS iron(II), whereas between 100 and 200 K an additional doublet due to the HS

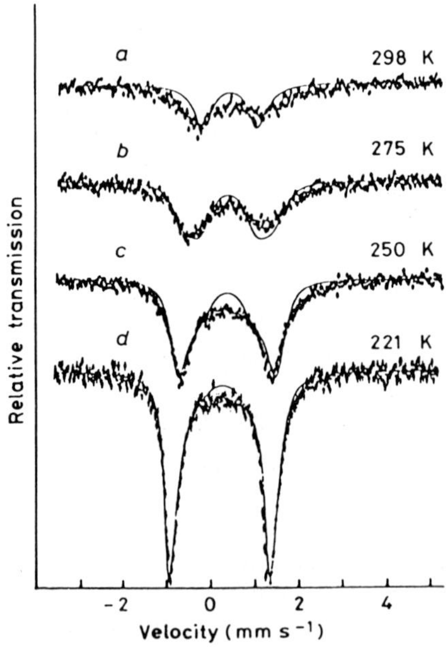

Fig. 27a–d. ^{57}Fe-Mössbauer spectra of [Fe(mim)$_2$(salacen)]PF$_6$ between 221 and 298 K. *Full curves* have been calculated on the basis of the stochastic theory of Blume and Tjon using the following parameter values: **a** $n_H = 0.75$, $\tau = 1.50 \times 10^{-7}$ s; **b** $n_H = 0.52$, $\tau = 2.3 \times 10^{-7}$ s; **c** $n_H = 0.33$, $\tau = 3.0 \times 10^{-7}$ s; **d** $n_H = 0.16$, $\tau = 3.2 \times 10^{-7}$ s. According to Ref. [166]

state is observed, its area fraction increasing with increasing temperature. Above 200 K, the LS absorption line and to some extent the HS lines show a significant broadening and an increasing asymmetry of the HS lines. Although distinct spectra for both spin states are observed, the behavior is characteristic for relaxation effects where $k_{LH} \simeq \omega_n$. The spectra above 200 K were therefore calculated on the basis of the stochastic theory of line shape of Blume [31]. The numerical fit employs six parameters: the effective thickness of the two states t^{H}_{eff}, t^{L}_{eff}, the isomer shifts δ^{IS}_{H}, δ^{IS}_{L}, the quadrupole splitting ΔE^{H}_{Q} and the relaxation rate k_{HL}. The rate constants k_{HL} and k_{LH} for the intersystem crossing process:

$$^{1}A_1(Fe^{2+}; LS) \underset{k_{HL}}{\overset{k_{LH}}{\rightleftharpoons}} {}^{5}T_2(Fe^{2+}; HS)$$

were identified with the transition probabilities W_{21} and W_{12} as outlined in Sect. 8.2 above. The measured spectra and their numerical simulation are displayed in Fig. 28, the resulting values of the rate constant k_{HL} are listed in Table 10. The rate constant k_{LH} for the reverse process can be obtained from the requirement of detailed balance Eq. (116). From the temperature dependence of k_{LH} and k_{HL}, the Arrhenius equation gives the activation energies $\Delta E_{LH} = 4.92 \pm 0.14\ \mathrm{kcal\,mol^{-1}}$ and $\Delta E_{HL} = 2.99 \pm 0.19\ \mathrm{kcal\,mol^{-1}}$, the frequency factors resulting as $A_{LH} \simeq 3 \times 10^{11}\ \mathrm{s^{-1}}$ and $A_{HL} \simeq 1 \times 10^{9}\ \mathrm{s^{-1}}$. The activation parameters for the intersystem crossing processes have been derived on the basis of absolute reaction rate theory. The activation enthalpy has been obtained as $\Delta H^{\ddagger}_{LH} = 4.44$ and $\Delta H^{\ddagger}_{HL} = 2.46\ \mathrm{kcal\,mol^{-1}}$, the activation entropy being $\Delta S^{\ddagger}_{LH} = -7.9$ and $\Delta S^{\ddagger}_{HL} = -18.6\ \mathrm{cal\,K^{-1}\,mol^{-1}}$. It should be observed that $\Delta H^{\ddagger}_{HL}$ is within the range of values encountered for spin conversion in solution, viz. 0.62 to 3.70 kcal mol^{-1}, the same being applicable to $\Delta S^{\ddagger}_{HL}$ where the solution values span the range between -13.9 and $-22.7\ \mathrm{cal\,K^{-1}\,mol^{-1}}$. The HS → LS conversion in the solid complex thus seems to be determined essentially by intramolecular factors. The smaller value of $\Delta H^{\ddagger}_{LH}$ for the reverse process may be due to the fact that this quantity contains the endothermic enthalpy difference ΔH° arising from the difference of potential energies of the HS and LS state.

The spin-state transition has been also studied for external pressures of 70 and 150 MPa [169], some spectra recorded at $p = 150$ MPa being displayed in Fig. 29. Again line broadening effects and an asymmetry of linewidths were found which are typical of relaxation. The solid lines in the figure have been obtained on the basis of the stochastic theory of lineshape. Since the quadrupole interaction occurs only in the HS state, specific assumptions concerning orientation of the efg's were not required. The parameters derived from the Arrhenius plot of the rate constants are listed in Table 11. Employing the results for $\Delta p = 70$ MPa and $\Delta p = 150$ MPa, the quantity δg of Eq. (36) has been plotted in Fig. 30 as a function of Δn_H. It is seen that within the limits of error, δg is independent of Δn_H. Thus, the interaction parameter $\Gamma \simeq 0$, whereas from the ordinate one obtains the volume change per complex molecule as $\Delta V = 6.8 \pm 0.9\ \text{Å}^3$ or $4.1\ \mathrm{cm^3\,mol^{-1}}$ from the series with $\Delta p = 70$ MPa and as

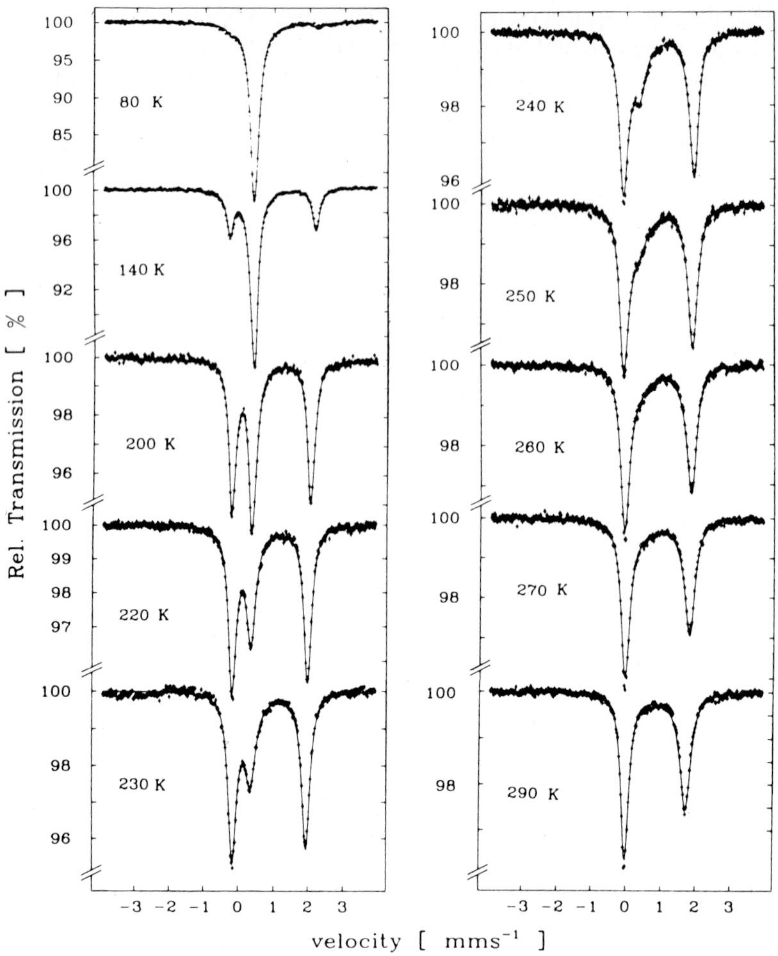

Fig. 28. ^{57}Fe-Mössbauer spectra of $[Fe(2\text{-pic})_3](PF_6)_2$ between 80 and 290 K. *Solid lines* are calculated on the basis of the stochastic theory of Blume using values for the rate constant k_{HL} listed in Table 10. According to Ref. [167]

$\Delta V = 8.0 \pm 0.7$ Å^3 or 4.8 $cm^3\,mol^{-1}$ from the series with $\Delta p = 150$ MPa. Both values are in good agreement.

$[Fe(6\text{-Mepy})_3tren](PF_6)_2$. Within the series of iron(II) complexes of the type $[Fe(6\text{-Mepy})_{3-x}(py)_x tren](PF_6)_2$ containing the hexadentate ligand tris[4-(6-R)-2-pyridyl]-3-aza-3-butenyl]amine, R = H or CH_3, the compounds with $x = 1$, 2 show a similar HS ↔ LS transition in solution as well as in the solid state [170]. In solution, spin-state conversion rates have been obtained as shown in Table 3. The complex with $x = 0$ is pure HS in solution, whereas in the solid state a gradual $^5T_2 \leftrightarrow {}^1A_1$ transition is observed extending over the range 180 to 300 K [165], cf. Fig. 31. Whereas the spectrum at 160 K is a pure LS

Table 10. Lifetimes of the $^5T_2$ state and rate constants k_{HL} for the intersystem crossing in $[Fe(2\text{-}pic)_3](PF_6)_2$ according to a numerical simulation of the Mössbauer spectra[a]

T K	$\tau(^5T_2)$ s	k_{HL} s^{-1}
200	1.6×10^{-6}	6.0×10^5
220	7.1×10^{-7}	1.4×10^6
240	3.6×10^{-7}	2.8×10^6
260	2.3×10^{-7}	4.2×10^6
280	1.7×10^{-7}	6.0×10^6
290	1.5×10^{-7}	6.6×10^6

[a] According to Ref. [167]

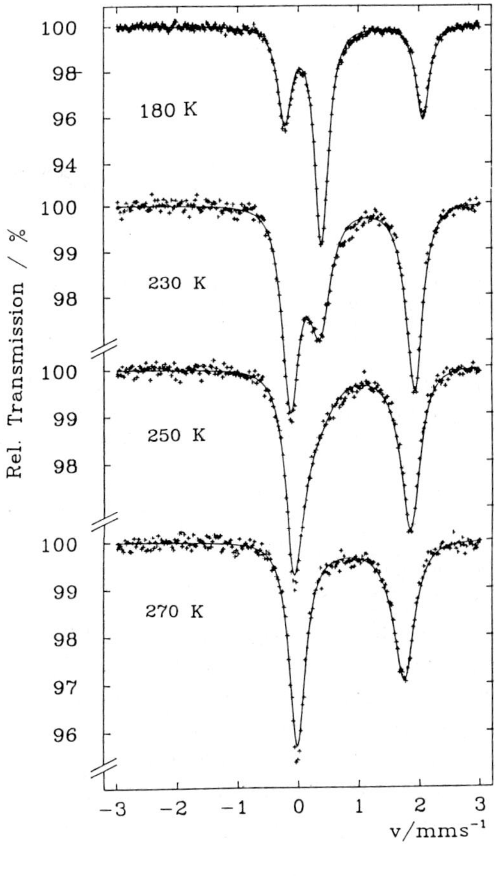

Fig. 29. ^{57}Fe-Mössbauer spectra of $[Fe(2\text{-}pic)_3](PF_6)_2$ for various temperatures recorded at $p = 150$ MPa. Spectra represented by *solid lines* were calculated using the stochastic theory of line shape by Blume. According to Ref. [169]

Table 11. Arrhenius parameters for the spin-state transition in $[Fe(2\text{-pic})_3](PF_6)_2$ at different pressures[a]

p MPa	ΔE_{HL} kJ mol^{-1}	ΔE_{LH} kJ mol^{-1}	ln A_{HL}	ln A_{LH}
0.1	14.3	23.2	22.0	27.9
70	13.3	25.0	21.5	28.8
150	14.0	24.1	22.0	28.2

[a] According to Ref. [169]

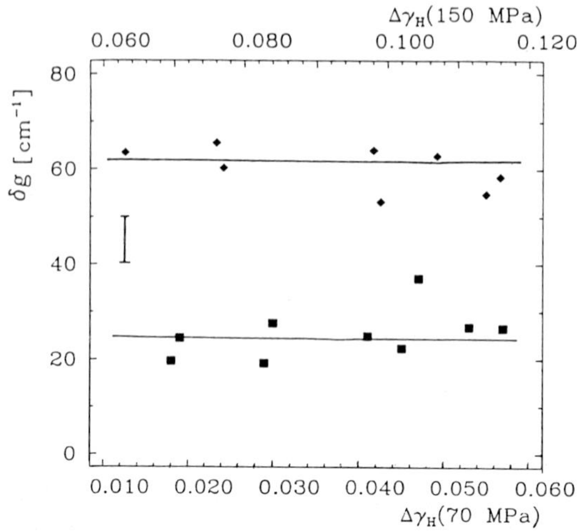

Fig. 30. Plot of the quantity δg of Eq. (35) as a function of $\Delta n_H(T)$ for $\Delta p = 70$ MPa (■) and $\Delta p = 150$ MPa (◆) for $[Fe(2\text{-pic})_3](PF_6)_2$. According to Ref. [169]. Notation in the figure uses Δg_H instead of Δn_H

doublet and, similarly, the spectrum at 320 K is a pure HS doublet, the spectra at 220 and 230 K show a superposition of both states. It is seen from the figure that the pure Mössbauer spectra are well reproduced by a sum of two Lorentzians, whereas deviations are observed for the composed spectra. These deviations are believed to arise from the dynamics of the HS ↔ LS transition. The spectra were therefore calculated employing the stochastic theory of lineshapes [31]. In the calculations, the same principal axes system for the efg in the HS and LS isomer was assumed, whereby opposite signs of V_{zz} were required. Figure 31 shows that the lineshapes of the Mössbauer spectra are well reproduced within the model, the activation energy resulting as $\Delta E_{HL} = 5.0 \pm 1.5$ kJ mol^{-1}.

The effect of a pressure of 80 and 150 MPa on the spin-state transition has been also studied [169], a series of spectra obtained at 150 MPa being shown in Fig. 32. The spectra show relaxation effects as line broadening and linewidth asymmetry. Calculated spectra were obtained in the same way as at ambient pressure. Rate constants for a number of temperatures are listed in Table 12, the parameter values resulting from an Arrhenius plot of the rate constants being listed in Table 13. In Fig. 33, the quantity δg of Eq. (36) has been plotted as a

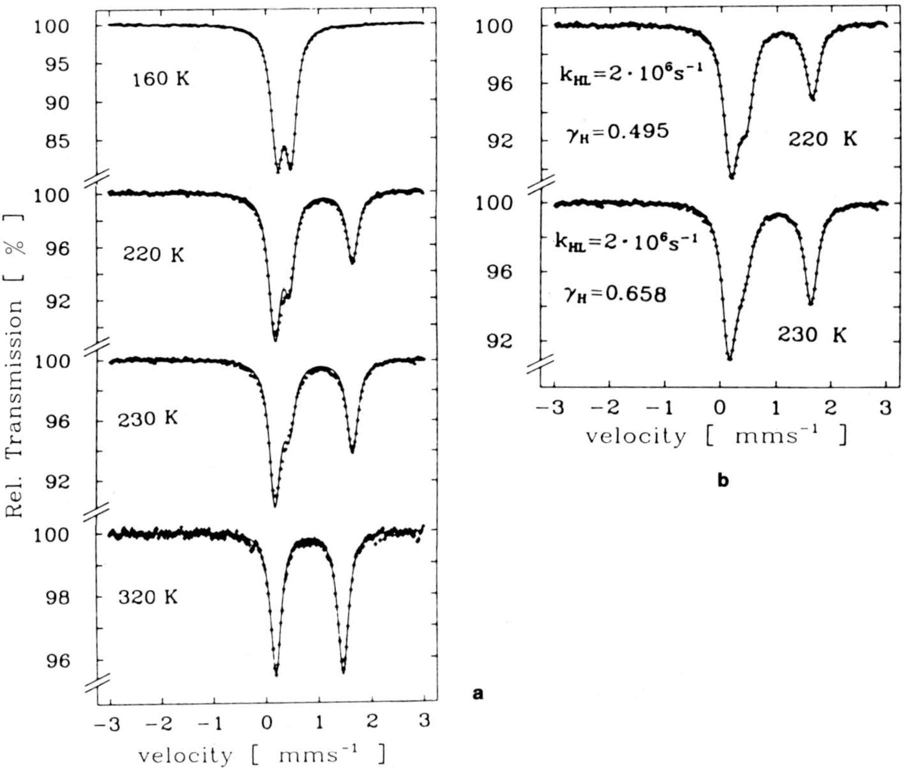

Fig. 31a. ^{57}Fe-Mössbauer spectra of $[Fe(6\text{-}Mepy)_3tren](PF_6)_2$ between 160 and 320 K. *Full lines* represent the best fit with a sum of Lorentzians; **b** ^{57}Fe-Mössbauer spectra at 220 and 230 K. Full lines are calculated by the stochastic theory of Blume using the rate constant $k_{HL} = 2 \times 10^6\ s^{-1}$, the values of the HS fraction being $n_H = 0.495$ at 220 and $n_H = 0.658$ at 230 K. According to Ref. [165]

function of Δn_H for $\Delta p = 70$ MPa. Thus δg increases linearly with increasing Δn_H giving the interaction parameter $\Gamma = 197\ cm^{-1}$ and the volume per complex molecule $\Delta V = 25 \pm 3$ Å^3 or $15\ cm^3\ mol^{-1}$. From Table 12 it may be seen that ΔE_{HL} for this complex is rather small. In fact, the total activation energy ΔE_{HL} consists of two contributions: the inner sphere reorganization energy ΔE^s_{HL} which takes account of the reorganization due to the spin-state change within the coordination sphere around the metal, and the lattice reorganization energy ΔE^{lat}_{HL} describing the response of the lattice. More detailed considerations suggest [169] that an exothermic ΔE^{lat}_{HL} partly compensates for the endothermic intramolecular contribution ΔE^s_{HL} and thus provides an explanation for the small value of ΔE_{HL}.

In order to determine the activation volume for the HS → LS conversion $\Delta V^{\ddagger}_{HL}$, Mössbauer spectra were recorded at 260 K for different pressures. A linear regression of $\ln k_{HL}$ versus pressure p following Eq. (55) gives $\Delta V^{\ddagger}_{HL} = -22 \pm 3$ Å^3. Thus $\Delta V^{\ddagger}_{HL}$ is negative and its absolute value is less than ΔV, the

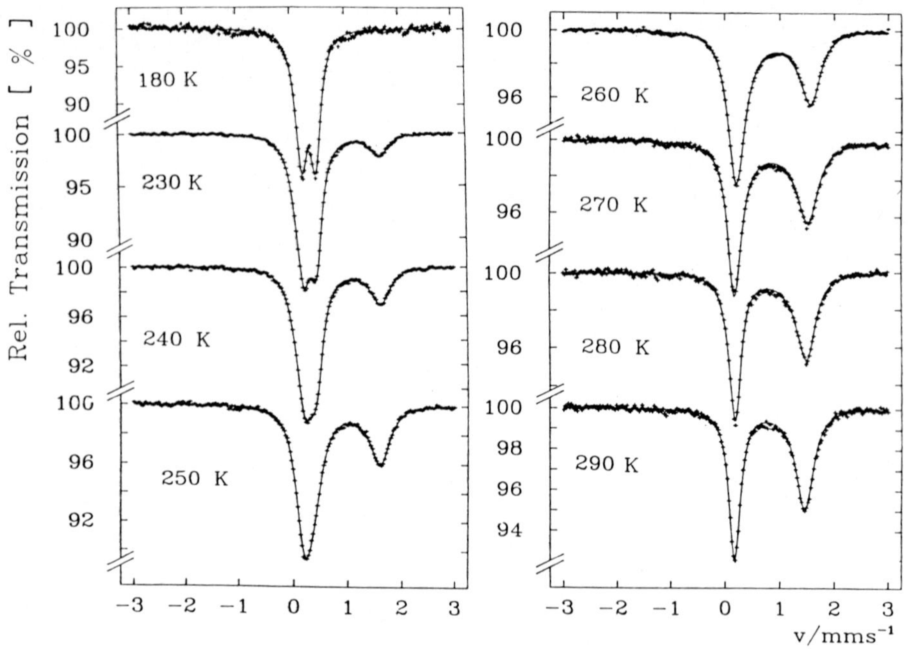

Fig. 32. ^{57}Fe-Mössbauer spectra of $[Fe(6\text{-}Mepy)_3tren](PF_6)_2$ for various temperatures recorded at $p = 150$ MPa. *Full lines* are spectra calculated using the stochastic theory of lineshape by Blume with $\mathrm{sgn}(V_{zz}^{H}) \neq \mathrm{sgn}(V_{zz}^{L})$. According to Ref. [169]

Table 12. Rate constants k_{HL} for the intersystem crossing in $[Fe(6\text{-}Mepy)_3tren](PF_6)_2$ under an external pressure of 150 MPa according to a numerical calculation of Mössbauer lineshape[a]

T K	n_H	k_{HL} s^{-1}
240	0.344	7.2×10^6
270	0.776	9×10^6
290	0.897	9×10^6

[a] According to Ref. [165]

Table 13. Arrhenius parameters for the spin-state transition in $[Fe(6\text{-}Mepy)_3tren](PF_6)_2$ at different pressures[a]

p MPa	ΔE_{HL} kJ mol^{-1}	ΔE_{LH} kJ mol^{-1}	ln A_{HL}	ln A_{LH}
0.1	5.0	31	17.1	31.3
80	4.2	32	17.5	32.0
150	2.4	32	17.4	31.6

[a] According to Ref. [169]

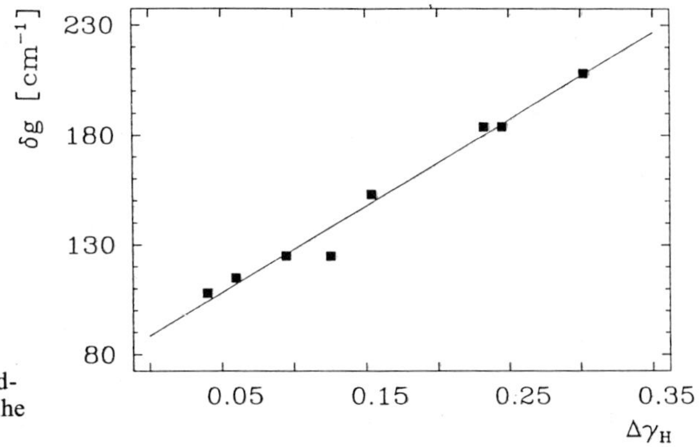

Fig. 33. Plot of the quantity δg of Eq. (35) as a function of Δn_H for $\Delta p = 70$ MPa for $[Fe(6\text{-}Mepy)_3tren](PF_6)_2$. According to Ref. [169]. Notation in the figure uses Δg_H instead of Δn_H

latter quantity corresponding to the change of the molal volume in solution, $\Delta V°$. This result indicates that the transition state in the present compound should be intermediate between the equilibrium configurations of the HS and LS state [169].

$[Fe(6\text{-}Mepy)_2(py)tren](ClO_4)_2$ Doped in PSS. Magnetic susceptibilities measured for a microcrystalline sample of the complex produce a magnetic moment value $\mu_{eff} = 0.36\ \mu_B$ at 10 K and 0.61 μ_B at 150 K, followed by a gradual increase to $\mu_{eff} = 2.80\ \mu_B$ at 311 K [138]. Thus ~ 26% of the complexes are in the HS state at 300 K if a magnetic moment of 5.1 μ_B is assumed for the pure HS compound. On the other hand, the complex doped into a polystyrene sulfonate (PSS) film does not provide any evidence for a thermal population of the HS state up to 340 K as demonstrated by variable-temperature UV-VIS and Mössbauer spectra. In fact, all the complexes doped into the PSS film are in the LS state at temperatures below 340 K. However, if irradiated by a single pulse of a Q-switched Nd/YAG laser (532 mμ), the complex is excited from the LS $^1A_1$ ground state to the HS $^5T_2$ states via an intermediate MLCT state and the metal $^3T_1$, $^3T_2$ states. The subsequent back relaxation from the excited $^5T_2$ state to the $^1A_1$ state has been followed between 300 and 4.2 K, the resulting relaxation rates being listed in Table 14. At temperatures above ~ 140 K, an Arrhenius plot of $\ln k$ versus $1/T$ gives a straight line with an apparent activation energy $\Delta E = 823\ \text{cm}^{-1}$, cf. Fig. 34. Below ~ 120 K, the relaxation rate k is relatively independent of temperature with $k = 1.4 \times 10^5\ \text{s}^{-1}$ in the 50–4.2 K range. The temperature independence of the rate is a direct evidence for quantum mechanical tunneling [171] which apparently occurs from the lowest vibrational level of the HS state through the barrier to the vibrational levels of the LS state.

The 300 – 4.2 K relaxation rate data for the PSS-doped complex were fitted to two different theoretical models. According to the model by Hopfield [172], electron tunneling between two molecular states is considered which are only very weakly interacting. The rate of tunneling between the two states a and b,

Table 14. Representative values of relaxation rate constants for $[Fe(6\text{-}Mepy)_2(py)tren](ClO_4)_2$ doped into PSS[a,c]

T K	$10^{-5}k^{b}$ s^{-1}	T K	$10^{-5}k^{b}$ s^{-1}
300	232.7	80	0.424
260	114.0	60	0.158
220	51.7	40	0.164
180	13.5	20	0.136
140	2.76	10	0.202
100	0.809	4.2	0.0875

[a] PSS = polystyrene sulfonate polymer
[b] $k = 1/\tau$, where τ is the relaxation time
[c] According to Ref. [138]

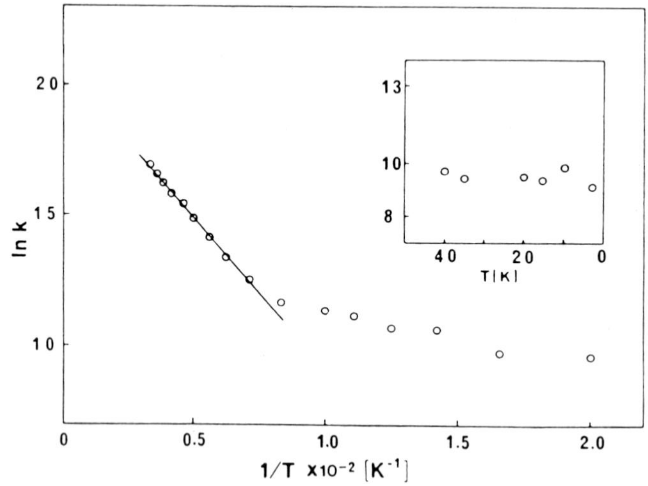

Fig. 34. Arrhenius plot of ln k versus $1/T$ for PSS-doped $[Fe(6\text{-}Mepy)_2(py)tren](ClO_4)_2$ for the temperature range 50 to 300 K. Here k is the relaxation rate constant, the *straight line* representing a least squares fit of the 150-300 K data producing $\Delta E = 823\ cm^{-1}$. The *insert* shows k versus T between 4.2 and 50 K. According to Ref. [138]

each with a single vibrational mode active, is determined by:

$$W_{ab} = \frac{2\pi}{\hbar}|T_{ab}|^2\left(\frac{1}{2\pi\sigma^2}\right)^{1/2}\exp[-(E_a - E_b - \Delta)^2/2\sigma^2] \tag{123}$$

where

$$\sigma^2 = (\tfrac{1}{2}k_a X_a^2)(k_B T_a)\coth\left(\frac{T_a}{2T}\right) \tag{124}$$

and

$$\Delta = \tfrac{1}{2}k_a X_a^2 + \tfrac{1}{2}k_b X_b^2 \tag{125}$$

Here, T_{ab} is the tunneling matrix element corresponding to the spin-orbit interaction between the $^1A_1$ and $^5T_2$ states, and $T_a = \hbar\omega_a/k_B$ is the temperature equivalent of the quantum of the active vibrational mode which is assumed to be the same in states a and b. Furthermore, $(E_a - E_b)$ is the difference of zero-point energies of the two states, $(E_a - E_b) = 400\ \text{cm}^{-1}$ being assumed, and $\frac{1}{2}k_a X_a^2$ reflects the reorganization energy for converting state a to state b along the potential energy curve for state a. The least squares fit of the relaxation data employs the parameter values [173] $T_{ab} = 6.4\ \text{cm}^{-1}$, $T_a = T_b = 287$ K and $\Delta = 0.790$ eV and is displayed in Fig. 35.

According to the model of Buhks et al. [117] the spin-state transition may be described as a radiationless nonadiabatic multiphonon process, the appropriate formalism having been reviewed in Sect. 6. The process involves two distinct spin states separated by a potential energy barrier which is large compared to $k_B T$, the resulting rate constant k being determined by Eq. (75). The vibrational overlap factor G involved in the expression for k is given by Eq. (79). In order to provide a numerical fit of the relaxation rates, the degeneracy change has been fixed to $g_f = 5$ and the reduced zero-point energy difference to $p = \Delta E_0/\hbar\omega = -1$. The fit employs the five parameters ζ, ΔE_1, ΔE_2, $\hbar\omega$, and S where ΔE_1 and ΔE_2 are defined by Eq. (81) and the coupling parameter S by Eq. (78). The parameter values actually used [173] are $\zeta = 59.6\ \text{cm}^{-1}$, $\Delta E_1 = 1.05 \times 10^4\ \text{cm}^{-1}$, $\Delta E_2 = 1.94 \times 10^4\ \text{cm}^{-1}$, $\hbar\omega = 312\ \text{cm}^{-1}$, and $S = 18.3$, the numerical fit being illustrated by Fig. 35.

Thus the relatively temperature independent relaxation rates for the PSS-doped complex at low temperature demonstrate that below ~120 K the prime

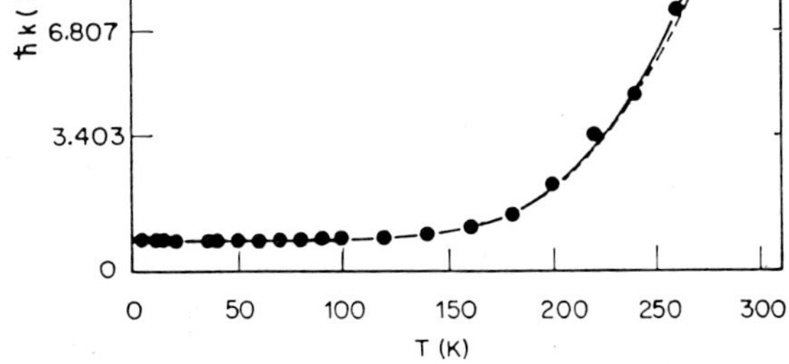

Fig. 35. Spin-state relaxation rate constant k versus temperature T for PSS-doped $[Fe(6\text{-}Mepy)_2(py)tren](ClO_4)_2$. Experimental data are indicated by *filled circles*. The *solid line* represents the fit to the tunneling model of Hopfield, the *dashed line* the fit to the quantum mechanical theory of Buhks et al. According to Ref. [138]

mechanism of ${}^5T_2 \rightarrow {}^1A_1$ interconversion is quantum mechanical tunneling. The analysis of the data over the range 300 – 4.2 K suggests that tunneling is also an important mechanism for spin-state conversion above ~140 K. The apparent activation energy of 823 cm^{-1} reflects a Boltzmann distribution of complexes over different vibrational levels. Apparently, each level is characterized by its intrinsic rate of tunneling.

[Fe(mtz)$_6$](PF$_6$)$_2$. The solid iron(II) complex hexafluorophosphate of the ligand mtz = 1-methyltetrazole shows a gradual ${}^5T_2 \leftrightarrow {}^1A_1$ transition which is centred at $T_c \simeq 204$ K [174]. The Mössbauer spectrum consists at 150 K of a sharp single line characteristic for LS iron(II), the spectra above 280 K showing a quadrupole doublet due to HS iron(II) which is also sharp. The spectra at intermediate temperatures are the superposition of LS and HS spectra which strongly overlap, texture effects being involved in addition, cf. Fig. 36. The spectra were therefore calculated employing a formalism which takes account of texture. The solid lines in the figure were obtained with a single texture component $t_{00}^2 = -1.57$ assuming an axially symmetric efg tensor for the HS state without considering relaxation effects. The pure LS spectrum is well reproduced as is the HS spectrum, whereas spectra consisting of the superposition of both species show differences which are attributed to the dynamics of the LS↔HS transition. If the numerical fit is based on the stochastic theory of Mössbauer lineshape, taking additionally account of texture, the spectra at 213 and 243 K are also reproduced well, cf. Fig. 37. The results from evaluation of

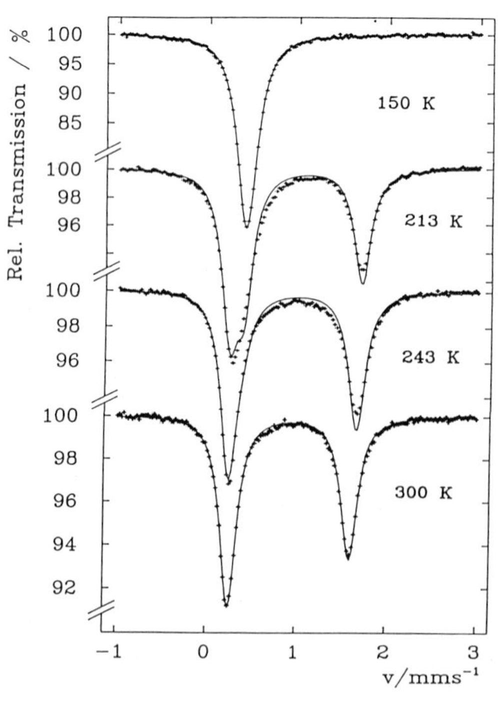

Fig. 36. ^{57}Fe-Mössbauer spectra of [Fe(mtz)$_6$](PF$_6$)$_2$ at the temperatures of 150, 213, 243, and 300 K. *Solid lines* are spectra calculated with account of texture but without consideration of relaxation effects. According to Ref. [174]

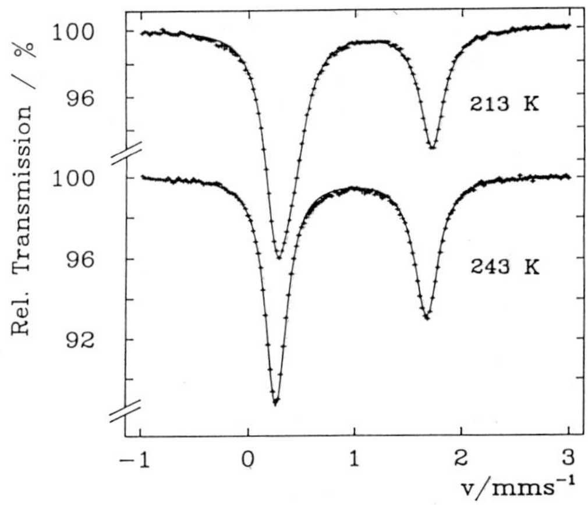

Fig. 37. ^{57}Fe-Mössbauer spectra of $[Fe(mtz)_6](PF_6)_2$ at 213 and 243 K. *Solid lines* represent spectra calculated on the basis of a relaxation model inclusive of texture. According to Ref. [174]

Table 15. Rate constants k_{HL} for the spin-state transition in $[Fe(mtz)_6](PF_6)_2$ derived on the basis of a relaxation model[a]

T K	n_H	k_{HL} $10^6\ s^{-1}$
150	0	
185.1	0.117	2.8
192.5	0.233	2.3
199.7	0.354	1.8
203.7	0.482	1.8
210.3	0.616	1.8
213.2	0.701	2.0
221.3	0.804	1.8
229.7	0.856	2.0
243.1	0.921	2.1
300.0	1	

[a] According to Ref. [174]

the spectra on the basis of the relaxation model are summarized in Table 15. The Arrhenius plot of the rate constants shows no temperature dependence of k_{HL} and thus a zero activation energy for the HS → LS conversion results, whereas from the values of k_{LH}, the activation energy $\Delta E_{LH} = 28 \pm 2\ kJ\,mol^{-1}$ and the frequency factor $A \simeq 3 \times 10^{13}\ s^{-1}$ are obtained. The lack of a temperature dependence of the rate constant for the exothermic HS → LS conversion could arise if the intramolecular contribution to ΔE_{HL} were compensated by a lattice contribution. The small value of ΔE_{HL} for the spin-state transformation of $[Fe(6\text{-}Mepy)_3tren](PF_6)_2$ was attributed to such a situation [169]. Alternatively, the temperature independent rate constant could be interpreted as evidence for a tunneling mechanism as found for $[Fe(6\text{-}Mepy)_2(py)tren](ClO_4)_2$ in a polymer matrix [201].

9 The Molecular Reorganization Associated with Spin-State Equilibria in Solution and the Structural Changes Accompanying Spin-State Transitions in Solid Metal Complexes

Spin-state equilibria of transition-metal complexes in solution are accompanied by a variation of the nuclear configuration which may be characterized, at least in part, by the detailed study of relaxation kinetics. It is evident that the ability of the coordination sphere to reorganize is determined by the initial geometry of the coordination core, the presence of Jahn-Teller distortions, the ligand rigidity, the existence of intra- or intermolecular steric interactions and other effects. Of great importance is the nature of the metal-ligand bonds, Me–N bonds being generally far more affected by the spin-state change than Me–O or Me–S bonds [62, 110, 175, 176].

It has been pointed out above that the molecular reorganization in a spin transition molecule is closely related to the transfer of electrons between the nonbonding t_{2g} and the formally antibonding e_g^* orbitals, the metal-ligand bond variations in spin transition complexes of cobalt(II) ($\Delta S = 1$) being expected to be smaller than in iron(II) or iron(III) complexes ($\Delta S = 2$). In addition, the configuration change involves modifications of bond angles between the metal and the ligand donor atoms, changes of intraligand bond lengths and angles and possibly also changes of vibrations and orientations of the surrounding solvent molecules.

In solution studies, the modification of the equilibrium nuclear configuration appears primarily as a change of the partial molar volume ΔV° of the two spin states. The presently available values of ΔV° for spin conversions in solution are collected in Table 16. There is no apparent difference between the values for iron(II) and iron(III) spin transition compounds, the variation being

Table 16. Differences of molar volume ΔV° between LS and HS isomers of spin transition complexes on the basis of solution relaxation measurements and pressure studies

Compound	Temp. Range K	$\Delta V^{\circ a}$ $cm^3\,mol^{-1}$	Ref.
$Fe^{II}[HB(pz)_3]_2$	278–313	23.6 ± 1	[29]
$[Fe^{II}(paptH)_2]Cl_2$	283–313	11 ± 0.3	[29]
$[Fe^{II}(pyim)_3](BPh_4)_2{}^b$	298	8.1 ± 1.3	[89, 91]
$[Fe^{II}(pybim)_3](BPh_4)_2{}^b$	298	9.6 ± 0.4	[91]
$[Fe^{II}(phenmethoxa)_2](BF_4)_2{}^b$	298	12.3 ± 1.0	[89]
$[Fe^{III}(acac_2trien)]NO_3$	278–298	10.3 ± 0.2	[94]
$[Fe^{III}(Sal_2trien)]NO_3 \cdot H_2O$	278–298	11.9 ± 0.1	[94]
$[Co^{II}(terpy)_2]Cl_2 \cdot 2H_2O$	298	10.1 ± 0.4	[177]

[a] Average values over temperature range stated
[b] In acetone

between 8.1 cm^3 mol^{-1} for $[Fe(pyim)_3](BPh_4)_2$ [89] and 23.6 cm^3 mol^{-1} for $Fe[HB(pz)_3]_2$ [29].

Interesting is a comparison of the volumes occupied by individual complexes in solution and in the solid state. The partial molal volumes can be obtained from precise measurements of the solution densities of the complexes as a function of concentration [177]. These values may be subsequently compared with the unit cell volumes per complex molecule derived from the crystal structure. For $Fe[HB(pz)_3]_2$, the apparent molal volume in tetrahydrofuran solution was determined as 340.9 cm^3 mol^{-1}. Taking into account that the complex in solution forms an equilibrium between 86% LS and 14% HS isomers and employing the volume difference between the two spin states $\Delta V^\circ = 23.6$ cm^3 mol^{-1}, the volume of the LS isomer was calculated as 337.6 cm^3 mol^{-1}. This value agrees closely with the volume of 337.3 cm^3 mol^{-1} for the completely LS complex in solid $Fe[HB(pz)_3]_2$ [105].

The partial molal volume of the cation $[Fe(acac_2trien)]^+$ in aqueous solution was derived from the measured value of 301.1 cm^3 mol^{-1} for the nitrate, subtracting therefrom the partial volume for the anion (34 cm^3 mol^{-1}), as 267 cm^3 mol^{-1}. Taking account of the equilibrium existing in solution between 67% LS and 33% HS isomers and the volume difference between the spin states of $\Delta V^\circ = 10.3$ cm^3 mol^{-1}, the partial molal volume for the HS isomer results as 274 cm^3 mol^{-1}. A crystal structure is available [110] for $[Fe(acac_2trien)]PF_6$ which is HS and which possesses a molal volume of 341.4 cm^3 mol^{-1}. Applying a correction for the PF_6 anion (66 cm^3 mol^{-1}) gives the volume of the cation in the solid as 275 cm^3 mol^{-1} which is again in excellent agreement with the volume in solution. Similarly, the calculated solution value for the LS cation $[Fe(Sal_2trien)]^+$ of 272 cm^3 mol^{-1} is in agreement with the solid state volume of 275 cm^3 mol^{-1} based on the crystal structure of the chloride salt of the complex [110]. Moreover, the partial molal volume of the LS cation $[Co(terpy)_2]^{2+}$ was calculated from solution studies as 368 cm^3 mol^{-1} and thus compares favorably with the crystal volume of 372 cm^3 mol^{-1} [132]. The close agreement between the volume occupied by the complex in the crystal and the partial molal volume in solution suggests that the intermolecular or interionic contacts in the crystal are similar to the solvation of the complexes or complex cations in solution.

A direct measurement of the change in bond length between the HS and LS isomers of a given complex in solution is not possible. An estimate may be obtained if the complexes are considered as simple spheres [177]. Thus the effective spherical radius of the LS complex $Fe[HB(pz)_3]_2$ may be calculated from its partial molal volume of 337.6 cm^3 mol^{-1} as $r_L = 5.115$ Å. From the relation:

$$\Delta V^\circ = \frac{4\pi}{3} N_L (r_H^3 - r_L^3) \tag{126}$$

for the volume difference and the known value of ΔV°, the spherical radius of the HS isomer results as $r_H = 5.232$ Å and thus the change in spherical radius due to the spin-state conversion is obtained as $\Delta r = 0.12$ Å. The difference between the Fe–N distances in the crystal structures of LS $Fe[HB(pz)_3]_2$ and HS $Fe[HB(3,5\text{-}(CH_3)_2pz)_3]_2$ is found to be $\overline{\Delta R} = 0.199$ Å [105], however. Apparently, the calculated value of 0.12 Å is a minimum value since the expansion of the core is directed mainly along the metal-ligand bonds and thus the complex is not a perfect sphere but rather of irregular periphery.

For the iron(III) complexes $[Fe(acac_2trien)]^+$ and $[Fe(Sal_2trien)]^+$, the partial molal volumes of 263.6 and 271.5 $cm^3\,mol^{-1}$ correspond to spherical radii of the LS complexes of $r_L = 4.710$ and 4.757 Å, respectively. Employing the volume differences ΔV°, the differences in spherical radii result as 0.061 and 0.069 Å, respectively. On the other hand, the average iron-donor atom bond length change between the LS $[Fe(Sal_2trien)]Cl_2 \cdot 2H_2O$ and the HS $[Fe(acac_2trien)]PF_6$ has been determined as $\overline{\Delta R} = 0.126$ Å [110]. The solution value for the change may be improved if it is assumed that the complex remains solvated in the same way in the two spin states. This seems to be a consequence of the open nature of these complexes which is revealed in their crystal structures. Taking the solvated radius of the metal-ligand core as the average metal-ligand distance of 1.939 Å plus the average van der Waals radius of the coordinated atoms of 1.467 Å, the volume differences correspond to metal-ligand bond length changes of 0.114 and 0.131 Å, respectively.

For $[Co(terpy)_2]^{2+}$, the partial molal volume of 368 $cm^3\,mol^{-1}$ corresponds to a spherical radius of 5.264 Å and in combination with ΔV°, the difference in spherical radii obtains as only $\Delta r = 0.048$ Å. This value has to be considered again as a lower limit since the bond length expansion is concentrated along a single axis of the complex. According to crystal structure data [132, 152], the spin-state change consists of a lengthening of the bond to the central N atom of each ligand by as much as 0.21 Å, whereas the bonds to the two distal N atoms change by only 0.07 Å.

In summary, the calculated changes of spherical radii between the spin states are much smaller than the actual variation of metal-ligand bond distances in the crystal. This suggests that the complex remains solvated to much the same extent in both spin states. For the ligands deriving from triethylenetetramine in which the coordinated atoms are accessible to hydrogen bonding with the solvent, good agreement is obtained if solvation of the first coordination sphere of the complex is assumed.

The structural changes associated with a spin-state transition in solid metal complexes are much more amenable to experimental studies. The largest amount of information derives from single crystal X-ray structure investigations, particularly if followed over a range of temperatures. A recent review concentrating on results of this kind is available [6]. In principle, neutron diffraction, EXAFS (extended X-ray absorption fine structure) and FIR spectroscopy as well as various other methods should contribute to the resulting conception, although the availability of such data is extremely limited.

The X-ray structure results clearly establish the volume expansion within the course of the LS → HS transition in these compounds. Unfortunately, there exist no multiple temperature structure studies of the particular compounds for which ΔV° values from solution measurements are available. Above we have compared partial molal volumes from solution with the volumes occupied by a single complex in the crystal. In the following, we compare unit cell volume changes referred to a single metal atom with ΔV° values from solution as listed in Table 16. Thus the unit cell volume of the iron(II) complex $[Fe(2\text{-pic})_3]Cl_2 \cdot C_2H_5OH$ where 2-pic = 2-picolylamine is found [178] to increase by 35.3 Å^3 per Fe atom or 21.3 cm^3 mol^{-1} and that of $[Fe(2\text{-pic})_3]Cl_2 \cdot CH_3OH$ [179] by 24.6 Å^3 per Fe atom or 14.8 cm^3 mol^{-1} in the course of a LS → HS conversion. It is evident that these values do not differ markedly from the ΔV° values of Table 16. In the recently reported [180] complex $[Fe(4\text{-mpt})_3](ClO_4)_2$ where 4-mpt = 4-methyl-2-(pyridin-2-yl)thiazole, the volume difference estimated for the complete LS ⇌ HS transformation is 32.1 Å^3 per Fe atom or 19.3 cm^3 mol^{-1} again consistent with the ΔV° values for iron(II). Many spin transitions are incomplete at the lower or higher temperature end of the measurements. Therefore, values of the unit cell volume for the pure HS and LS state V_H and V_L have been obtained assuming a linear variation with n_H, the HS fraction, and $n_L = 1 - n_H$, the LS fraction, respectively. The HS fraction n_H, on the other hand, has been estimated from the temperature dependence of the effective magnetic moment squared, μ_{eff}^2, which is the weighted sum of $(\mu_{HS})^2$ and $(\mu_{LS})^2$, the squares of the limiting values for the pure HS and the pure LS state, respectively:

$$[\mu_{eff}(T)]^2 = n_H(\mu_{HS})^2 + (1 - n_H)(\mu_{LS})^2 \tag{127}$$

An interesting example is the iron(III) dimethyldithiocarbamato complex [181] $[Fe((CH_3)_2NC(S)S)_3]$ where the unit cell amounts to 44.4 Å^3 per Fe atom or 26.7 cm^3 mol^{-1}. This value is somewhat higher than the ΔV° values of Table 16. It should be noted, however, that the dithiocarbamato complex is of $[Fe^{III}\text{–}S_6]$ type, whereas the iron(III) complexes in Table 16 are examples of the type $[Fe^{III}\text{–}N_4O_2]$. As far as cobalt(II) is concerned, for the complex $[Co(nnp)(NCS)_2]$ where nnp = *N*-[(diphenylphosphino)ethyl]-*N*′-diethylethylenediamine [182, 183], the volume change has been obtained as 21.5 Å^3 per Co atom or 13.0 cm^3 mol^{-1}, whereas for $[Co(terpy)_2]I_2 \cdot 2H_2O$ [132] where terpy = 2,2′,2″-terpyridine, the corresponding results are 33.7 Å^3 per Co atom or 20.3 cm^3 mol^{-1}. Again this value is higher than the ΔV° value for the closely related $[Co(terpy)_2]Cl_2$ complex of Table 16.

The overall volume expansion which occurs in the course of the LS → HS transformation of a solid metal complex is the result of various more specific changes within the individual complex molecules. The results of X-ray structure investigations show [6] that these changes comprise an increase of metal-ligand bond length, a decrease of the donor atom-metal-donor atom bite angle for bidentate ligands, sometimes represented by a decrease of the corresponding ligand bite (e.g., the S–C–S angle in dithiocarbamato ligands), and various modifications of bond distances and angles within the ligands that are dependent on the detailed geometry of the ligand, the disposition of the ligands about the central metal atom, and the crystal packing. Of these changes, the variation of metal-ligand bond length ΔR appears to be the most significant. Thus the volume difference between the HS and LS isomers considered above depends to a large extent on ΔR. The magnitude of the molecular expansion or contraction is involved directly or indirectly in several theoretical models of spin transition [40, 44, 74, 76, 78]. The vibrational contribution to the entropy change associated with the spin transition mainly depends on the frequency variation of the metal-ligand stretching mode and thus on ΔR [41]. Spin-state lifetimes or equally spin interconversion rates in solid metal complexes are largely influenced by the extent of molecular reorganization, in particular by the magnitude of ΔR [134]. Various other properties such as cooperativity and the width of hysteresis [24, 54] show a volume or ΔR dependence.

Bond length differences between HS and LS isomers have been determined for a number of iron(II), iron(III) and cobalt(II) complexes on the basis of multiple temperature X-ray diffraction structure studies [6]. The available results have been collected in Table 17. Average values for the bond length changes characteristic for a particular transition-metal ion have been extracted from these data and are obtained as $\overline{\Delta R} \simeq 0.17$ Å for iron(II) complexes, $\overline{\Delta R} \simeq 0.13$ Å for iron(III) complexes, and $\overline{\Delta R} \simeq 0.06$ Å for cobalt(II) complexes. These values may be compared with the differences of ionic radii between the HS and LS forms of iron(II), iron(III) and cobalt(II) which were estimated some time ago [184] as 0.16, 0.095, and 0.085 Å, respectively.

The $\overline{\Delta R}$ value for cobalt(II) resulting from the data of Table 17 is not representative. Firstly, data are available for only two structures. Secondly, the complex $[Co(nnp)(NCS)_2]$ is five-coordinate, while for $[Co(terpy)_2]I_2 \cdot 2H_2O$ the spin transition is rather incomplete at the higher temperature limit of the measurements. Since, in addition, the magnetism of $[Co(terpy)_2]^{2+}$ complexes is dependent on subtle properties of the sample such as the amount of hydration [152], extrapolations of $\bar{R}_{HS}$ values on the basis of the magnetic moment are somewhat uncertain. In order to obtain a better estimate of the magnitude of metal-ligand bond length change for cobalt(II), we present in Table 18 an adaptation of a table of Thuéry et al. [8] containing bond lengths for a number of related cobalt(II) complexes from single crystal X-ray structure studies. According to these results, the average bond length for five-coordinate square-pyramidal cobalt(II) complexes is 2.03 Å for the HS case and 1.94 Å for the LS case, thus giving an average bond length change $\overline{\Delta R} \simeq 0.09$ Å. The variation of bond length due to a spin-state transition depends to a high extent on the nature of the ligand donor atoms. Thus for Co–N bonds one finds $\overline{\Delta R}_{eq} \simeq 0.15$–$0.21$ Å, whereas for Co–O bonds it is $\overline{\Delta R}_{eq} \simeq 0.04$–$0.05$ Å. Similar results have been obtained for iron(III) compounds with a N_4O_2 donor atom set [110, 176] as well as for iron compounds with salen as ligand [62, 175]. In addition, the cobalt(II) compounds of Table 18 being mostly five-coordinate, differences arise for equatorial versus axial bonds. In general, the bond length variations are more pronounced for equatorial bonds where $\overline{\Delta R}_{eq} \simeq 0.10$–$0.14$ Å while for the

Table 17. Average metal-ligand bond lengths $\bar{R}$ for HS and LS isomers and bond length variation $\overline{\Delta R}$ for complete HS ↔ LS conversion of spin transition compounds

Compound[a]	Donor atoms	$\bar{R}$(HT)[b] Å	$\bar{R}$(LT)[b] Å	$\overline{\Delta R}$[c] Å	Ref.
$[Fe(2\text{-pic})_3]Cl_2 \cdot C_2H_5OH$	N_6	2.195	2.013	0.182	[178]
$[Fe(2\text{-pic})_3]Cl_2 \cdot CH_3OH$	N_6	2.198[d]	2.016	0.23[e]	[179]
$[Fe(phen)_2(NCS)_2]$, polymorph II	N_6	2.156	1.992	0.164	[11]
$[Fe(bpy)_2(NCS)_2]$	N_6	2.14[f]	2.02[f]	0.15	[12]
$[Fe(4\text{-mpt})_3](ClO_4)_2$	N_6	2.191[d]	2.034	0.17[e]	[180]
$[Fe(6\text{-Mepy})_3tren](PF_6)_2$	N_6	2.213	2.093[g]	0.16[e]	[170]
$[Fe_3(Ettrz)_6(H_2O)_3](CF_3SO_3)_6$	N_6[h]	2.174	2.031	0.143	[185]
$[Fe(dppen)_2Cl_2] \cdot 2(CH_3)_2CO$	P_4Cl_2	2.510	2.310	0.2000	[103]
$[Fe(hptpd)Br]BPh_4 \cdot CH_2Cl_2$	P_4Br	2.261[d]	2.229[g]	0.13[e]	[186]
$[Fe(OEP)(3\text{-Clpy})_2]ClO_4$	N_6	2.074[d]	2.007	0.13[e]	[187]
$[Fe(TPP)(NCS)py]$[i]	N_6	2.040[d]	1.978	0.12[e]	[188]
$[Fe(salen)(imd)_2]ClO_4$	N_4O_2	2.038	1.936	0.102	[189]
$[Fe(3\text{-OEt-SalAPA})_2]ClO_4 \cdot C_6H_6$	N_4O_2	2.060	1.946	0.114	[100]
$[Fe(Me_2dtc)_3]$	S_6	2.415[d]	2.303	0.18[e]	[181, 190]
$[Fe(Et_2dtc)_3]$	S_6	2.358[d]	2.306[g]	0.12[e]	[131]
$[Fe(4\text{-mpdtc})_3] \cdot CH_2Cl_2$	S_6	2.427[d]	2.357[g]	0.13[e]	[191]
$[Fe(2\text{-HOEt}_2dtc)_3]$	S_6	2.390[d]	2.331[g]	0.15[e]	[192]
$[Fe(NCPr_2dtc)_3]$	S_6	2.324[d]	2.308[g]	0.10[e]	[193]
$[Fe(Bzdtc)_3]$	S_6	2.349[d]	2.314[g]	0.18[e]	[194]
$NH_4[Fe(3,5\text{-Cltsa})_2] \cdot 1.5H_2O$[j]	$N_2O_2S_2$	2.144[d]	2.069[g]	0.13[e]	[195]
$K_4[Fe(3,5\text{-Cltsa})_2] \cdot 1.5H_2O$[j]	$N_2O_2S_2$	2.092[d]	2.022[g]	0.23[e]	[196]
		2.137[d]	2.048[g]	0.16[e]	
$[Fe(salen)NO]$	N_3O_2	1.950	1.910	0.05	[62, 175]
$[Co(nnp)(NCS)_2]$	N_4P	2.09[d]	2.06	0.06[e]	[182]
$[Co(terpy)_2]I_2 \cdot 2H_2O$	N_6	2.023[d]	1.998	0.05[e]	[132]

[a] Ligand abbreviations: 2-pic = 2-aminomethylpyridine; phen = 1,10-phenanthroline; bpy = 2,2'-bipyridine; 4-mpt = 4-methyl-2-(pyridin-2-yl)thiazole; (6-mepy)$_3$tren = tris[4-(6-methyl-2-pyridyl)-3-aza-3-butenyl]amine; Ettrz = 4-ethyl-1,2,4-triazole; dppen = *cis*-1,2-bis(diphenylphosphino)-ethylene; hptpd = hexaphenyl-1,4,7,10-tetraphosphadecane; OEP = octaethylporphyrinato dianion; TPP = *meso*-tetraphenylporphinato dianion; salen = *N,N'*-ethylenebis(salicylaldiminato) dianion; imid = imidazole; 3-OEt-SalAPA = monoanion of Schiff base condensate of 3-ethoxysalicylaldehyde and *N*-(3-aminopropyl)aziridine; Me$_2$dtc = *N,N'*-dimethyldithiocarbamato anion; Et$_2$dtc = *N,N'*-diethyldithiocarbamato anion; 4-mpdtc = 4-morpholinecarbodithiolato anion; 2-HOEt$_2$dtc = *N,N'*-bis(2-hydroxyethyl)dithiocarbamato anion; NCPr$_2$dtc = *N,N'*-dipropionitriledithiocarbamato anion; Bzdtc = *N,N'*-dibenzyldithiocarbamato anion; 3,5-Cltsa = 3,5-dichlorosalicylaldehyde thiosemicarbazonato anion; nnp = *N*-[2-(diphenylphosphino)ethyl]-*N'*-diethyl ethylenediamine; terpy = 2,2',2''-terpyridine

[b] Highest temperature (HT) and lowest temperature (LT) data, not necessarily representing the pure HS and LS isomers

[c] Average metal-ligand bond length difference between pure HS and LS isomers

[d] Lower limit to R_{HS}. Value of $\bar{R}$(HT) corresponds to incomplete LS → HS transformation

[e] Value estimated for a complete HS ⇌ LS transformation, by assuming a linear dependence of $\bar{R}$ on n_H, the latter being obtained from the effective magnetic moment employing Eq. (127)

[f] Low accuracy data

[g] Upper limit to R_{LS}. Value of $\bar{R}$(LT) corresponds to incomplete HS → LS transformation

[h] Donor atom set of central Fe atom, the only one showing a spin-state transition

[i] Two crystallographically independent iron(III) atoms involved, only one showing a spin-state transition

[j] Two crystallographically independent iron(III) atoms involved, both showing a spin-state transition with different n_H values in the potassium salt, whereas only one shows a spin-state transition in the ammonium salt

Table 18. Bond lengths $\bar{R}$ for equatorial and axial bonds in HS and LS Cobalt(II) complexes with salen and salen-type ligands

Compound[a]	Spin state	$(Co{-}N)_{eq}$[b] Å	$(Co{-}O)_{eq}$[b] Å	$(Co{-}N)_{ax}$[c] Å	$(Co{-}O)_{ax}$[c] Å	$\bar{R}$[d]	Ref.
[Co(3-MeOsalen)(H_2O)]	HS	2.05	1.95		2.12	2.02	[197]
[Co(salen)(py)]	LS	1.90	1.90	2.10		1.94	[198]
[Co(salbn)(py)]	LS	1.87	1.90	2.14		1.94	[199]
[Co(salphen)(2-MeImd)]	HS	2.08	1.97	2.07		2.04	[200]
[Co(H_2fsa$_2$en)(py)$_2$]	HS	2.135	2.023	2.262		2.14	[201]

[a] Ligand abbreviations: salen = *N,N′*-ethylenebis(salicylaldimine) dianion; salbn = *N,N′*-butylenebis(salicylaldimine) dianion; salphen = *N,N′-o*-phenylenebis(salicylaldimine) dianion; 2-MeImd = 2-methylimidazole; H_4fsa$_2$en = *N,N′*-ethylenebis(3-carboxysalicylaldimine)
[b] eq = equatorial
[c] ax = axial
[d] Average metal-ligand bond lengths

axial cobalt-ligand bonds the variation of $\overline{\Delta R}_{ax}$ is from + 0.02 to − 0.07 Å for the LS → HS transformation. Since the axial bond lengths are not appreciably modified upon spin conversion, it is expected that the overall $\overline{\Delta R}$ values should be rather small as are the changes of molecular volume. For [Co(H_2fsa$_2$en)(py)$_2$] e.g. where $Z = 4$, the unit cell volume changes from 2491 Å^3 at 128 K to 2568 Å^3 at 294 K [201], i.e., by 19.25 Å^3 per Co atom or 11.6 cm^3 mol^{-1}. The volume change for the complete spin conversion will be even somewhat larger since the value at 128 K does not correspond to the pure LS state. Recently, EXAFS measurements have been employed in order to obtain estimates for the average Co-ligand bond length change in [Co(H_2fsa$_2$en)(H_2O)$_2$] and [Co(H_2fsa$_2$en)(py)$_2$] [8]. According to these results, $\overline{\Delta R}$ has been extrapolated for a complete spin conversion in the first compound to be 0.09 Å, whereas it was shown not to differ significantly from the experimental value of 0.12 Å in the second compound. These $\overline{\Delta R}$ values are in good agreement with the average bond length changes obtained on the basis of X-ray structure studies.

The observation of different metal-ligand bond lengths, bite angles and various minor features of the ligands implies that discrete structures are associated with each of the two spin states of a complex. This conclusion may be arrived at in a number of ways and it has in fact been invoked more than once in the course of this study. However, this situation has been most impressively demonstrated by the crystallographic resolution of the spin isomers of the iron(II) complex [Fe(2-pic)$_3$]Cl$_2$ · CH$_3$OH over a range of temperatures by Katz and Strouse [179]. The gradual spin-state transition in a mononuclear complex may thus be described by the variation of the fractional occupancy of the two discrete structures corresponding to the two spin isomers of the compound.

It should be noted that the structural changes which occur in the course of a spin-state transition are concentrated in the central region of the molecule and these changes are considerably larger than any modification encountered in peripheral areas. The central part of the molecule is directly influenced by electronic changes on the metal atom and thus the changes of molecular structure are directly related to the rearrangement of t_{2g} and e_g electrons of the metal. In view of these facts, it is justified to consider the change of spin-state and the concurrent modification of molecular structure as the *driving force* of the macroscopic change of crystal structure. It should be noted that other authors have often argued the other way around. Thus order-disorder phenomena [178], changes in hydrogen bonding and structural phase transitions were frequently considered as the primary phenomenon. This concept seemed to be supported by the entropy changes involved. The spin change $\Delta S = 2$, i.e. from $S = 0$ to $S = 2$, e.g. in an octahedral iron(II) complex, results in an entropy change $\Delta S = -Nk_B[\ln(2S+1)_{HS} - \ln(2S+1)_{LS}] = -Nk_B \ln 5 = 13.38\ \mathrm{J\,K^{-1}\,mol^{-1}}$, whereas for a $\Delta S = 1$ change one obtains $\Delta S = -Nk_B[\ln 4 - \ln 2] = -5.76\ \mathrm{J\,K^{-1}\,mol^{-1}}$. These values should be compared with values for the total entropy change of between 27.5 and 65 $\mathrm{J\,K^{-1}\,mol^{-1}}$ for iron(II) and 10 to 18 $\mathrm{J\,K^{-1}\,mol^{-1}}$ for cobalt(II) complexes which were derived from absolute calorimetry measurements or DSC studies. These data are summarized in Table 19. It has been demonstrated [41, 202] that the major portion of the entropy change ΔS originates from a variation of vibrational frequencies rather than the spin change proper, the entropy acquisition due to the Fe–N stretching mode alone being 17.4 and 22.5 $\mathrm{J\,K^{-1}\,mol^{-1}}$ for $[Fe(phen)_2(NCS)_2]$ and $[Fe(phen)_2(NCSe)_2]$, respectively. Obviously, the change of vibrational frequencies is a consequence of the variation in bond distances and bond angles, and these are a part of the rearrangement of the molecule which occurs in the course of the spin-state transition.

The volume expansion accompanying the spin-state transition has been demonstrated, at least in quantitative terms, on the basis of multi-temperature crystal structure studies [6]. An alternative method starts from the determination of the lattice constants as a function of temperature and therefrom, the volume of the unit cell may be calculated at each temperature. It has been shown [70] that the variation of the unit cell volume $V(T)$ may be reproduced by:

$$V(T) = n_H V_H + (1 - n_H) V_L + \alpha_v T V_L \tag{128}$$

Here, V_H and V_L are the unit cell volumes of the pure HS and LS isomer at 0 K, respectively, and α_v is the coefficient of thermal expansion which is assumed to be equal in both lattices. The expression of Eq. (128) may be rewritten as:

$$V(T) = V_{LS}\left[1 + \alpha_v T + \frac{(V_H - V_L)}{V_L} n_H\right] \tag{129}$$

In order to reproduce the temperature variation of the lattice constants, the anisotropy of the lattice expansion has to be taken into account. For this purpose, the tensor of thermal expansion $\boldsymbol{\alpha}$ is introduced instead of the scalar α_v, and the tensor of deformation due to the HS $\leftrightarrow$ LS transition $\boldsymbol{\varepsilon}$ is employed instead of the dilation $(V_H - V_L)/V_L$. Each lattice vector $\mathbf{x}(T)$ can now be

Table 19. Results of calorimetric and DSC measurements on spin-state transitions in solid transition-metal complexes

Compound[a]	T_c K	ΔH kJ mol^{-1}	ΔS J K^{-1} mol^{-1}	Method[b]	Ref.
$[Fe(phen)_2(NCS)_2]$	176.29	8.60 ± 0.14	48.78 ± 0.71	cal	[41]
$[Fe(phen)_2(NCSe)_2]$	231.26	11.60 ± 0.44	51.22 ± 2.33	cal	[41]
$[Fe(bpy)_2(NCS)_2]$	212	10.1 ± 0.5	48 ± 4	DSC	[203]
$[Fe(phy)_2](ClO_4)_2$	248.2	15.8 ± 1	64.6 ± 4	DSC	[204]
$[Fe(phy)_2](BF_4)_2$	283.4	15.6	55.5	DSC	[204]
$[Fe(2\text{-pic})_3]Cl_2 \cdot C_2H_5OH$	114.04, 122.21	6.14	50.59	cal	[202]
$[Fe(phen)(bpy)(NCS)_2]$	180	~ 1	~ 5.5	DSC	[205]
$[Fe(bt)_2(NCS)_2]$	176.0	9.54 ± 0.5	54.4 ± 3	DSC	[206, 207]
$[Fe(bt)_2(NCSe)_2]$	219.5	9.11 ± 0.5	41.5 ± 3.0	DSC	[206]
$[Co(H_2fsa_2en)(py)_2]$	121	1.4	10	DSC	[208]
$[Co(H_2fsa_2en)(4\text{-t-Bupy})_2]$	146	2.4	18	DSC	[208]

[a] Ligand abbreviations: phen = 1,10-phenanthroline; bpy = 2,2′-bipyridine; phy = 1,10-phenanthroline-2-carbaldehyde phenylhydrazone; 2-pic = 2-aminomethylpyridine; bt = 2,2′-bi-2-thiazoline; bts = 2,2′-bi(5-methyl-2-thiazoline); H_4fsa_2en = *N,N′*-ethylenebis (3-carboxysalicylaldimine)

[b] cal = absolute calorimetry; DSC = differential scanning calorimetry

expressed as:

$$\boldsymbol{x}(T) = [1 + \boldsymbol{\alpha}T + \boldsymbol{\varepsilon}n_H(T)]\boldsymbol{x}(T=0) \quad (130)$$

In the principal axes system, Eq. (130) may be written in component notation as:

$$x_i(T) = (1 + \alpha_i T + \varepsilon_i n_H)x_{io} \quad (131)$$

In Eq. (131) α_i and ε_i are the diagonal elements of the two tensors, the scalars α_v and $(V_H - V_L)/V_L$ being given by the traces of the tensors $\boldsymbol{\alpha}$ and $\boldsymbol{\varepsilon}$. The temperature dependence of the lattice constants may then be accounted for on the basis of Eq. (131), that of the unit cell volume by Eq. (129). Figure 38 shows experimental data for lattice constants *a*, *b*, *c* and for the unit cell volume *V* for the iron(II) complex $[Fe(2\text{-pic})_3]Cl_2 \cdot CH_3OH$ [39]. Included are curves representing the numerical fit to the measured data. The components α_i and ε_i of the tensors in the principal axes system resulting from the fit are listed in Table 20. The measured temperature dependence of lattice constants *a*, *b*, *c* and of the unit cell volume *V* for $[Fe(2\text{-pic})_3]Br_2 \cdot C_2H_5OH$ and

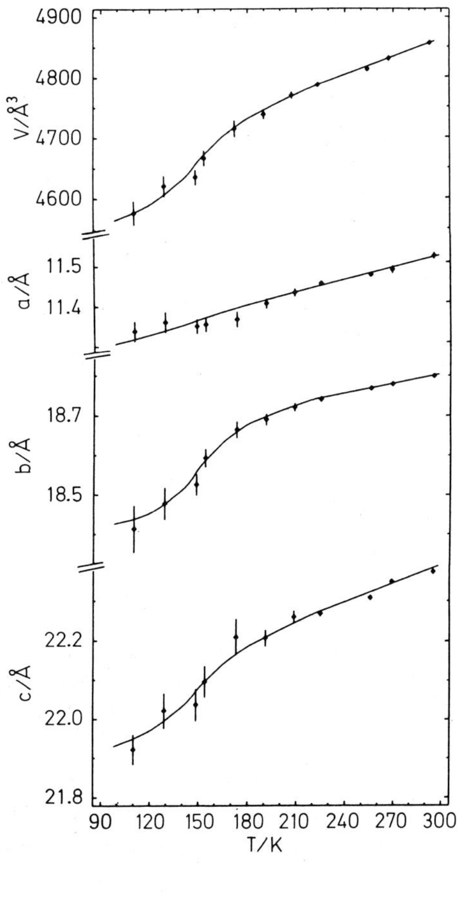

Fig. 38. Temperature dependence of the lattice constants *a*, *b*, *c* and the unit cell volume *V* of $[Fe(2\text{-pic})_3]Cl_2 \cdot CH_3OH$. The *solid lines* are calculated using the parameter values of Table 20. According to Ref. [39]

Table 20. Tensor elements referred to the principal axes of the tensor of thermal expansion $\boldsymbol{\alpha}$ and the deformation tensor $\boldsymbol{\varepsilon}$ due to the HS↔LS transition in $[Fe(2\text{-pic})_3]Cl_2 \cdot CH_3OH$; according to Ref. [39]

i	α_i $10^{-4}\,K^{-1}$	ε_i 10^{-2}
1	0.85	0.18
2	0.33	1.43
3	0.64	0.83
Trace	1.82	2.44

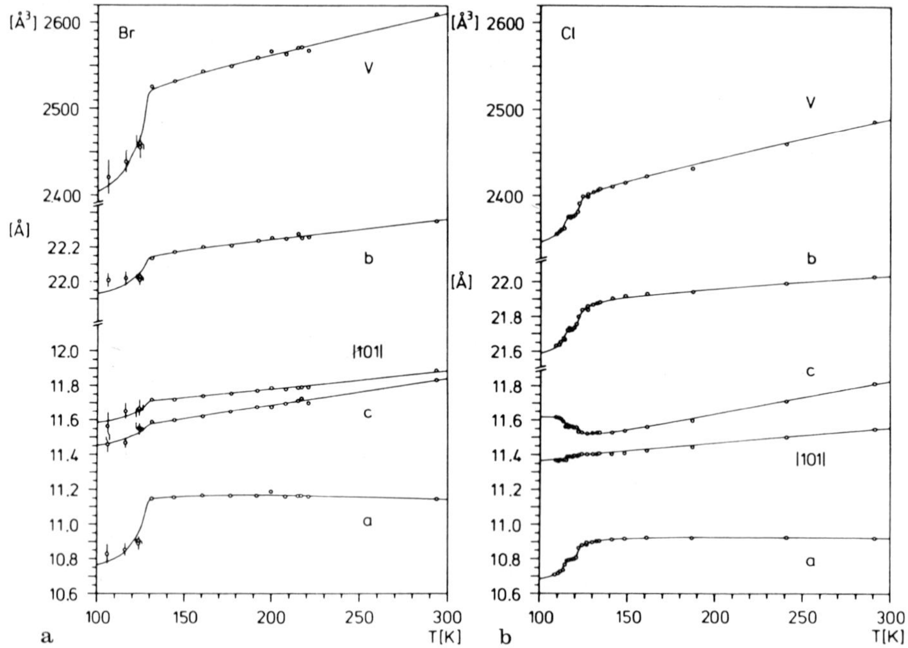

Fig. 39. Temperature dependence of the lattice constants *a*, *b*, *c* and the unit cell volume *V* of $[Fe(2\text{-pic})_3]Br_2 \cdot C_2H_5OH$ (*a*) and $[Fe(2\text{-pic})_3]Cl_2 \cdot C_2H_5OH$ (*b*). *Solid lines* are calculated using the parameter values of Table 21. According to Ref. [70]

$[Fe(2\text{-pic})_3]Cl_2 \cdot C_2H_5OH$ is displayed in Fig. 39 [70]. The figure also includes the curves calculated using the parameter values collected in Table 21. For $[Fe(2\text{-pic})_3]Cl_2 \cdot CH_3OH$, the HS fractions $n_H(T)$ were taken from the Mössbauer data, for the ethanolate complex from the magnetic susceptibility $\chi(T)$ which is the weighted sum of χ_{HS} and χ_{LS}. $[Fe(2\text{-pic})_3]Cl_2 \cdot CH_3OH$ crystallizes in a rhombic cell [179], the principal axes of the tensors are therefore oriented along the crystal axes *a*, *b*, *c*. The ethanolate complex crystallizes in a monoclinic

Table 21. Elements of the tensor of thermal expansion $\boldsymbol{\alpha}$ and the deformation tensor $\boldsymbol{\varepsilon}$ due to the HS$\leftrightarrow$LS transition in $[Fe(2\text{-pic})_3]Br_2 \cdot C_2H_5OH$ and $[Fe(2\text{-pic})_3]Cl_2 \cdot C_2H_5OH$ referred to their principal axes[a]; according to Ref. [70]

	Br^-	Cl^-
α_1	1.54	1.86
α_2	0.53	0.31
α_3	−0.30	−0.31
$Tr(\alpha)$	1.77	1.86
$\delta(\alpha)$	97	107
ε_1	4.45	2.93
ε_2	1.03	1.57
ε_3	−0.19	−2.07
$Tr(\varepsilon)$	5.29	2.43
$\delta(\varepsilon)$	−1	14
η	0.45	1.72

[a] The quantity δ is the (clockwise) angle between the crystallographic $\boldsymbol{a}$ axis and the $\mathbf{x}$ axis of the principal axis system. The anisotropy parameter η of the tensor $\boldsymbol{\varepsilon}$ is defined by $\eta = (\varepsilon_2 - \varepsilon_3)/(\varepsilon_1 - Tr(\varepsilon)/3)$. Units: δ, deg; ε_i, 10^{-2}; α_i, $10^{-4}\ K^{-1}$

cell [70, 178], the relation of the tensors $\boldsymbol{\alpha}$ and $\boldsymbol{\varepsilon}$ to the crystal axes is somewhat more complex. One axis which is chosen as the y axis is defined by the 2-fold axis b of the crystal. In addition, the temperature dependence of the lengths of the three axes a, c and $a + c$ determines the principal axes of the tensors in the plane normal to the b axis, the a axis forming angle δ with the axis x of the principal axes system of the tensors $\boldsymbol{\alpha}$ and $\boldsymbol{\varepsilon}$.

The studies show that the observed crystal volume is in fact composed of the fractional contributions from the unit cell volumes of the HS and LS isomers of the compound and a linear volume change with temperature as expressed in Eq. (128). Similarly, the observed lattice constants are formed from a deformation contribution proportional to the HS fraction n_H and a contribution from thermal expansion following Eq. (131). This is a convincing demonstration that it is the internal variation of the molecular units occurring in the course of the spin-state transition which determines, at least in principle, the observed crystal properties.

10 Conclusions

Spin-state transitions in metal complexes are always associated with a considerable reorganization of molecular geometry. The reorganization involves changes of metal-ligand bond lengths and angles, possibly a modified ligand orientation as well as more specific changes of bond distances and angles within the ligands. Of these changes, the variation of the metal-donor atom bond length ΔR is always the most significant, average values being $\overline{\Delta R} \simeq 0.17$ Å for iron(II), $\simeq 0.13$ Å for iron(III) and $\simeq 0.09$ Å for cobalt(II) [6]. The bond length change is associated with the transfer of electrons from the nonbonding t_{2g} to the σ anti-bonding e_g^* orbitals of the metal as the LS state transforms to the HS state. The transfer of a single electron such as in cobalt(II) ($\Delta S = 1$) results in a smaller value of $\overline{\Delta R}$ than the transfer of two electrons such as in iron(II) or iron(III) ($\Delta S = 2$). A significant change is also observed for the donor atom-metal-donor atom bite angle for bidentate ligands, the variation of which is consistent with the change of metal-ligand bond length. Thus, in $[Fe(2\text{-pic})_3]Cl_2 \cdot C_2H_5OH$, the N–Fe–N angle shows a decrease by 6.1° on transformation of the LS to the HS state [178]. Additional changes of molecular geometry seem to be dictated by space requirements of the ligands, the disposition of ligands about the metal atom, individual contacts between peripheral ligand atoms, anions and solvent molecules, and the crystal packing. The elongation of the average metal-ligand bond by as much as up to ~0.20 Å or ~10% invariably produces a concomitant expansion of molecular volume. This volume change ΔV assumes typical values of about 25 Å^3 per metal atom or 3.5% for an iron(II) complex. The determination of the bond length change between HS and LS isomers of a complex in solution is not possible. The change of the partial molal volume $\Delta V°$ of the two spin states may be obtained from ultrasonic measurements or pressure studies. The available results indicate that $\Delta V°$ values and the changes of unit cell volume as referred to a single complex molecule are almost identical. In the solid complex, the observed crystal structure may be described by the fractional occupation of the individual structures of the HS and LS isomers as a function of temperature [179]. Similarly, the variation of the unit cell volume with temperature is reproduced by the sum of the weighted contributions of V_H and V_L, the unit cell volumes of the HS and LS isomer, respectively, an additional term describing the expansion of the lattice with increasing temperature. Also, electronic, vibrational (IR and Raman) and EPR spectra are a superposition of the spectra of the HS and LS isomer. The analysis of these data provides interesting conclusions with regard to the physical nature and the mechanism of spin-state transitions.

The observed changes of physical properties may be reasonably well described employing a number of theoretical models which are based on the familiar Bragg and Williams approximation. Accordingly, the properties such as Gibbs free energy G are expressed as the sum of fractional contributions of the properties of the HS and LS state, e.g. G_H and G_L, an additional term taking account of the interaction between the two terms. The models differ only by the specific form of the interaction term. It follows that the peculiar variation of crystal properties with temperature is a manifestation of the change of n_H, the HS fraction. The variation thus represents the gradual spin-state conversion of the individual molecules which is mediated and slightly modified by the elastic and/or phonon interactions of the lattice.

It has been pointed out above that, at least for a number of suitable systems, the partial molal volume in solution and the volume occupied by a single molecule in the crystal are practically identical [177]. This finding provides additional support for the applicability of the theoretical models considered. In addition, the intermolecular contacts in the crystal should be similar to the solvation of the complexes in solution.

As far as the dynamics of spin-state conversions is concerned, these are in general rapid, both in solution and in the solid state. In solution, the fastest dynamics is represented by the $\Delta S = 1$ transitions of cobalt(II) complexes which proceed in less than a nanosecond, i.e., with rate constants $k \geq 10^9\ s^{-1}$. The $\Delta S = 2$ transitions in six-coordinate iron(II) and iron(III) complexes are slower by a factor of 10 to 10^3, rate constants between $k = 4 \times 10^5$ and $2 \times 10^7\ s^{-1}$ or between $k = 6.1 \times 10^7$ and $3.2 \times 10^8\ s^{-1}$, respectively, having been determined. The lower rates for $\Delta S = 2$ transitions seem to be associated with the larger variation of nuclear configuration such as a more pronounced metal-ligand bond length change. In fact, the bond length difference for iron(III) is somewhat smaller than that for iron(II) with an average of $\overline{\Delta R} \simeq 0.15$ Å for complexes with dithiocarbamate ligands and $\simeq 0.13$ Å for an N_4O_2 donor atom core [6]. This seems to account for the somewhat faster rates for spin-state conversions, viz. $k = 6 \times 10^7$ to $2.2 \times 10^8\ s^{-1}$. If an even larger reorganization of the coordination sphere is involved such as for the planar-tetrahedral equilibria of nickel(II), the rates are slower again with values of k between 10^3 and $10^6\ s^{-1}$ and resulting lifetimes in the microsecond to millisecond range [30].

In the solid state, dynamics of spin-state transitions have been studied only for complexes of iron(II) and iron(III). At room temperature, the available spin-state conversion rates for iron(II) complexes are $k_{HL} = 6.6 \times 10^6\ s^{-1}$ for $[Fe(2\text{-pic})_3](PF_6)_2$, $k_{HL} = 2.1 \times 10^6\ s^{-1}$ for $[Fe(mtz)_6](PF_6)_2$, $k_{HL} = 9 \times 10^6\ s^{-1}$ for $[Fe(6\text{-Mepy})_3tren](PF_6)_2$ and $k' = 2.3 \times 10^7\ s^{-1}$ for $[Fe(6\text{-Mepy})_2(py)tren](ClO_4)_2$ in PSS, $k' = 1/\tau$ being the relaxation rate [138, 165, 167]. For iron(III) complexes, the values $k_{LH} = 2.9 \times 10^6\ s^{-1}$ for $[Fe(acpa)_2]BPh_4 \cdot H_2O$ and $k' = 6.7 \times 10^6\ s^{-1}$ for $[Fe(mim)_2(salacen)]PF_6$ have been obtained [156, 166]. The rate constants derived from the line shape analysis of Mössbauer spectra thus vary between 2.1×10^6 and $2.3 \times 10^7\ s^{-1}$ at room temperature, no significant difference between iron(II) and iron(III) being apparent. In addition, it is evident that the rates of spin-state conversion in solution and in the crystalline solid are almost the same. For iron(II) complexes, for example, the solution rates vary between $k_{HL} = 5 \times 10^6$ and $2 \times 10^7\ s^{-1}$, whereas in solid compounds values between $k_{HL} = 6.6 \times 10^6$ and $2.3 \times 10^7\ s^{-1}$ have been obtained. Rates resulting from the relaxation of thermally quenched spin transition systems are considerably slower, since they have been measured only over a small range of relatively low temperatures. Extrapolation of the kinetic data to room temperature is, however, of uncertain validity.

The dynamics of spin-state transitions may be rationalized in various ways. Thus, if the geometry of complexes does not depart too far from octahedral, one-dimensional cross sections through Gibbs free energy surfaces $G = G(R)$ along the totally symmetric metal-ligand stretching coordinate may be considered. These potential energy diagrams may serve, in particular, to illustrate the magnitude of activation free energy, $\Delta G^{\ddagger}_{LH}$ and $\Delta G^{\ddagger}_{HL}$, i.e., the barrier height of the conversion processes. Thus the coordination sphere reorganization barrier is decreased in cobalt(II) and possibly also in iron(III) as compared to iron(II) complexes. Of particular interest is the magnitude of the transmission coefficient κ which determines the probability of intersystem crossing. A direct evaluation of κ from experimental data is not possible since it appears with the activation entropy in the expression for the rate constant k of absolute reaction rate theory. Studies of the volumes of activation based on the pressure dependence of spin conversion rates indicate that the transition state is about midway between the HS and the LS state [89, 91]. The values of κ are then estimated to not less than about 10^{-2} [30]. The entropy of activation $\Delta S^{\ddagger}$ includes contributions from increased solvation and decreased vibrational partition functions which arise due to the compression of the HS state to give the transition state. Obviously, the above discussion implies that the radial compression of the metal-ligand bonds is the predominant geometric change associated with the spin-state conversion.

A quantum-mechanical description of spin-state equilibria has been proposed on the basis of a radiationless nonadiabatic multiphonon process [117]. Calculated rate constants of, e.g., $k \simeq 10^9\ s^{-1}$ for iron(II) and iron(III) are in reasonable agreement with the observed values between 10^7 and $10^8\ s^{-1}$. Here again the quantity of largest influence is the metal-ligand bond length change ΔR and the consequent variation of stretching vibrations.

As far as the solid complexes are concerned, the above conclusions are generally valid for gradual spin-state transitions, whereas additional features such as hysteresis effects are observed for transitions which show abrupt changes of physical properties. In fact, abrupt transitions seem to be formed if the volume change ΔV associated with the spin-state conversion of the molecules cannot be conveniently accommodated by the lattice.

Acknowledgements. The author appreciates helpful comments on the manuscript by H.A. Goodwin (Sydney) and J. Dengler (Erlangen). Special thanks are due to G. Ritter (Erlangen) for inspiring discussions throughout the preparation of this review.

11 References

1. Holm RH, Everett GW, Chakravorty A (1966) Progr Inorg Chem 7: 83
2. Holm RH, O'Connor MJ (1973) Progr Inorg Chem 14: 241
3. Barefield EK, Busch DH, Nelson SM (1968) Quart Rev Chem Soc 22: 457
4. Chisholm MH, Kober EM, Ironmonger DJ, Thornton P (1985) Polyhedron 4: 1869
5. Imoto H, Simon A (1982) Inorg Chem 21: 308

6. König E (1987) Progr Inorg Chem 35: 527
7. Chandrasekhar K, Bürgi HB (1984) Acta Crystallogr B40: 387
8. Thuéry P, Zarembowitch J, Michalowicz A, Kahn O (1987) Inorg Chem 26: 851
9. Tanabe Y, Sugano S (1954) J Phys Soc Japan 9: 766
10. König E, Madeja K (1967) Inorg Chem 6: 48
11. Gallois B, Real JA, Hauw C, Zarembowitch J (1990) Inorg Chem 29: 1152
12. König E, Watson KJ (1970) Chem Phys Lett 6: 457
13. Gerloch M, Slade RC (1973) Ligand-field parameters, Cambridge Univ Press, Cambridge
14. Bragg WL, Williams EJ (1934) Proc Roy Soc (London) A 145: 699
15. Guggenheim EA (1952) Mixtures, Oxford Univ Press, London
16. König E, Kremer S (1971) Theor Chim Acta 23: 12
17. König E, Kremer S (1977) Ligand field energy diagrams, Plenum Press, New York
18. Coronel P, Barraud A, Claude R, Kahn O, Ruaudel-Teixier A, Zarembowitch J (1989) J Chem Soc Chem Commun 193
19. Martin RL, White AH (1969) Trans Metal Chem 5: 113
20. Goodwin HA (1976) Coord Chem Rev 18: 293
21. Gütlich P (1981) Struct Bonding 44: 83
22. Gütlich P (1981) Adv Chem Ser 194: 405
23. Gütlich P (1984) in Long GJ (Ed) Mössbauer spectroscopy applied to inorganic chemistry, Vol 1: 287, Plenum Press, New York
24. König E, Ritter G, Kulshreshtha SK (1985) Chem Rev 85: 219
25. Toftlund H (1989) Coord Chem Rev 94: 67
26. Bacci M (1988) Coord Chem Rev 86: 245
27. Rao CNR (1985) Int Rev Phys Chem 4: 19
28. Beattie JK, Sutin N, Turner DH, Flynn GW (1973) J Am Chem Soc 95: 2052
29. Beattie JK, Binstead RA, West RJ (1978) J Am Chem Soc 100: 3044
30. Beattie JK (1988) Adv Inorg Chem 32: 1
31. Blume M (1968) Phys Rev 174: 351
32. Blume M, Tjon JA (1968) Phys Rev 165: 446
33. Tjon JA, Blume M (1968) Phys Rev 165: 456
34. Dose EV, Hoselton MA, Sutin N, Tweedle MF, Wilson LJ (1978) J Am Chem Soc 100: 1141
35. Zimmermann R, König E (1977) J Phys Chem Solids 38: 779
36. Rao PS, Ganguli P, McGarvey BR (1981) Inorg Chem 20: 3682
37. Drickamer HG, Frank CW (1973) Electronic transitions and the high pressure chemistry and physics of solids, Chapman and Hall, London
38. Slichter CP, Drickamer HG (1972) J Chem Phys 56: 2142
39. Adler P, Wiehl L, Meissner E, Köhler CP, Spiering H, Gütlich P (1987) J Phys Chem Solids 48: 517
40. Spiering H, Meissner E, Köppen H, Müller EW, Gütlich P (1982) Chem Phys 68: 65
41. Sorai M, Seki S (1974) J Phys Chem Solids 35: 555
42. Gütlich P, Köppen H, Link R, Steinhäuser HG (1979) J Chem Phys 70: 3977
43. Chestnut DB (1964) J Chem Phys 40: 405
44. Wajnflacz J (1970) Phys Stat Solidi 40: 537
45. Bari RA, Sivardière J (1972) Phys Rev B5: 4466
46. Zimmermann R (1983) J Phys Chem Solids 44: 151
47. Eshelby JD (1956) Solid State Phys 3: 79
48. König E, Ritter G, Kulshreshtha SK, Nelson SM (1984) Inorg Chem 21: 3022
49. König E, Ritter G, Irler W, Nelson SM (1979) Inorg Chim Acta 37: 169
50. König E, Ritter G, Dengler J, Nelson J (1989) Inorg Chem 28: 611
51. Müller EW, Spiering H, Gütlich P (1982) Chem Phys Lett 93: 567
52. König E, Ritter G (1976) Solid State Commun. 18: 279
53. König E, Ritter G, Kulshreshtha SK, Csatary N (1984) Inorg Chem 23: 1903
54. König E, Ritter G, Irler W, Goodwin HA (1980) J Am Chem Soc 102: 4681
55. König E, Ritter G, Kulshreshtha SK, Waigel J, Goodwin HA (1984) Inorg Chem 23: 1896
56. König E, Kanellakopulos B, Powietzka B, Goodwin HA (1989) Inorg Chem 28: 3993
57. Ritter G, König E, Irler W, Goodwin HA (1978) Inorg Chem 17: 224
58. Sorai M, Ensling J, Hasselbach KM, Gütlich P (1977) Chem Phys 20: 197
59. König E, Ritter G, Dengler J, Nelson SM (1987) Inorg Chem 26: 3582
60. Timken MD, Scott RW, Hendrickson DN (1985) Inorg Chem 24: 3450
61. Mohan M, Madhuranath PH, Kumar A, Kumar M, Jha NK (1989) Inorg Chem 28: 96

62. Wells FV, McCann SW, Wickman HH, Kessel SL, Hendrickson DN, Feltham RD (1982) Inorg Chem 21: 2306
63. König E, Ritter G, Dengler J, Thuéry P, Zarembowitch J (1989) Inorg Chem 28: 1757
64. Zarembowitch J, Kahn O (1984) Inorg Chem 23: 589
65. Thuéry P, Zarembowitch J (1986) Inorg Chem 25: 2001
66. Zarembowitch J, Claude R, Kahn O (1985) Inorg Chem 24: 1576
67. Meissner E, Köppen H, Spiering H, Gütlich P (1983) Chem Phys Lett 95: 163
68. Jacobi R, Spiering H, Wiehl L, Gmelin E, Gütlich P (1988) Inorg Chem 27: 1823
69. Sanner I, Meissner E, Köppen H, Spiering H, Gütlich P (1984) Chem Phys 86: 227
70. Wiehl L, Kiel G, Köhler CP, Spiering H, Gütlich P (1986) Inorg Chem 25: 1565
71. Willenbacher N, Spiering H (1988) J Phys C 21: 1423
72. Spiering H, Willenbacher N (1989) J Phys Condens Matter 1: 10089
73. Kambara T (1979) J Chem Phys 70: 4199
74. Kambara T (1981) J Chem Phys 74: 4557
75. Ogata F, Kambara T, Sasaki N, Gondaira GI, J Phys C 16: 1391
76. Sasaki N, Kambara T (1981) J Chem Phys 74: 3472
77. Kambara T (1980) J Phys Soc Japan 49: 1806
78. Ohnishi S, Sugano S (1981) J Phys C 14: 39
79. Turner DH, Flynn GW, Sutin N, Beitz JV (1972) J Am Chem Soc 94: 1554
80. Petty RH, Dose EV, Tweedle MF, Wilson LJ (1978) Inorg Chem 17: 1064
81. Brody SS (1957) Rev Sci Instrum 28: 1021
82. Demas JN, Crosby GA (1970) Anal Chem 42: 1010
83. Crawford TH, Swanson J (1971) J Chen Educ 48: 382
84. Evans DF (1959) J Chem Soc 2003
85. Laidler KJ (1965) Chemical Kinetics, McGraw-Hill, New York
86. Lown DA, Thirsk HR, Wynne-Jones L (1968) Trans Faraday Soc 64: 2073
87. Creutz S, Chan M, Netzel TL, Okumura M, Sutin N (1980) J Am Chem Soc 102: 1309
88. Hoselton MA, Drago RS, Wilson LJ, Sutin N (1976) J Am Chem Soc 98: 6967
89. DiBenedetto J, Arkle V, Goodwin HA, Ford PC (1985) Inorg Chem 24: 455
90. McGarvey JJ, Lawthers I, Hermans K, Toftlund H (1990) Inorg Chem 29: 252
91. McGarvey JJ, Lawthers I, Heremans K, Toftlund H (1984) J Chem Soc Chem Commun 1575
92. McGarvey JJ, Lawthers I (1982) J Chem Soc Chem Commun 906
93. Purcell KF (1979) J Am Chem Soc 101: 5147
94. Binstead RA, Beattie JK, Dewey TG, Turner DH (1980) J Am Chem Soc 102: 6442
95. Lawthers I, McGarvey JJ (1984) J Am Chem Soc 106: 4280
96. Dose EV, Murphy KMM, Wilson LJ (1976) Inorg Chem 15: 2622
97. Tweedle MF, Wilson LJ (1976) J Am Chem Soc 98: 4824
98. Binstead RA, Beattie JK, Dose EV, Tweedle MF, Wilson LJ (1978) J Am Chem Soc 100: 5609
99. Simmons MG, Wilson LJ (1977) Inorg Chem 16: 126
100. Timken MD, Strouse CE, Soltis SM, Deverio SA, Hendrickson DN, Abdel-Mawgound AM, Wilson SR (1986) J Am Chem Soc 108: 395
101. Ewald AH, Martin RL, Ross IG, White AH (1964) Proc Roy Soc (London) A 280: 235
102. Ewald AH, Martin RL, Sinn E, White AH (1969) Inorg Chem 8: 1837
103. Cecconi F, Di Vaira M, Midollini S, Orlandini A, Sacconi L (1981) Inorg Chem 20: 3423
104. Hitchman MA (1982) Inorg Chem 21: 821
105. Oliver JD, Mullica DF, Hutchinson BB, Milligan WO (1980) Inorg Chem 19: 165
106. Hutchinson B, Hoffbauer M, Takemoto J (1976) Spectrochim Acta 32A: 1785
107. Baker AT, Goodwin HA, Rae AD (1984) Aust J Chem 37: 443
108. Takemoto JH, Hutchinson B (1972) Inorg Nucl Chem Lett 8: 769
109. Takemoto JH, Hutchinson B (1973) Inorg Chem 12: 705
110. Sinn E, Sim G, Dose EV, Tweedle MF, Wilson LJ (1978) J Am Chem Soc 100: 3375
111. Grandjean F, Long GJ, Hutchinson BB, Ohlhausen L, Neill P, Holcomb JD (1989) Inorg Chem 28: 4406
112. Franke PL, Haasnoot JP, Zuur AP (1982) Inorg Chim Acta 59: 5
113. Müller EW, Ensling J, Spiering H, Gütlich P (1983) Inorg Chem 22: 2074
114. Hutchinson B, Neill P, Finkelstein A, Takemoto J (1981) Inorg Chem 20: 2000
115. Eyring H, Walter J, Kimball GE (1954) Quantum chemistry, Wiley, New York
116. Sutin N (1973) in Eichhorn GI (Ed) Inorganic biochemistry, Vol 2, Elsevier, Amsterdam
117. Buhks E, Navon G, Bixon M, Jortner J (1980) J Am Chem Soc 102: 2918
118. Vanquickenborne LG, Pierloot K (1981) Inorg Chem 20: 3673

119. Springer CS, Sievers RE (1967) Inorg Chem 6: 852
120. Ray P, Dutt NK (1943) J Indian Chem Soc 20: 81
121. Lawrance GA, Stranks DR (1977) Inorg Chem 16: 929
122. Lawrance GA, Stranks DR (1978) Inorg Chem 17: 1804
123. Buhks E, Bixon M, Jortner J, Navon G (1979) Inorg Chem 18: 2014
124. Kestner NR, Logan J, Jortner J (1974) J Phys Chem 78: 2148
125. Levich VG (1970) In: Eyring H, Henderson D, Jost V (eds) Physical chemistry: An advanced treatise, Vol 9B, Academic Press, New York
126. Griffith JS (1961) The theory of transition metal ions, Cambridge Univ Press, London
127. Jørgensen CK (1962) Absorption spectra and chemical bonding in complexes, Pergamon Oxford
128. Jørgensen CK (1962) Progr Inorg Chem 4: 73
129. Dunn TM (1961) Trans Faraday Soc 57: 1441
130. Ferraro JR (1971) Low frequency vibrations of inorganic and coordination compounds, Plenum, New York
131. Leipoldt JG, Coppens P (1973) Inorg Chem 12: 2269
132. Figgis BN, Kucharski ES, White AH (1983) Aust J Chem 36: 1527
133. Goodwin HA, Sugiyarto KH (1987) Chem Phys Lett 139: 470
134. Hauser A, Gütlich P, Spiering H (1986) Inorg Chem 25: 4245
135. Hauser A (1986) Chem Phys Lett 124: 543
135a. Hauser A, private communication
136. König E, Ritter G, Goodwin HA (1976) Chem Phys Lett 44: 100
137. Decurtins S, Gütlich P, Hasselbach KH, Hauser A, Spiering H (1985) Inorg Chem 24: 2174
138. Xie CL, Hendrickson DN (1987) J Am Chem Soc 109: 6981
139. Batchelet WH, Rose NJ (1983) Inorg Chem 22: 2083
140. Herber RH (1987) Inorg Chem 26: 173
141. Herber RH, Casson LM (1986) Inorg Chem 25: 847
142. Sorai M (1978) J Inorg Nucl Chem 40: 1031
143. Flick C, Gelerinter E (1973) Chem Phys Lett 23: 422
144. Pandeya KB, Singh R (1988) Inorg Chim Acta 147: 5
145. Flick C, Gelerinter E, Zimmermann JB, Duffy NV (1976) Inorg Chem 15: 2945
146. Hall GR, Hendrickson DN (1976) Inorg Chem 15: 607
147. Haddad MS, Lynch MW, Federer WD, Hendrickson DN (1981) Inorg Chem 20: 123
148. Federer WD, Hendrickson DN (1984) Inorg Chem 23: 3861
149. Ohshio H, Maeda Y, Takashima Y (1983) Inorg Chem 22: 2684
150. Tinken MD, Abdel-Mawgoud AM, Hendrickson DN (1986) Inorg Chem 25: 160
151. Tinken MD, Hendrickson DN, Sinn E (1985) Inorg Chem 24: 3947
152. Kremer S, Henke W, Reinen D (1982) Inorg Chem 21: 3013
153. Maiti B, McGarvey BR, Rao PS, Stubbs LC (1983) J Magn Res 54: 99
154. Gutowski HS, McCall DW, Slichter CP (1953) J Chem Phys 21: 279
155. McConnell HM (1958) J Chem Phys 28: 430
156. Maeda Y, Tsutsumi N, Takashima Y (1984) Inorg Chem 23: 2440
157. Anderson PW, Weiss PR (1953) Rev Mod Phys 25: 269
158. Kubo R (1954) J Phys Soc Japan 9: 935
159. Kubo R, Tomita K (1954) J Phys Soc Japan 9: 888
160. Messiah A (1970) Quantum mechanics, Vol 2, North-Holland Publ, Amsterdam
161. Maeda Y, Takashima Y (1983) Mem Fac Sci Kyushu Univ Ser C 14: 107
162. Abragam A (1961) The Theory of Nuclear Magnetism, Oxford University Press, London
163. Numata Y, Kubokura K, Nonaka Y, Okawa H, Kida S (1980) Inorg Chim Acta 43: 193
164. Dengler J, König E, Larkworthy LF, Ritter G, Sengupta SK (1990) Hyperfine Interact 56: 1443
165. Adler P, Spiering H, Gütlich P (1988) Hyperfine Interact 42: 1035
166. Maeda Y, Takashima Y, Matsumoto N, Ohyoshi A (1986) J Chem Soc Dalton Trans 1115
167. Adler P, Spiering H, Gütlich P (1987) Inorg Chem 26: 3840
168. Maeda Y, Ohshio H, Takashima Y, Mikuriya M, Hidaka M (1986) Inorg Chem 25: 2958
169. Adler P, Spiering H, Gütlich P (1989) J Phys Chem Solids 50: 587
170. Hoselton MA, Wilson LJ, Drago RS (1975) J Am Chem Soc 97: 1722
171. DeVault D (1984) Quantum-mechanical tunneling in biological systems, Cambridge Univ Press, Cambridge
172. Hopfield JJ (1974) Proc Natl Acad Sci USA 71: 3640
173. Conti AJ, Xie CL, Hendrickson DN (1989) J Am Chem Soc 111: 1171

174. Adler P, Poganiuch P, Spiering H (1989) Hyperfine Interact 52: 47
175. Haller KJ, Johnson PL, Feltham RD, Enemark JH, Ferraro JR, Basile LJ (1979) Inorg Chim Acta 33: 119
176. Sim PG, Sinn E, Petty RH, Merrill CL, Wilson LJ (1981) Inorg Chem 20: 1213
177. Binstead RA, Beattie JK (1986) Inorg Chem 25: 1481
178. Mikami M, Konno M, Saito Y (1980) Acta Crystallogr B 36: 275
179. Katz BA, Strouse CE (1979) J Am Chem Soc 101: 6214
180. Baker AT, Goodwin HA, Rae AD (1987) Inorg Chem 26: 3513
181. Albertsson J, Oskarsson A, Stahl K, Svensson C, Ymen I (1981) Acta Crystallogr B 37: 50
182. Gatteschi D, Ghilardi CA, Orlandini A, Sacconi L (1978) Inorg Chem 17: 3023
183. Orlandini AB, Calabresi C, Ghilardi CA, Orioli PL, Sacconi L (1973) J Chem Soc Dalton Trans 1383
184. Shannon RD, Prewitt CT (1969) Acta Crystallogr B 25: 925
185. Vos G, DeGraaff RAG, Haasnoot JG, Van Der Kraan AM, De Vaal P, Reedijk J (1984) Inorg Chem 23: 2905
186. Bacci M, Ghilardi CA, Orlandini A (1984) Inorg Chem 23: 2798
187. Scheidt WR, Geiger DK, Haller KJ (1982) J Am Chem Soc 104: 495
188. Geiger DK, Chunplang V, Scheidt WR (1985) Inorg Chem 24: 4736
189. Kennedy BJ, McGrath AC, Murray KS, Skelton BW, White AH, Inorg Chem 26: 483
190. Albertsson J, Oskarsson A (1977) Acta Crystallogr B 33: 1871
191. Stahl K (1983) Acta Crystallogr B 39: 612
192. Albertsson J, Oskarsson A, Nygren M (1979) Acta Crystallogr B 35: 1473
193. Albertsson J, Oskarsson A, Stahl K (1982) Acta Chem Scand A 36: 783
194. Albertsson J, Elding I, Oskarsson A (1979) Acta Chem Scand A 33: 703
195. Ryabova NA, Ponomarev VI, Zelentsov VV, Atovmyan LO (1982) Kristallografiya 27: 279; Sov Phys Crystallogr 27: 171
196. Ryabova NA, Ponomarev VI, Zelentsov VV, Atovmyan LO (1982) Kristallografiya 27: 81; Sov Phys Crystallogr 27: 46
197. Calligaris M, Nardin G, Randaccio L (1974) J Chem Soc Dalton Trans 1903
198. Calligaris M, Minichelli D, Nardin G, Randaccio L (1970) J Chem Soc A 2411
199. Bresciani N, Calligaris M, Nardin G, Randaccio L (1974) J Chem Soc Dalton Trans 498
200. Kennedy BJ, Fallon GD, Gatehouse BMKC, Murray KS (1984) Inorg Chem 23: 580
201. Charpin P, Nierlich M, Vigner D, Lance M, Thuéry P, Zarembowitch J, d'Yvoire FJ (1988) J Crystallogr Spectrosc Res 18: 429
202. Kaji K, Sorai M (1985) Thermochim Acta 88: 185
203. Kulshreshtha SK, Iyer RM (1984) Chem Phys Lett 108: 501
204. Kulshreshtha SK, Iyer RM (1987) Chem Phys Lett 134: 239
205. Kulshreshtha SK, Maiti B (1987) Inorg Chim Acta 134: 131
206. Kulshreshtha SK, Iyer RM, König E, to be published
207. Kulshreshtha SK, Sasikala R, König E (1986) Chem Phys Lett 123: 215
208. Zarembowitch J, Thuéry P, Dworkin A, Michalowicz A (1987) J Chem Res (S) 146

Optical Electron Transfer Between Metal Ions and its Consequences

G. Blasse

Debye Research Institute, University of Utrecht, P.O. Box 80 000, 3508 TA Utrecht, The Netherlands

This paper surveys several aspects of metal-to-metal charge-transfer transitions. Species of interest originate from non-molecular and molecular solids and from solutions. The parallel in the different approaches is stressed. In addition to the spectroscopy of these transitions, their influence or role in other phenomena is also discussed.

1 Introduction 154

2 Metal-to-Metal Charge Transfer (MMCT) Involving One Closed-Shell Transition-Metal Ion 156
2.1 General 156
2.2 Survey of Other Cases 158

3 Other Examples of Optical MMCT Transitions 167
3.1 Molecular Complexes 167
3.2 $M(d^n)M(d^m)CT$ in Solids 168
3.3 Other examples in solids 170

4 The Nature of MMCT Transitions 172

5 Mixed-Valence Compounds 175

6 The Role of MMCT in the Field of Semiconductors 177

7 The Role of CT States in Other Phenomena 180

8 Conclusion 184

9 References 185

1 Introduction

In this paper we will describe and discuss the metal-to-metal charge-transfer transitions as observed in optical spectroscopy. Their spectroscopic properties are of large importance with regard to photoredox processes [1–4]. However, these transitions are also responsible for the color of many inorganic compounds and minerals [5, 6], for different types of processes in semiconductors [7], and for the presence or absence of certain luminescence processes [8].

An early appreciation (Wherry ET (1918) Am. Min. 3: 161) states the following:

"Since ferrous iron usually colors minerals green, and ferric iron yellow or brown, it may seem rather remarkable that the presence of both together should give rise to a blue color, as in the case of vivianite. It may be pointed out, however, that this is by no means a unique instance of such an effect. Even apart from the artificial substances, "Prussian" and "Turnbull's" blues, which are complex cyanides containing both ferric and ferrous iron, there are several blue minerals in which the color seems explainable only on this basis. The most noteworthy of these are: crocidolite and related amphiboles; iolite; and the blue tourmaline or indicolite. Other instances may perhaps be discovered, should this subject ever be investigated as it deserves to be".

Let us consider a species ML_n (metal-ligands). This species may be present in a nonmolecular solid or in a solution, it may be a molecule or a coordination complex ion.

In a species ML_n (metal-ligands) we can in principle distinguish a large number of excited states [1]:

— metal-centered (MC) excited states for which the electronic excitation transitions are localized on the metal
— ligand-centered (LC) excited states for which the electronic excitation transitions are localized on the ligand
— charge-transfer (CT) excited states for which the electronic excitation transitions transfer an electron from metal to ligand (MLCT) or from ligand to metal (LMCT)
— metal-to-metal charge transfer (MMCT) for complexes with more than one type of metal; the excitation transitions transfer an electron from one metal to the other, they are sometimes called intervalence charge-transfer transitions.
— ligand-to-ligand charge transfer (LLCT) for complexes with more than one type of ligand.

The ML_n species may interact with a species in its second coordination sphere. Therefore one distinguishes inner-sphere charge-transfer and outer-sphere charge-transfer states.

The distinction between all these types of excited states is by no means sharp. For example, if the amount of admixture of ligand orbitals to the two metal orbitals involved in a metal-centered transition is different, this transition involves a certain amount of charge-transfer. The famous example of a CT transition in inorganic chemistry, the MnO_4^- ion, has, according to calculations [9], to be mainly described as a rearrangement of electronic charge on the metal ion and on the oxygen ions. Nevertheless, the description of several types of excited states as given above appears to be very useful.

This paper does not aim to give a full account of all types and all consequences of MMCT. We try to present data and phenomena which originate mainly from the solid-state field, but which can be connected to and may be of use in other, related fields. Unfortunately this exchange of ideas on MMCT between the several fields in which they are of importance is not very common.

Here we mention as an example that in the coordination-chemistry field optical MMCT transitions between weakly coupled species are usually evaluated using the Hush theory [10, 11]. The energy of the MMCT transition is given by $\nu_{max} = \Delta E + \chi$. Here ΔE is the difference between the potentials of both redox couples involved in the CT process. The reorganizational energy χ is the sum of inner-sphere and outer-sphere contributions. The former depends on structural changes after the MMCT excitation transition, the latter depends on solvent polarity and the distance between the redox centres. However, similar approaches are also known in the solid state field since long [12].

For cobalt ferrite, $CoFe_2O_4$, the energy effect in the valency reaction Co(II) + Fe(III) → Co(III) + Fe(II) has been calculated to be 1.35 eV from the ionization potentials and crystal-field stabilization effects. The combination Co(II) + Fe(III) is the ground state. For titanium in Fe_2O_3 the following reaction is of importance

$$\text{Fe(III)} + \text{Ti(III)} \rightarrow \text{Fe(II)} + \text{Ti(IV)}.$$

In this case polarization effects must also be taken into account. By several arguments it was possible to show that the Fe(II) + Ti(IV) combination has the lower energy [12].

The organization of this paper is as follows. In Sect. 2 we will discuss at length MMCT for those cases in which the metal ion which is reduced in the MMCT transition (the electron acceptor) is an ion with "d^0" configuration (e.g. Ti^{4+}, V^{5+}, Nb^{5+}, W^{6+}). In Sect. 3 we will mention several other cases without aiming at completeness. In Sect. 4 we will review the work by McGlynn et al. [5], since it bears a clear relation to the phenomena described in the foregoing sections. In Sect. 5 we will shortly enter into the problem of the mixed valence compounds which is in this aspect undoubtedly in order. In Sect. 6 the importance of MMCT transitions for semiconductors will be mentioned. Finally Sect. 7 will present the consequences of MMCT excited states for radiative and especially nonradiative processes between levels which are at first sight not connected to the MMCT states. Section 8 gives a short conclusion.

2 Metal-to-Metal Charge Transfer Involving One Closed-Shell Transition-Metal Ion

2.1 General

By way of introduction we start with the color of mixed metal oxides containing Cr(III). No doubt the most famous and most thoroughly investigated example is ruby, Al_2O_3:Cr(III). It is a pink to red gemstone which shows a deep-red luminescence. A short historical review of ruby research can be found in Ref. [13]. Crystal-field theory was able to explain the spectroscopical properties of Cr(III) in Al_2O_3. It has also been found that the red crystal color of ruby is exceptional. Isomorphous Cr_2O_3 is green. This is also the color of Cr(III) in other gemstones and/or mixed metal oxides. The red color is usually ascribed to a strong crystal-field at the site of the Cr(III) ion. The absorption bands responsible for the color of the composition concerned, viz. $^4A_2 \rightarrow {}^4T_2$ and $^4T_1$ shift to higher energy if the crystal-field strength increases [14].

As a contradistinction to the relatively simple case of Al_2O_3:Cr(III) where the color is due to a metal-centred electronic transition, we mention now on one hand the fact that the Cr(III) ion colors many transition-metal oxides brown (e.g. rutile TiO_2 or the perovskite $SrTiO_3$ [15]), and on the other hand the fact that the color of blue sapphire (Al_2O_3: Fe, Ti [16]) is not simply due to a metal-centred transition. By way of illustration Fig. 1 shows the diffuse reflection spectrum of $SrTiO_3$ and $SrTiO_3$:Cr(III) [17], and Fig. 2 the absorption spectrum of Al_2O_3:Ti(III) and Al_2O_3:Ti(III), Fe(III) [18]. It has been shown that these colors are due to MMCT transitions and cannot simply be described by metal-centred transitions [19].

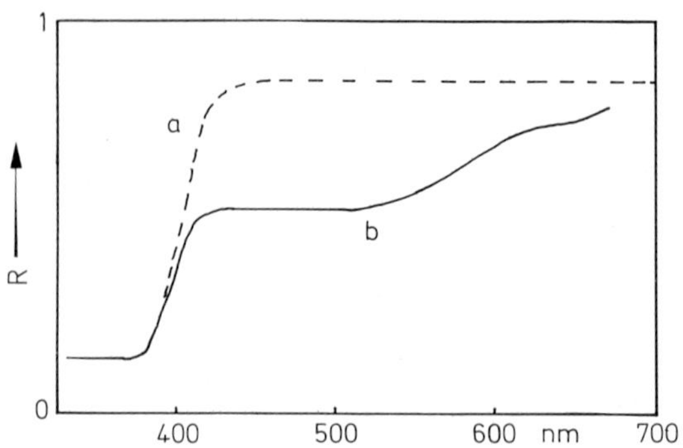

Fig. 1. Diffuse reflection spectrum of $SrTiO_3$ (*a*) and $SrTiO_3$:$LaCrO_3$ (0.5 mole %) (*b*). The extra absorption in (b) is due to Cr (III) → Ti(IV) MMCT

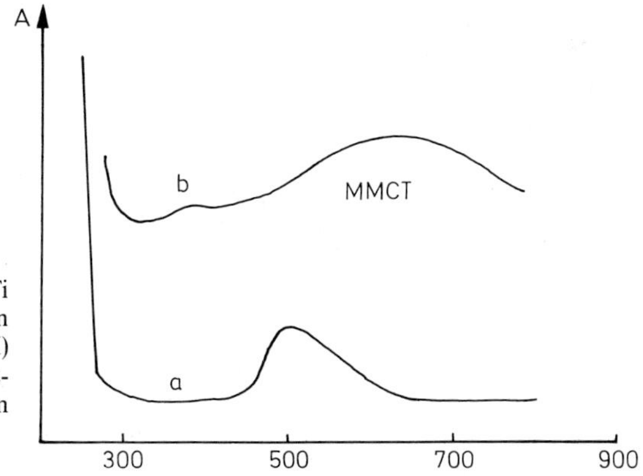

Fig. 2. Absorption spectrum of Al_2O_3:Ti (*a*) and Al_2O_3: Fe, Ti (*b*). The absorption band in (a) at about 500 nm is the Ti(III) crystalfield transition. The MMCT transition in (b) is indicated (modified from Ref. [18])

Consider first blue sapphire Al_2O_3: Ti(III), Fe(III) (Fig. 2). In the absence of Fe(III) the absorption spectrum is easy to interpret. The weak band with a maximum at about 500 nm is due to the $t_2 \rightarrow e$ crystal-field transition on Ti(III) ($3d^1$), the strong band at $\lambda < 280$ nm is due to a Ti(III)–O(−II) LMCT transition. The absorption band in the region around 700 nm in the case of the codoped crystal cannot be due to Fe(III). It has been ascribed to MMCT, i.e. to a transition within an iron-titanium pair:

$$\text{Ti(IV)} + \text{Fe(II)} \xrightarrow{h\nu} \text{Ti(III)} + \text{Fe(III)}.$$

This assignment was given for the first time by Townsend long ago [20]; it has been confirmed by other authors [16, 18].

Let us now turn to Cr^{3+}-doped titanates. Here it has been proposed that chromium-titanium MMCT transitions are responsible for the brown colors obtained [15, 17, 19]:

$$\text{Ti(IV)} + \text{Cr(III)} \xrightarrow{h\nu} \text{Ti(III)} + \text{Cr(IV)}.$$

In the isoelectronic zirconates this absorption band is not observed [17]. The spectral position of these MMCT bands has been interpreted in terms of the relevant ionization potentials [17], an approach which runs parallel with the Hush theory [10]. The fact that the MMCT transition is at higher energy in the Cr(III)–Ti(IV) pair than in the Fe(II)–Ti(IV) pair is due to the more than 10 eV higher ionization potentials of the trivalent transition-metal ions compared to the divalent transition-metal ions. The fact that the MMCT absorption band is not observed in the zirconates in contradiction to the titanates is due to the higher ionization potential of the Ti(III) species:

$$I_4(\text{Ti}) \simeq 43\text{ eV},\ I_4(\text{Zr}) \simeq 34\text{ eV [21]}.$$

Having introduced the MMCT transitions, two remarks are in order before continuing the discussion.

1. For every couple of metal ions of which one is easily oxidized whereas the other is easily reduced, the MMCT excited state should be at relatively low energy. This is not only easily predicted from the Hush theory [10], or a consideration of ionization potentials [17], but is straight forward. Nevertheless, this is not always generally realized. As an example, the red color of Cu_2WO_4 crystals [22] may seem enigmatical at first sight, because neither the Cu(I) ion nor the WO_4^{2-} ion absorbs in the visible. Since the Cu(II)–W(V) MMCT excited state cannot be far out, it is obvious to ascribe this intense absorption to a MMCT transition [23]. Also the black color of ilmenite $FeTiO_3$ is due to MMCT absorption. Actually this is precisely the same transition as mentioned above for blue sapphire, since the crystal structure of $FeTiO_3$ is a superstructure of corundum (Al_2O_3).
2. Utmost care has to be taken when preparing samples for study of MMCT transitions. The reason for this is obvious. Every deviation from stoichiometry, either in the metal-to-ligand ratio or in the metal-to-metal ratio will be translated into different valencies of one and the same metal. This configuration will also result in MMCT absorption. Let us take $FeTiO_3$ as an example. If the Fe/Ti ratio is slightly less than one, Ti(III) will be present resulting in Ti(III)–Ti(IV) MMCT. A slight deficiency of oxygen will also result in the presence of Ti(III). This type of equilibria has been discussed extensively [24].

2.2 Survey of Other Cases

Having defined a MMCT transition and having shown its presence in a few illustrative examples, we now turn to a more or less systematic discussion of the case in which one of the metal ions is a closed-shell transition metal ion, i.e. an ion which has lost its *d* electrons ("d^0" configuration). Typical examples are Ti^{4+}, V^{5+}, Cr^{6+}, Mn^{7+}, Nb^{5+}, Mo^{6+}, Ta^{5+}, W^{6+}. In solids they usually occur as complex anions like VO_4^{3-}, CrO_4^{2-}, MoO_4^{2-}, WO_6^{6-}. The electronic absorption spectra of these anions are of the LMCT type. In many cases return from the excited LMCT state to the ground state occurs radiatively. The spectra consist of broad bands and the Stokes shift of the emission is large ($\geq 10\,000\ cm^{-1}$). The compound $CaWO_4$ has been in use as a luminescent material for far over 50 years [25], YVO_4 has been for years the host lattice for efficient luminescent materials [26], and $YTaO_4$-based materials are nowadays used as X-ray phosphors [25]. The spectroscopy of the LMCT transitions in these complexes has been reviewed by us earlier [27].

How complicated the nature of these excited states can be, has been shown convincingly by Van der Waals et al., who studied these states by resonance techniques. A strong Jahn-Teller distortion appears to be present [28–30].

The spectroscopic properties of these closed-shell complex anions are drastically changed if metal ions which are easily oxidized enter the second coordination sphere. Electron transfer occurs to the empty *d* orbital of the closed-shell transition-metal ion.

Consider first transition-metal ions with an incompletely filled *d* shell as electron donors. Two examples have already been given above. Reynolds [31] observed these MMCT transitions not only in solids $MSO_4 . Ti(SO_4)_2$, but also in M(II) and Ti(IV) containing solutions (M is a divalent transition metal ion). Other authors reported these absorption transitions, but did not comment on them. Reinen [32], for example, studied the Ni^{2+} spectra of the perovskites $La_2Ni_{0.4}Mg_{0.6}TiO_6$ and $La_2Ni_{0.4}Mg_{0.6}ZrO_6$ and observed a strong absorption band starting at about 20 000 cm^{-1} in the titanate case. We have assigned this to a Ni(II)–Ti(IV) MMCT transition [17].

In agreement with the ionization potential involved, it has been observed that the MMCT transition between ions $M(d^n)$ and $M(d^0)$ shifts to lower energy if the valency of the former (the donor) ion decreases [17, 19]. This transition is for M(IV) donor ions hardly distinguishable from the CT transition in the $M(d^0)O_n$ complex, because they overlap in the spectra. For M(III) donor ions (like Cr(III), see above) the MMCT is observed below the CT transition and the sample concerned is strongly colored. For M(II) donor ions (like Fe(II), see above) the MMCT is even at still lower energies.

An interesting case is the optical absorption of M(II)-doped $MgTi_2O_5$ [33]. The spectra of interest are given in Fig. 3. The undoped $MgTi_2O_5$ shows a strong optical absorption which starts at about 320 nm. This is due to the O(– II)–Ti(IV) LMCT transition. The spectra of $MgTi_2O_5$ doped with Mn(II), Fe(II), Co(II) and Ni(II) show considerable additional absorption in the visible. Only Co(II) and Ni(II) are expected to show spin-allowed crystal-field transitions in this spectral range [14]. These are in fact observed (see Fig. 3): $^4T_1 \rightarrow {}^4T_1$ (P) for Co(II) at 590 nm, and $^3A_2 \rightarrow {}^3T_1$ (P) for Ni(II) at 450 nm.

The interesting observation is the presence of a broad, intense absorption region covering a part of (Mn(II), Co(II), Ni(II)) or even the whole visible region (Fe(II)). These absorptions are ascribed to M(II)–Ti(IV) MMCT. The energy of the onset of this absorption increases in the sequence Fe(II), Mn(II), Co(II), Ni(II). This is in good agreement with the increasing third ionization potential of these ions.

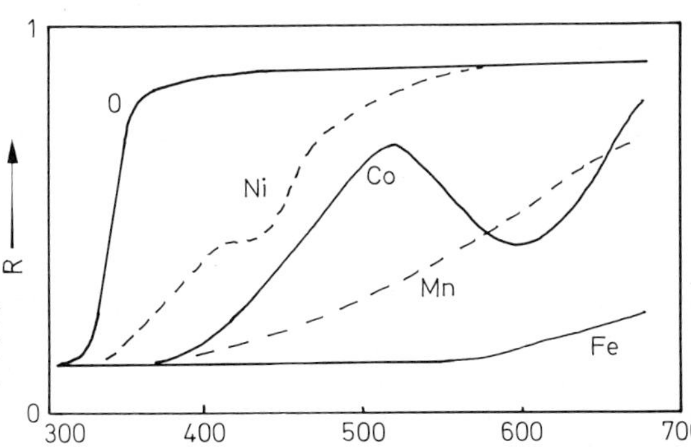

Fig. 3. Diffuse reflection spectra of nominally pure $MgTi_2O_5$ (O) and of $MgTi_2O_5$:M(II). The nature of M is indicated

In the case of $MgTi_2O_5$:Fe(III) the Fe(III)–Ti(IV) MMCT occurs at much higher energy than the Fe(II)–Ti(IV) MMCT mentioned above [33]. This agrees with the agruments used above.

Rossman et al. [34] have studied the problem of the color of nickel(II) oxides. The well-known green color of Ni(II) is due to crystal-field transitions. However, many nickel(II) compounds are intensively yellow or even brown to red colored. For example, $NiNb_2O_6$ and $NiWO_4$ are brown, and NiV_2O_6 red. This color is due to MMCT between Ni(II) and the $M(d^0)$ ion concerned. Figure 4 shows several reflection spectra reported in Ref. [34]. Note the onset of an intense, broad absorption band which extends to far in the visible. It is clear that Ni(II)–V(V) MMCT is situated at the lowest energy in agreement with the high 5th ionization potential of vanadium. Those for Ti(IV), W(VI) and Nb(V) are at higher energy. Before leaving the topic of $M(d^n)M(d^0)$CT it may be remarked that in this paragraph we have neglected the degree of delocalization of the electrons involved. This will be discussed more in detail in Sect. 6. However, there is no doubt that the MMCT states in the case of, for example, $SrTiO_3$ and TiO_2 are strongly delocalized since the transferred electron will occupy the conduction band built up for a substantial part from the titanium 3*d* orbitals. This is not the case for, for example, $MgTi_2O_5$ or $CaWO_4$, which are much more localized [19, 23]. For certain applications a high degree of delocalization of the MMCT excited state is essential, as will be shown below.

Let us now consider MMCT for the case in which the donating ion is a lanthanide ion with a partly filled 4*f* shell: $M(f^n)M(d^0)$CT. The trivalent lanthanide ions with a low fourth ionization potential are Ce^{3+}, Pr^{3+}, Tb^{3+}. Their optical absorption spectra show usually allowed 4*f*–5*d* transitions in the ultraviolet part of the spectrum [6, 35]. These are considered as MC transitions, although they will undoubtedly have a certain CT character due to the higher admixture of ligand orbitals into the *d* orbitals. In combination with $M(d^0)$ ions these $M(f^n)$ ions show MMCT transitions. An early example has been given by Paul [36] for Ce(III)–Ti(IV) MMCT in borosilicate glasses. The absorption maximum was at about 30 000 cm^{-1}.

The Ce(III) ion gives in titanates an additional optical absorption. Figure 5 gives an example, viz. $SrTiO_3$:Ce(III). This absorption, extending down to 600 nm, has been ascribed to MMCT between Ce(III) and Ti(IV) [37].

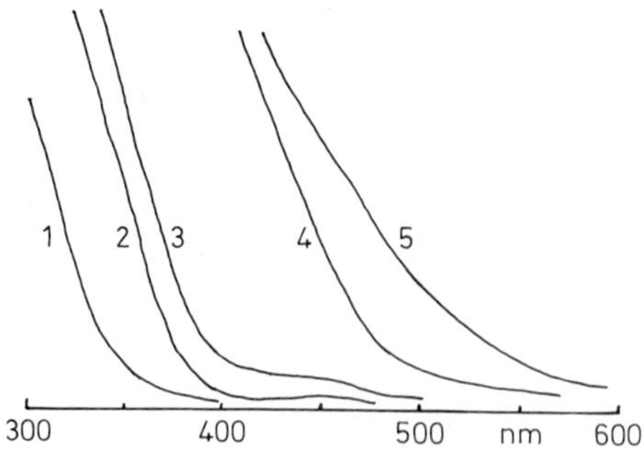

Fig. 4. Absorption spectra of some nickel compounds: (*1*) $NiNb_2O_6$, (*2*) $NiWO_4$, (*3*) $NiTiO_3$, (*4*) $Ni_3V_2O_8$ (*5*) NiV_2O_6 (modified from Ref. [34])

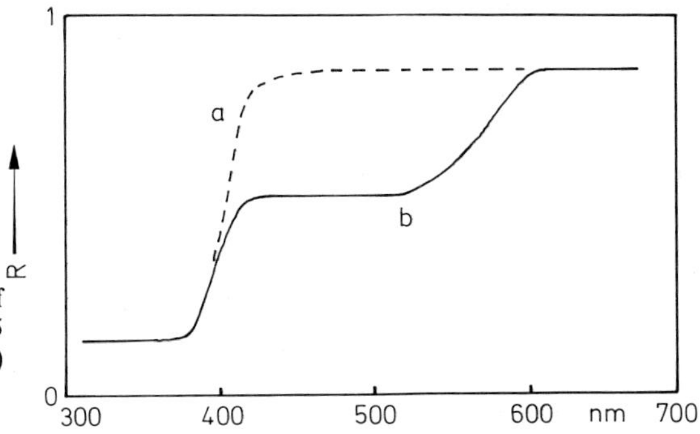

Fig. 5. Diffuse reflection spectrum of $SrTiO_3$ (*a*) and $SrTiO_3$:$CeAlO_3$ (0.5 mole %) (*b*). The extra absorption in (b) is due to Ce(III) → Ti(IV) MMCT

Transitions of this type play a detrimental role in the field of luminescent materials (see also Sect. 7). The Tb(III) and Ce(III) ions in YVO_4, for example, spoil all possible luminescence due to the presence of a MMCT excited state at low energy. This state is ascribed to Tb(III)/Ce(III)–V(V) MMCT resulting in an excited state with character Tb(IV)/Ce(IV)–V(IV) [38, 39]. Very similar are the excited MMCT states in the lanthanide decatungstates $Ln[W_{10}O_{36}]^{9-}$ [40] in the cases Ln = Ce, Pr, Tb. Here the excited state has Ln(IV)–W(V) character.

These MMCT excited states $M(4f^{n-1})M(d^1)$ are not always observable in the spectra. We mention two examples in which a different physical factor is responsible for the high energy position. The first is $YTaO_4$:Tb(III) [41]. The MMCT state Tb(IV)–Ta(IV) is at too high energy to be observed. This is related to the high value of the 5th ionization potential of tantalum. The tantalate CT transition in $YTaO_4$ is also at very high energy, viz. $\sim 45\,000\ cm^{-1}$.

More intriguing is the case of codoped $CaSO_4$:Tb(III), V(V). In contrast to YVO_4:Tb(III) the Tb(IV)–V(IV) MMCT state appears to be at very high energy [41]. This time the reason has to be sought in the reorganization energy of the surroundings. The Tb(III) ion in $CaSO_4$ occupies calcium sites, the V(V) ion sulfur sites. Therefore they carry an effective charge relative to the host lattice constituents. The attraction between the two species results in pair formation, which can be written as $(Tb^{\bullet}_{Ca}\,V'_{s})^{x}$, according to the Kröger notation [24]. A MMCT transition would result in a pair $(Tb^{\bullet\bullet}_{Ce}\,V''_{S})^{x}$ and obviously this takes more energy than in the situation without effective charges.

Gruehn et al. [42] have reported on the isomorphy between modifications of $LaTa_3O_9$ and $CeTa_3O_9$. Both contain trivalent lanthanide ions. The former is white, the latter is orange (P-modification), brass (O-modification) or yellow (M-modification). These colors are due to Ce(III)–Ta(V) MMCT absorptions, the Ce(III) and tantalate chromophores absorbing only in the ultraviolet. No doubt there are many more examples of this type in the literature.

In view of this the divalent lanthanide ions are expected to show intense optical absorption in the whole visible region in compounds containing $M(d^0)$ ions. Since these compounds are not easy to prepare, our earlier warning is relevant. Only results on well-prepared and characterized compounds may be taken seriously. Let us mention one example, viz. the compounds in the system $EuO–TiO_2$, viz. $EuTiO_3$, Eu_2TiO_4 and $Eu_3Ti_2O_7$ [43]. These compounds are isostructural with the analogous strontium compounds. This is not surprising, the Eu(II) and Sr(II) ions being of equal size. These europium titanates are all black. This shows the MMCT transition Eu(II) + Ti(IV) → Eu(III) + Ti(III) to be at low energy. The reason for the black color is the same as for e.g. $FeTiO_3$ with low-lying MMCT transitions Fe(II) + Ti(IV) → Fe(III) + Ti(III).

Reference [43] cites that $EuTiO_3$ at 1150 °C contains 5% Eu(III)–Ti(III). Using a simple Boltzmann distribution this yields for the MMCT transition an energy of only 3000 cm^{-1}. All other lanthanide titanates $LnTiO_3$ have all cations trivalent in the ground state. The different valency state in the europium titanates has interesting consequences: $EuTiO_3$ is an antiferromagnet (T_N 5.1 K) and Eu_2TiO_4 and $Eu_3Ti_2O_7$ are ferromagnets (T_c 9 and 8.5 K, respectively).

Metal ions with d^{10} configurations can also show MMCT transitions with M(d^0) ions. We mentioned already above the case of Cu_2WO_4 with a MMCT transition of the type Cu(I) + W(VI) → Cu(II) + W(V) resulting in the red color of this compound. Similar are $CuNbO_3$ (44) and $CuLn(WO_4)_2$ [19]. Returning to red Cu_2WO_4, note that Ag_2WO_4 is yellow; Na_2WO_4 is colorless with a tungstate charge-transfer transition at about 40 000 cm^{-1}. The change in colors follows closely the second ionization potential of the monovalent ion: Cu 20.3, Ag 21.5, Na 47.3 eV [21]. To this class belong also the cases mentioned before by Reisfeld and Jørgensen [6]: $AgMnO_4$ is blue, and Ag_2CrO_4 red (compare K_2CrO_4 yellow).

A careful magnetic and optical study of $AgMnO_4$ has been made by Sperka and Fritzer [45]. They find the Ag(I)–Mn(VII) MMCT transition at about 10 000 cm^{-1} (see Fig. 6). The ground state Ag(I)Mn(VII)O_4 is confirmed by magnetic susceptibility measurements.

The higher-charged ions with d^{10} configuration do not show pronounced MMCT transitions with the d^0 ions. Only an additional absorption at the onset of the M(d^0)–O(−II) CT transition has been noted. Examples are $HgWO_4$ and $HgMoO_4$ (46), Cd(II) in $CaMoO_4$, $CaWO_4$ and $MgWO_4$ [47]. A more pro-

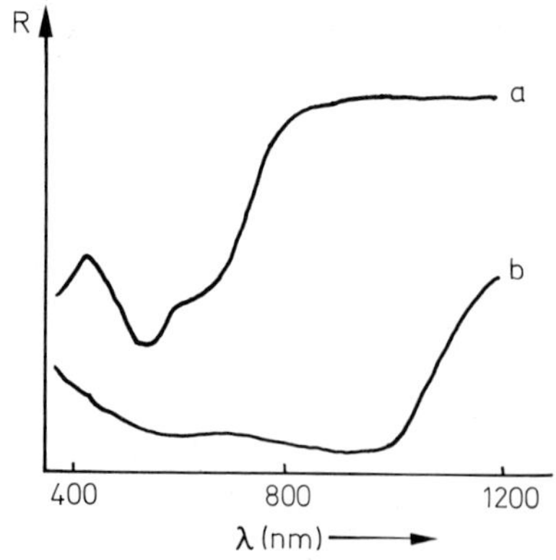

Fig. 6. Diffuse reflection spectrum of $KMnO_4$ (*a*) and $AgMnO_4$ (*b*) (modified from Ref. [45])

nounced effect is given by In(III) [47, 48]. Obviously the MMCT-character of the excited state is restricted.

Interestingly enough, it is possible to study these systems also by emission spectroscopy. The results for In(III) are conspicious (see Table 1). Figure 7 gives the luminescence spectra of $La_3TaO_4Cl_6$:In(III) to illustrate the type of spectra [48]: we are dealing with broad bands; the emission is strongly Stokes-shifted relative to the absorption transition.

It is not unexpected that M(II) ions with d^{10} configuration give hardly observable MMCT transition in combination with M(d^0) ions. Since the ionization potential of the donor ion increases with its charge, the consequence of this observation must be that M(III) (d^{10})–M(d^0) CT will always be at higher energy than the M(d^0)O_n CT transition. Therefore the absorptions induced by In(III) as mentioned in Table 1 are probably not due to MMCT. It seems more obvious to assign this to a complicated type of CT in which an oxygen electron is promoted to an orbital of a mixed metal character, viz. *d* of M(d^0) and $(n+1)s$ and $(n+1)p$ of M(nd^{10}). This implies that the band structure of $InVO_4$ is in first

Table 1. Some optical properties of In(III)-doped host lattices [47, 48]

Composition	Maximum absorption band (10^3 cm^{-1})	Maximum emission band (10^3 cm^{-1})	T_q(K)[a]	q(%)[b]
YVO_4	32.5	24.1	300	5
YVO_4:Sc	—	22.0	370	30
YVO_4:In	—	20.8	400	30
$InPO_4$:V	33.7	17.9	475	55
Y_2WO_6	—	22		—
Y_2WO_6:In	—	19.5		—
$La_3TaO_4Cl_6$	43.5	29.5	300	—
$La_3TaO_4Cl_6$:In	37	20	> 300	60

[a] Quenching temperature of the emission; [b] quantum efficiency of the emission at 300 K

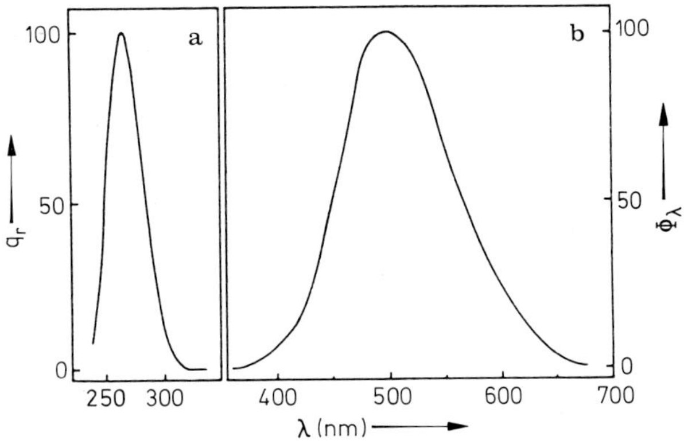

Fig. 7. Excitation (*a*) and emission (*b*) spectra of the luminescence of $La_3TaO_4Cl_6$: In at 4.2 K. After MJJ Lammers, thesis, Utrecht, 1986

approximation as follows: the valence band consists of oxygen orbitals, the conduction band of 5*s* and 5*p* orbitals of indium and 3*d* orbitals of vanadium. This example shows that one cannot simply assign every additional absorption band to a MMCT transition.

Considerably more simple is the case of s^2 ions as electron-donating ions to $M(d^0)$ ions. Examples of s^2 ions are Sn^{2+}, Sb^{3+}, Tl^{+}, Pb^{2+}, Bi^{3+}. In the pair $M(s^2)M(d^0)$ there is usually a low-lying MMCT state with strong $M(s^1)M(d^1)$ character. This has been discussed by us before [27]. In many cases this MMCT state shows efficient luminescence.

Let us illustrate this with a few examples. Table 2 gives some data on YVO_4, YVO_4:Bi(III), $BiPO_4$ and $BiPO_4$:V(V) [47, 49]. The undoped compounds show a blue luminescence which is due to a CT transition in the vanadate group in case of YVO_4, and to a MC ($s^2 \rightarrow sp$) transition on Bi^{3+} in $BiPO_4$. The doped materials, however, show a green luminescence with an excitation band at lower energies than the luminescence of the undoped compounds (see Fig. 8). This transition is ascribed to a MMCT transition between Bi(III) and V(V).

In the comparable set of compositions $CaWO_4$, $CaWO_4$:Pb(II), $PbSO_4$ and $PbSO_4$:W(VI) very similar phenomena have been observed [47]. The extra absorption band which appears as a result of doping is ascribed to a MMCT transition between Pb(II) and W(VI). The optical properties of $PbWO_4$ have

Table 2. Some spectroscopical data on undoped and doped YVO_4 and $BiPO_4$ [47, 49]

Composition	Maximum absorption band	Maximum emission band
YVO_4	32 500	24 100
$YV_{0.95}Bi_{0.05}O_4$	29 500	17 900
$BiPO_4$	~ 35 000	23 500
$BiP_{0.95}V_{0.05}O_4$	~ 30 000	18 500

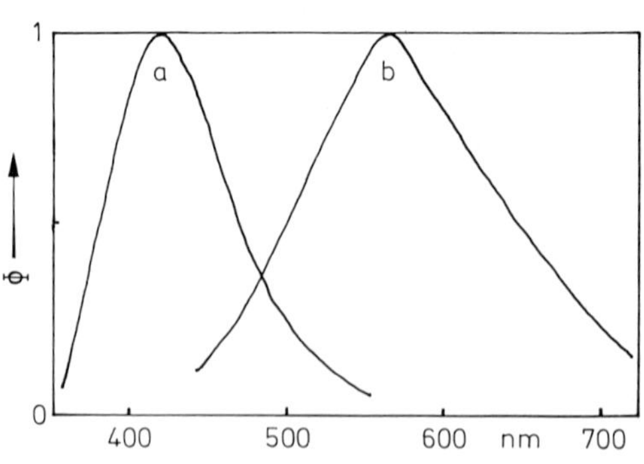

Fig. 8. Emission spectra of YVO_4 (*a*) and YVO_4:Bi (*b*) (modified from Ref. [47])

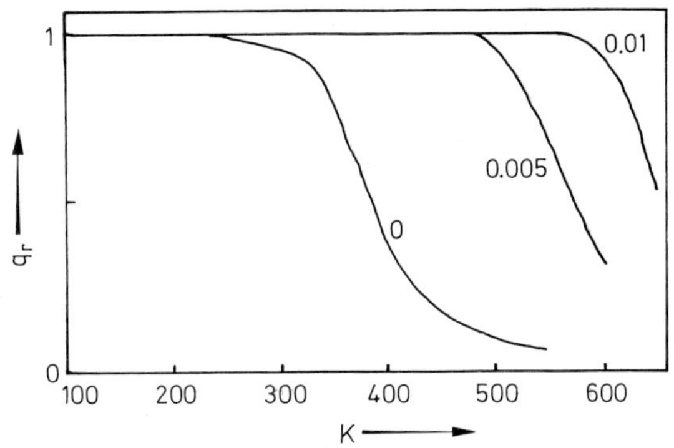

Fig. 9. Temperatures dependence of the luminescence efficiency of the system $Ca_{1-x}Pb_xWO_4$ for three values of x (modified from JA Groenink, thesis, Utrecht, 1979)

also been interpreted along this line [50]. This MMCT state shows efficient luminescence with surprisingly high quenching temperature. In Fig. 9 the temperature dependence of the luminescence intensity of the system $Ca_{1-x}Pb_xWO_4$ has been given as a function of x. Note the drastic shift of the quenching range to high temperatures for low Pb(II) concentrations. For higher values of x there is again a decrease due to concentration quenching [50].

The drastic increase of the luminescence quenching temperature of $CaWO_4$ upon introduction of Pb(II) has been interpreted as an indication for a change in the nature of the transition. Whereas the CT transition in the tungstate group occurs between bonding and antibonding orbitals, the MMCT transition between Pb(II) and W(VI) occurs between two antibonding orbitals. This description is undoubtedly an oversimplification, but it explains the higher quenching temperature as well as the smaller Stokes shift in case of the MMCT transition [50].

These considerations can be easily extended. Actually ions like Pb(II) and Bi(III) show an additional absorption band in many niobates, molybdates and tungstates [47, 51]. Excitation into this band gives an emission band which belongs specifically to the new centre $M(s^2)M(d^0)$ [47].

This type of centre has also been observed in chlorides [35]. Whereas Cs_2ZrCl_6 and Cs_2HfCl_6 show blue luminescence due to a CT transition in the $ZrCl_6^{2-}$ and $HfCl_6^{2-}$ octahedra, the introduction of Pb(II) yields a red luminescence due to MMCT between Pb(II) and Zr(IV)/Hf(IV).

The red color of Ag_2CrO_4 becomes even more deeply red in $AgBi(CrO_4)_2$ [52]. This may point to a contribution of Bi(III)–Cr(VI) MMCT in the transition involved. Unfortunately the spectra of many of these colored compounds are not given in the literature.

$SnWO_4$ with Sn(II) and W(VI) ions shows an absorption edge “in the near UV region” [53]. This might well be the MMCT transition between Sn(II) and W(VI), the Sn(II) ion and the tungstate group having only absorptions in the short-wavelength ultraviolet.

Another special case is the yellow compound $LiVTeO_5$. The charge distribution in the ground state is $Li(I)Te(IV)V(V)O_5$ [54]. This suggests that the yellow color is due to a Te(IV)($5s^2$)–V(V)($3d^0$) MMCT transition.

A CT transition which is very similar to the $s^2 \rightarrow d^0$ MMCT transition has been observed by Vogler et al. [55] for complexes [M(2,2′-bipyridyl)X_3] with X = Cl, Br, I and M = Sb, Bi. These authors report MLCT transitions involving the promotion of an electron from the s^2 lone pair to the π^* orbital of the bipyridyl ligand. For example, for M = Sb and X = Br they observe an orange color for the complex due to an absorption band with a maximum at 435 nm. In the complexes considered by us the transition is to an antibonding π^* orbital (with pronounced *d* character) on the filled-shell transition-metal complex ion.

Let us close this paragraph with MMCT transitions between two ions of the same element, one with d^0, the other with d^1 configuration. Jørgensen [56] showed long ago that the simultaneous presence in 12 M hydrochloric acid of pale yellow Ti(IV) and blue Ti(III) yields a dark brownish-purple 1:1 complex with a stronger absorption band than the single constituents have. The broad absorption band has a maximum at 482 nm. The new absorption band is ascribed to a Ti(III)–Ti(IV) MMCT transition. A near-infrared absorption band in titanium-doped Al_2O_3 seems to be due to the same transition [57].

Another example originates from the absorption spectra of WO_3 and MoO_3 [58]. Consider WO_3. At room temperature the optical bandgap is situated at 22 200 cm^{-1} (O–W charge transfer). The charge carriers are large polarons and consequently the near-infrared spectral region shows free-carrier absorption. In a low-temperature modification of WO_3, however, the edge shifts to 25 600 cm^{-1}, the free-carrier absorption vanishes and a distinct peak appears in the near-infrared spectral region (around 6000 cm^{-1}). This phenomenon is due to strong electron-phonon coupling. Charge carriers are trapped at tungsten positions corresponding to the formation of W(V) and consequent polarization of the surrounding structure (small polaron formation). This near-infrared absorption can be considered as an intervalence charge transfer (or MMCT) between W(V) and W(VI). The introduction of H stabilizes the W(V) state: H(0) + W(VI) → H(I) + W(V) and the W(V)–W(VI) intervalence charge transfer is now at about 10 000 cm^{-1}. Further, the optical transition Mo(V) to W(VI) was found at about 7500 cm^{-1}. For $MoO_{2.99}$ a Mo(V)–Mo(VI) absorption transition has been observed at 10 000 cm^{-1}.

Consider now Na_xWO_3 or $LiTi_2O_4$. One might expect to find W(V)–W(VI) and Ti(III)–Ti(IV) MMCT. However, in these compounds all metal ions are equivalent and the *d* electrons available are spread out in a conduction band. The bronzes Na_xWO_3 are metallic. They become superconducting at 6 K, whereas $LiTi_2O_4$ becomes superconducting at even 13 K [59]. Here we meet the central problem of mixed-valence compounds [60] which we will postpone till Sect. 5.

Having discussed at length the case of MMCT with d^0 ions as electron acceptors, the following section will discuss examples of optical MMCT involving other acceptor ions.

3 Other Examples of Optical MMCT Transitions

3.1 Molecular Complexes

After an extensive review of MMCT transitions involving d^0 ions in solids, it seems wise to start this paragraph with some molecular species, because many of these have been investigated in much more detail than their counterparts in non-molecular solids. It is suitable to make a distinction between outer-sphere charge-transfer (OSCT) and inner-sphere charge-transfer (ISCT) transitions [1]. In the former the metal ions do not have ligands in common, in the latter they are connected by a common ligand. Studies are usually performed on metal-ion pairs in solution.

An impressive example of a detailed study of optical MMCT of the OSCT type has been given by Curtis and Meyer [61]. The subject of their investigations is the ion pair $[M(II)(CN)_6, Ru(III)(NH_3)_5L]^-$, where M = Fe, Ru, Os and L is pyridine or a substituted pyridine. These ion pairs show broad absorption bands in the red and near-infrared part of the spectrum. These are assigned to OSCT (or MMCT) of the type M(II) + Ru(III) → M(III) + Ru(II). The absorption maxima are in the range 600–940 nm, ε_{max} is about 30 M^{-1} cm^{-1} and the halfwidth is some 7000 cm^{-1}. From this the order of magnitude of the oscillator strength is estimated to be 10^{-3}. The authors were able to derive the potential energy surfaces of the excited MMCT state. From this they can show that the return from the excited state to the ground state is a nonradiative process (see also Sect. 7).

Vogler and Kunkely [62] have reported outer-sphere MMCT transitions in the ion pairs $[Pt(NH_3)_5Cl]^{3+}$ $[M(CN)_6]^{4-}$ where M = Fe, Ru, Os (Fig. 10). Whereas aqueous solutions of $[Pt(NH_3)_5Cl](ClO_4)_3$ and $K_4[M(CN)_6]$ are colorless, their mixture is yellow due to a MMCT transition M(II) + Pt(IV) → M(III) + Pt(III). For M = Fe, λ_{max} = 418 nm and ε_{max} = 88

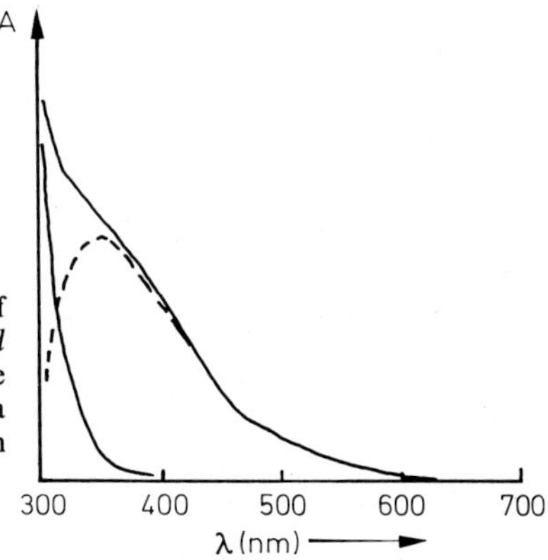

Fig. 10. Absorption spectra of aqueous solutions of $[Pt(NH_3)_5Cl](ClO_4)_3$ and $K_4[Os(CN)_6]$. *Left-hand curve*: separate solutions; *right-hand curve*: mixture. The *broken curve* gives the difference between the two spectra and corresponds to Os(II)–Pt(IV) MMCT (after data in Ref. [62])

$M^{-1}\,cm^{-1}$. For M = Ru λ_{max} = 332 nm, for M = Os λ_{max} = 352 nm. This shift to shorter wavelengths can be related to the change in redox couple when varying M.

In the ion pair $[Pt(NH)_3)_5Cl]^{3+}$ $[Pt(CN)_4]^{2-}$ these authors observed a MMCT transition with a maximum at 295 nm: Pt(II) + Pt(IV) → Pt(III) + Pt(III). It is interesting to note that in Wolffram's red, $[Pt(etn)_4]$-$[Pt(etn)_4Cl_2]Cl_4 . 4H_2O$ (etn = ethylamine) this MMCT transition has been observed at 500 nm, i.e. at much lower energy [63]. Vogler and Kunkely ascribe this effect to a smaller reorganization energy.

Examples of inner-sphere MMCT are the following. Ludi et al. [3] have reported the spectra of the $[(CN)_5Fe(III)NC\ Fe(II)(CN)_5]^{6-}$ ion in aqueous solution and in KBr pellets. The MMCT transition [III + II] → [II + III]* is situated at 1300 nm with $\varepsilon_{max} \simeq 10^3\ M^{-1}\,cm^{-1}$. The oscillator strength is one order of magnitude larger than for the outer-sphere MMCT transitions. The NC bridge plays a role in the CT mechanism. For $[(CN)_5Fe(III)pyr\ Fe(II)(CN)_5]^{5-}$ the absorption maximum has shifted to 1180 nm (pyr = pyrazine).

Another inner-sphere MMCT transition is observed if aqueous solutions of $[Co(NH_3)_5H_2O]^{3-}$ and $[Ru(CN)_6]^{4-}$ are added together. The following complex is formed: $[(NH_3)_5Co(III)NC\ Ru(II)\ (CN)_5]^{-1}$. It shows a MMCT transition with a maximum at 375 nm and ε_{max} = 690 $M^{-1}\,cm^{-1}$ [63] due to Co(III) + Ru(II) → Co(II) + Ru(III).

From a tabulated list in Lever's book [65] we derive the following examples of MMCT in ion pairs:

$Fe(III)/[Mo(IV)(CN)_8]^{4-}$ (12 500 cm^{-1}),

$Cu(II)/[Mo(IV)(CN)_8]^{4-}$ (20 200 cm^{-1}),

$U(VI)O_2/[Mo(IV)(CN)_8]^{4-}$ (24 650 cm^{-1}),

$V(IV)O/[Mo(IV)(CN)_8]^{4-}$ (29 000 cm^{-1}),

and the same electron acceptors with $[Ru(II)(CN)_6]^{4-}$ as donor (absorption maxima at 18 000, 25 200, 29 100 and 31 500 cm^{-1}, respectively. Let us now turn to the situation in solids.

3.2 $M(d^n)M(d^m)CT$ in Solids

There is not much doubt that the most intriguing and best studied example in this field is Prussian blue $Fe_4[Fe(CN)_6]_3$ [66, 67]. It is the first synthetic coordination compound and has been used as a pigment. The deep blue color is due to a MMCT transition Fe(II) + Fe(III) → Fe(III) + Fe(II) (Fig. 11). The maximum of the absorption band is situated at 14 700 cm^{-1} (ε_{max} $10^3\ M^{-1}\,cm^{-1}$). Figure 12 gives the energy diagram of Prussian blue as proposed by Ludi [66]. The optical transition is indicated by E_{op}; further E_o gives the energy difference between the ground state and the MMCT state and

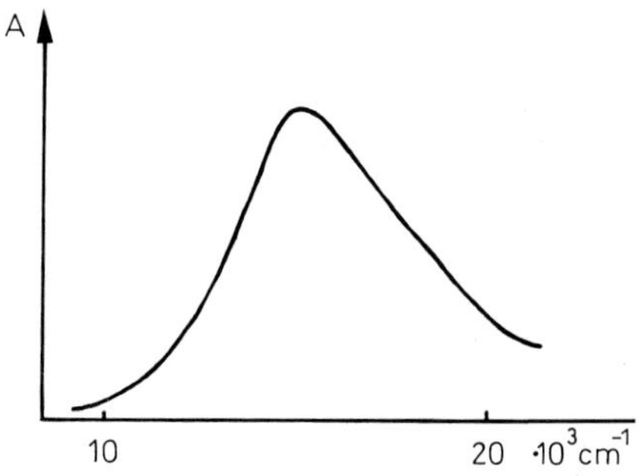

Fig. 11. Absorption spectrum of a Prussian-blue sol showing the MMCT transition (after data in Ref. [66])

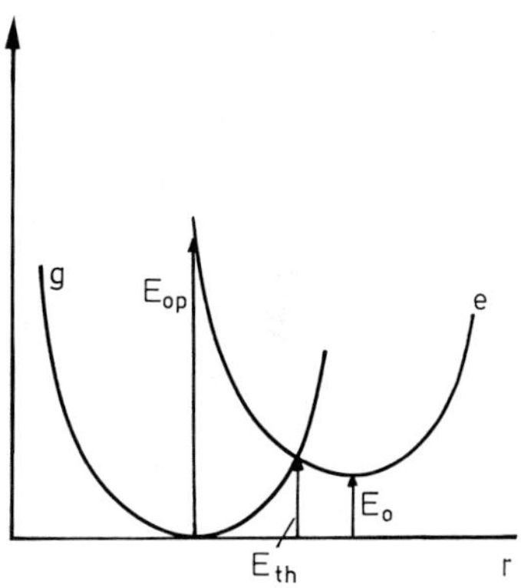

Fig. 12. Configurational coordinate diagram of Prussian blue. Curve *g* gives the ground state Fe(III)–NC–Fe(II); Curve *e* gives the MMCT state Fe(II)–NC–Fe(III). The optical transition is indicated by E_{op}, whereas E_0 gives the energy difference between the two states. See also text (after data in Ref. [66])

E_{th} the activation energy for the transition between these two states along the coordinate.

The ideal composition of Prussian blue is $Fe(III)_4[Fe(II)(CN)_6]_3 \cdot 15H_2O$. The crystal structure is cubic. All Fe(III) lattice sites are occupied, whereas those of Fe(II) are only 75% occupied. At low temperatures the paramagnetic Fe(III) ions order ferromagnetically ($T_c = 5.6$ K). Finally we note that upon replacing Fe(II) by Ru(II) or Os(II) the color properties are not drastically influenced.

Unfortunately considerably less is known about MMCT in other solids. This does not mean that it is not of importance. Consider for example, the ferrites. Those with only Fe(III) are brown-red colored like Fe_2O_3, $LiFe_5O_8$, $MgFe_2O_4$. However, $MnFe_2O_4$, $CoFe_2O_4$ and $NiFe_2O_4$ are black. This is undoubtedly due to a MMCT transition of the type M(II) + Fe(III) → M(III) + Fe(II) in the near infrared.

In the same way green NiO becomes black if small amounts of Li(I) ions are introduced. The Li(I) ions are charge compensated by Ni(III) (see, however, below): $Ni(II)_{1-2x}Li_xNi(III)_xO$ [12]. The black color is due to Ni(II) + Ni(III) → Ni(III) + Ni(II) MMCT. Simultaneously the electrical resistance drops drastically.

The importance of MMCT transitions for minerals has been clearly indicated by Burns et al. [68]. After blue sapphire (see above) another famous case

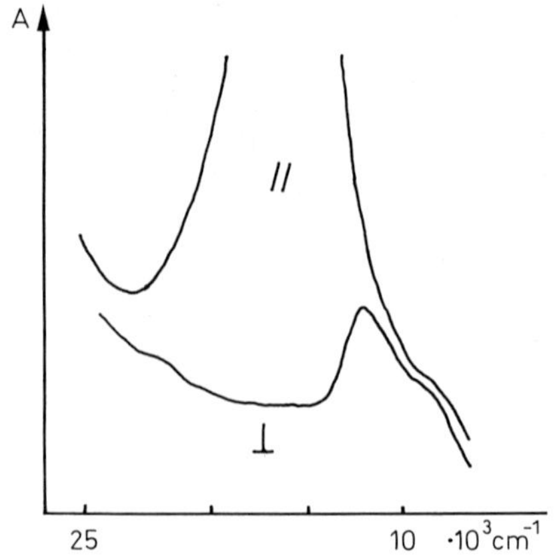

Fig. 13. Absorption spectrum of vivianite ($Fe_3(PO_4)_2 \cdot 8H_2O$) showing the MMCT transition. This transition shows a strong polarization dependence (simplified from Ref. [68])

is vivianite, $Fe_3(PO_4)_2 . 8H_2O$. The crystal structure contains two crystallographic sites for iron. Freshly mined crystals are pale green (Fe(II)), but turn blue upon exposure to air. The Fe(II) ions are partly oxidized and Fe(II)–Fe(III) MMCT transitions become possible. The absorption band shows a strong polarization (Fig. 13) which can be explained from the structural data. The absorption maximum is at 15 200 cm^{-1} (close to the value for Prussian blue). Reference [68] gives many other examples of minerals with MMCT transitions. Here we mention still Mn(II) → Mn(III) in yoderite. There are two absorption maxima, viz. 16 500 and 21 000 cm^{-1} depending on the crystallographic sites of the manganese ions.

Recently Oshio et al. [137] have reported on the physical properties of new mixed-valence compounds with one-dimensional chain structures. The chemical composition is $[Cu(en)_2][PtX_2(en)_2][ClO_4]_4$ (en = ethylenediamine, X = Cl or Br). The structure contains chains with composition Cl–Pt(IV)–Cl– –Cu(II)– –Cl–Pt(IV)–Cl– –Cu(II)– – along the b axis. The optical absorption spectrum is highly anisotropic. For polarisation along the b axis there is a strong and broad absorption band with a maximum at 22 000 cm^{-1}, which is ascribed to a metal-to-metal charge transfer transition from Cu(II) to Pt(IV). For polarisation perpendicular to the b axis this band is lacking and only weak bands at 20 000 cm^{-1} and 33 600 cm^{-1} are observed. These are due to d–d crystal-field transitions on Cu(II) and Pt(IV). Resonance Raman spectra using charge-transfer excitation show a progression in the totally symmetric Cl–Pt(IV)–Cl stretching vibration.

3.3 *Other Examples in Solids*

Other examples are the following [6, 133]. The compound Ag_2ReCl_6 is orange whereas salts of $ReCl_6^{2-}$ are pale green [134]. This indicates Ag(I)–Re(IV) MMCT. In the same way Tl_2OsBr_6 is black, although $OsBr_6^{2-}$ is tomato-red

[134]: Tl(I)–Os(IV) MMCT. Whereas $IrCl_6^{2-}$ is orange-brown, Ag_2IrCl_6 is blue and Tl_2IrCl_6 dark green [135]. Another intriguing color is the fox-red of $K_2[Fe(II)(CN)_6Cu(II)]$ and $K_2[Fe(II)(CN)_6UO_2]$ with MMCT from Fe(II) to Cu(II) and $U(VI)O_2$, respectively [69].

An extended set of measurements have been reported on compounds containing $Sb(V)X_6^-$ and $Sb(III)X_6^{3-}$ where X is a halogen [70]. These compounds are colored due to a broad absorption band in the visible. Its origin is a MMCT transition of the type Sb(III) + Sb(V) → Sb(IV) + Sb(IV). Examples of these compounds are Cs_2SbCl_6 and $(NH_4)_2SbBr_6$. Their crystal structures contain clearly two different types of SbX_6 octahedra. The bond length difference is ~ 0.25 Å. Several physical properties indicate the presence of $Sb(III)X_6^{3-}$ and $Sb(V)X_6^-$ octahedra.

In solid solution series like $Cs_2Sb_{1-x}Sn_xCl_6$ and $Cs_2Sb_{1-x}In_xCl_6$ the concentration of these octahedra can be varied. In the latter one the Sb(III) concentration decreases with increasing value of x, in the former the concentration of both antimony ions decreases with increasing x. The intensity of the MMCT transition (see Fig. 14) appears to be proportional to the product of the Sb(III) and the Sb(V) concentrations. Figure 14 shows also an intense absorption band in the ultraviolet which is due to a $5s^2 \rightarrow 5s5p$ transition on Sb(III) [71, 72].

It is interesting in this aspect to note that the $5s^2 \rightarrow 5s5p$ transitions of Sb(III) in $SbCl_6^{3-}$ are very similar in the solid model compounds $Cs_2NaLnCl_6$:Sb(III) (Ln = La, Y, Sc) [72], in the mixed valence compounds of the type Cs_2SbCl_6, and in aqueous solutions [73].

Figure 14 shows that the intensity of the Sb(III)–Sb(V) MMCT transition is relatively weak. This follows from comparison with the Sb(III) transition in the ultraviolet which is spin forbidden. The intensity of the MMCT transition is determined by the amount of Sb(IV)Sb(IV) admixture in the Sb(III)Sb(V) ground state. From the intensity of the MMCT transition the amount of admixture into the ground state is found to be less than 1%.

MMCT transitions between the lanthanide ions have not been thoroughly investigated as far as we know. That they exist, and even in the visible spectral

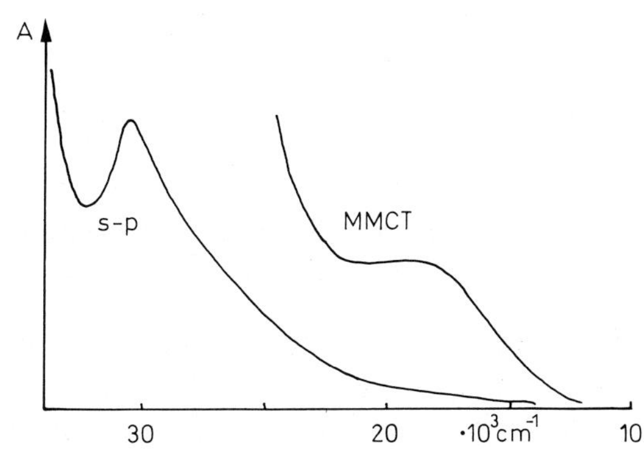

Fig. 14. Absorption spectrum of $(CH_3NH_3)_2Sb_xSn_{1-x}Cl_6$. The MMCT and *s–p* transitions are indicated. See also text (simplified from Ref. [70])

region, is immediately clear from the dark colors of the commercial oxides Pr_6O_{11} and Tb_7O_{12}. This color is due to a Ln(III)–Ln(IV) MMCT transition. The binary lanthanide oxides have been extensively reviewed by Eyring [74]. From his compilation of data we cite $CeO_{1.50}$ (white), $CeO_{1.80}$ (dark blue), CeO_2 (pale yellow); $PrO_{1.5}$ (light green), $PrO_{1.80}$ (black), PrO_2 (black); $TbO_{1.50}$ (white), $TbO_{1.714}$ (brown), TbO_2 (brown). The colors are given in parentheses. It should be realized that the LnO_2 oxides are colored by Ln(IV)–O(−II) CT transitions [75]. It is interesting to note that $Ce(OH)_{3+x}$ in aqueous suspension has been reported to be pale purple [136].

Similar MMCT transitions occur for phases in certain systems Ln_2O_3–$M(IV)O_2$. As early as 1915 the name "Ceruranblau" (cerium-uranium blue) was given to a dark blue phase with approximate composition $2CeO_2 . UO_2$ [76]. The electron exchange equilibrium seems to be [6, 77]: U(IV) + Ce(IV) $\rightleftarrows$ U(V) + Ce(III).

This brings us to U(VI) as an electron acceptor in MMCT transitions. A few examples were mentioned above already. Krol et al. have shown and used the U(V)–U(VI) MMCT transition in oxygen-deficient uranates [78]. It is situated in the near-infrared. This transition plays an important role in the quenching of the luminescence of uranates. The luminescence of $U(VI)O_2^{2+}$ in solution is quenched by Mn(II) and it has been shown by flash photolysis that this is due to an excited MMCT state Mn(III)U(V) [79].

It has been shown by Meyer [80] that reduced halides of the lanthanides can be prepared relatively easily. Many of these are strongly colored as are the reduced oxides. As examples we mention first $NaNd_2Cl_6$ (green-black) and $NaPr_2Cl_6$ (blue-black) [81]. In $NaNd_2Cl_6$ there are Nd(II) and Nd(III) ions statistically distributed over the neodymium sites of Nd(III) in $NdCl_3$: $NaNd(II)Nd(III)Cl_6$. In isomorphous $NaPr_2Cl_6$ there seems to be only one type of praseodymium ions; it has been formulated as $Na(I)\ (Pr(III))_2(e^-)Cl(-I)_6$ and should be metallic in accordance with its metallic lustre. The other example is the $EuCl_2/EuCl_3$ system [82]. Three compounds exist: Eu_3Cl_7, Eu_4Cl_9 and Eu_5Cl_{11}. They are distinguishable by their colors.

Summarizing we conclude that optical MMCT transitions have been widely observed, but only in a few cases thoroughly studied and analyzed. In the next chapter we wish to discuss in how far MMCT transitions can be really considered as pure MMCT transitions using an approach by McGlynn et al. [5] and the example of Cu(I) [23].

4 The Nature of MMCT Transitions

The term MMCT transition is in our opinion more suitable for classification than for characterizing its nature. The same is true for the seemingly more simple LMCT and MLCT transitions. To illustrate this we have tabulated in

Table 3. Some spectral data on Cu(I) complexes (all in values in 10^3 cm^{-1})

Composition	Maximum absorption band	Maximum emission band	Ref.
Cu(I) in SiO_2 glass	38.5	20	[83]
$LiSrPO_4$:Cu(I)	~ 37	24	[84]
NaCl:Cu(I)	36	29	[85]
$Sr_3(PO_4)_2$:Cu(I)	~ 33	22	[84]
Zeolite-Y:Cu(I)	~ 32	18.5	[86, 87]
$[Cu(I)py\ I]_4$	29.5	14	[88]
Cu(I) in $(CuS)_3$ cluster	28.5	21	[89]
$[Cu(I)\ (PPh_3)_2(phen)]^+$	25	18.5	[90]
$CuLaO_2$	25	18	[91]
Cu_2O	17.5	—	[92]
Cu_2WO_4	~ 17	—	[93]
$CuNbO_3$	~ 17	—	[94]
Cu(I) catenates	16.1	13.8	[95]
Cu_2S	10	—	[96]

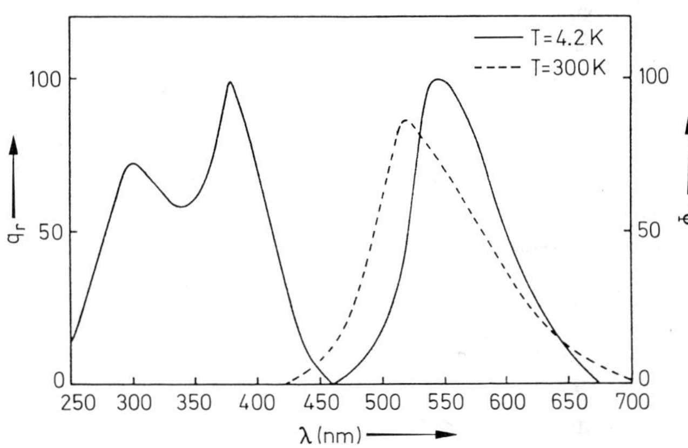

Fig. 15. Emission and excitation spectra of $[Cu(PPh_2CH_3)\ dmp]^+$ at 4.2 and 300 K (from PA Breddels, thesis, Utrecht, 1983)

Table 3 a number of spectral data on Cu(I) complexes. Figure 15 gives an example of a spectrum. These are all characterized by a broad absorption band in the ultraviolet or visible. Many of these show luminescence with a large Stokes shift and high quantum efficiency, even at room temperature.

The absorption band shifts to lower energy from top to bottom in the table. Roughly speaking the degree of covalency increases also in this sequence. Therefore it may be thought that we are dealing with one and the same transition. This, however, is unlikely. The literature contains many different, and sometimes not firmly based, assignments.

The ionic case, for example NaCl:Cu(I), is discussed in terms of transitions corresponding to an interconfigurational MC transition $d^{10} \rightarrow d^9s$ and d^9p

on the copper ion. Quantitative calculations are available [85]. For $[Cu(PPh_3)_2(phen)]^+$ and the Cu(I) catenates there is strong evidence that we are dealing with a MLCT transition [90, 95]. Actually the $d^{10} \rightarrow d^9s$ description contains already some MLCT character due to the diffuseness of the 4*s* orbital. For the sulfur ligands a LMCT transition is obvious. For Cu_2WO_4 and $CuNbO_3$ it has been discussed above that a MMCT transition is responsible for the deep colors of these compounds.

Other assignments made in the literature might be incorrect. The absorption band of $CuLaO_2$ was assigned to a $d^{10} \rightarrow d^9s$ transition [91], but in the light of Table 2 this seems doubtful. Texter et al. [86] give the same assignment for Cu(I) in zeolite Y. However, extended-Hückel-MO calculations show the highest-occupied MO to be concentrated on the oxygens [87], so that the transition must be of the charge-transfer type. Vogler and Kunkely [88] ascribed the strongly Stokes-shifted emission from the tetrameric copper(I) iodide complexes to a transition from a metal-centered $3d^9 4s^1$ excited state of copper(I) which is strongly modified by copper-copper interaction in the tetramer. However, it seems to us that copper(I)-iodide charge-transfer in the $[CuI]_4$ cubane core cannot be neglected in this description.

These examples show that the lowest excited state in the Cu(I) complexes in Table 3 is different from complex to complex. An interesting general consideration on CT transitions which takes this problem into account has been given by McGlynn et al. [5]. This relates immediately to the color of a large number of inorganic salts. The authors consider configurational interaction of several excited states in the following way. Consider the simple system M^+A^-. The ground state configuration is written $\psi_G = \psi(M^+A^-)$. Two possible excited configurations are the LMCT configuration $\psi_{CT} = \psi(MA)$ and the ligand-excited (LE) configuration $\psi_{LE} = \psi[M^+(A^-)^*]$. The excited state can be written as $\psi_E = a\psi_{LE} + b\psi_{CT}$. Figure 16 shows how the energy of the excited levels varies as a function of the ratio a^2/b^2.

The lowest absorption (and the emission) transition is ligand localized on the right-hand side, and CT on the left-hand side of Fig. 16. In between is an intermediate region. Since the excited states are open shell, spin-singlets and spin-triplets appear. This approximation can be extended by including other excited configurations, for example a back-CT state (or MLCT state), i.e. $\psi(M^{2+}A^{2-})$, a MC excited state, i.e. $\psi[(M^+)^*A^-]$, or a MMCT state, i.e. $\psi(MA^-M^{2+})$ taking another M species into account (which is not necessarily equivalent with the original M). It is not so an easy task to determine the coefficients a, b, etc. which give a characterization of the nature of the excited state.

McGlynn et al. [5] use their approach to discuss color phenomena in heavy-metal salts. Also here it is observed that compounds of colorless anions and cations are colored. As an example we mention lead nitrite. Although the Pb(II) and the $[NO_2]^-$ ions are colorless when isolated, $Pb(NO_2)_2$ is strongly orange colored. The interpretation of McGlynn et al. [5] is as follows. The singlet-triplet absorption transition of the nitrite ion is in the visible region, but it has a very low oscillator strength. However, by mixing the LE nitrite excited state

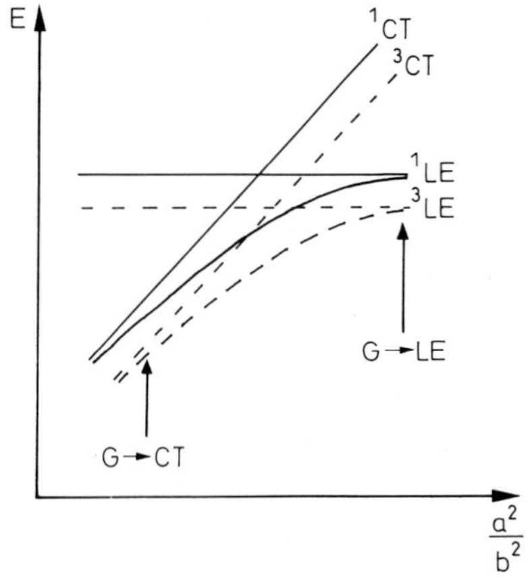

Fig. 16. Interaction between $^1CT/^1LE$ and $^3CT/^3LE$ states leading to extreme and intermediate mixing cases (see also text)

with the LMCT state $[Pb(I).(NO_2)]^+$, the weak absorption intensity increases by a factor 100 if one compares $NaNO_2$ and $Pb(NO_2)_2$. This is mainly due to the large spin-orbit coupling of the Pb(II) ion.

A description as a MMCT transition is not very obvious for this case. However, there is no essential difference between the physical origin of the colors of $Pb(NO_2)_2$ and, for example, Cu_2WO_4. Unfortunately the literature shows sometimes discussions on the nature of their excited states in terms of either MMCT or metal-ion-induced CT transitions. To us, such a discussion does not seem to be very fruitful. In the classification it is a matter of taste which nomenclature is used, in the (more difficult) characterization it is essential to determine the coefficients which indicate the amount of configuration interaction. The latter describe the nature of the excited state.

Very recently Kyle et al. [138] reported ab initio calculations on the tetrahedral $Cu_4I_4A_4$ molecules (A = pyridine or substituted pyridine) and come to very similar conclusions as presented above.

5 Mixed-Valence Compounds

After a consideration of optical transitions in which MMCT plays a role, and after a characterization of the excited states involved, a short review of mixed-valence compounds and their spectroscopy is in order. For more extended reviews we refer to Refs. [60, 97]. At least 40 elements of the periodic table form mixed-valence species which are of importance in solid state physics and chemistry, inorganic chemistry, materials science, geology and bioinorganic chemistry. It is usually their colors which are their most striking property (see also above), but they have more intriguing properties, for example electrical and magnetic properties.

Mixed-valence compounds are divided into three classes following proposals by Robin and Day [98] and Hush [99]. The basis for this classification is the degree to which the two sites occupied by elements of different valence can be distinguished in the ground state. This is related to the ease with which an electron can be transferred from one site to the other (MMCT).

In class I compounds (or complexes) the two sites are very different from each other and the valences are strongly localized. The properties of the complex are the sum of the properties of the constituting ions. The optical MMCT transitions are at high energy. The compounds are insulators. Here are some examples [60, 97]. In $GaCl_2$, or Ga(I)[Ga(III)Cl_4] there are dodecahedrally coordinated Ga(I) ions with Ga–Cl distances of 3.2–3.3 Å and tetrahedrally coordinated Ga(III) ions with Ga–Cl distance 2.2 Å. In $[Co(III)(NH_3)_6]_2$-$\{Co(II)Cl_4)_3\}$ there are low-spin, octahedrally coordinated Co(III) ions and high-spin, tetrahedrally coordinated Co(II) ions. For our purpose this class is not the most interesting one.

In class II compounds the two sites are more similar to each other, but nevertheless distinguishable (e.g. crystallographically). However, the valences can exchange easily, so that MMCT transitions are situated in the visible or near-infrared spectral range. A classic example is Prussian blue mentioned already above. Another one is $(NH_4)_2SbBr_6$ with octahedrally coordinated Sb(III) with Sb–Br distances 2.79 Å and octahedrally coordinated Sb(V) with Sb–Br distance 2.56 Å. Also $BaBiO_3$, nowadays well known for its superconductivity when doped, falls into this category. It should be written Ba_2Bi(III)Bi(V)–O_6 with Bi(V)–O 2.14 Å and Bi(III)–O 2.26 Å [100]. The linear chain compound Wolffram's Red with Pt(II) and Pt(IV) was mentioned above. This material shows an electrical conductance in the chain direction which is more than two orders higher than perpendicular to the chains.

The wavenumbers of the MMCT transition between Pt(II) and Pt(IV) are related to the Pt(II)–Pt(IV) distance in the chain, as follows from a study of these compounds using different types of ligands. This relation is linear [97]. If the linear curve is extrapolated to a MMCT transition energy of zero, we obtain a distance equal to twice the Pt(IV)-ligand distance. This is the situation in which the ligand, for example Cl^- or Br^-, is placed in between the two platinum ions, i.e. in which the difference between the platinum ions vanishes.

In this way we come to class III complexes, i.e. complexes in which the two sites are indistinguishable and the element has a non-integral oxidation state (delocalized valence). Usually one divides this class in two subclasses. In class IIIA the delocalization of the valence electrons takes place within a cluster of equivalent metal ions only. An example is the $[Nb_6Cl_{12}]^{2+}$ ion in which there are six equivalent metal ions with oxidation state + 2.33. In class IIIB the delocalization is over the whole lattice. Examples are the linear chain compound $K_2Pt(CN)_4 . Br_{0.30} . 3H_2O$ with a final oxidation state for platinum of 2.30, and three-dimensional bronzes like Na_xWO_3.

A special representative of class IIIA is the Creutz-Taube complex [101]. This is a binuclear complex $[(NH_3)_5Ru(pyrazine)Ru(NH_3)_5]^{5+}$. Various experimental results led to conflicting conclusions concerning its electronic structure. A combined attack [101] has shown, however, that we are not dealing with a

Ru(II)–Ru(III) valence distribution, but that there is complete delocalization with formal ruthenium valence 2.5. The ruthenium ions are crystallographically indistinguishable. The intervalence (MMCT) absorption band is at 6400 cm^{-1} (ε_{max} = 5500 M^{-1} cm^{-1}).

A related complex is $[(NH_3)_5Ru(bqd)Ru(NH_3)_5]^{5+}$ (bqd = *p*-benzoquinone diimine) [102]. Also here there is complete delocalization. The absorption spectrum of this ion is given in Fig. 17. It consists of an extremely intense band in the visible (ε_{max} = 108.000 M^{-1} cm^{-1}) and a much weaker one in the near infrared ($\varepsilon_{max} \simeq$ 1000 M^{-1} cm^{-1}). The latter has been assigned to the MMCT transition (like in the Creutz-Taube ion), the former to a MLCT transition. These assignments are confirmed by a Hückel MO calculation for which the reader is referred to the original literature [102].

Whereas the relation between MMCT and mixed-valence compounds is immediately clear, this is not the case for MMCT and semiconductors. Their relation will be discussed in the next paragraph.

6 The Role of MMCT in the Field of Semiconductors

First we consider the origin of band gaps and characters of the valence and conduction electron states in 3*d* transition-metal compounds [104]. Band structure calculations using effective one-particle potentials predict often either metallic behavior or gaps which are much too small. This is due to the fact that the electron-electron interactions are underestimated. In the Mott-Hubbard theory excited states which are essentially MMCT states are taken into account: $d_i^n d_j^n \rightarrow d_i^{n-1} d_j^{n+1}$. The subscripts i and j label the transition-metal sites. These transitions involve the *d-d* Coulomb and exchange interaction and involve an energy U. The value of U is up to 10 eV. The semiconductor gap in this case is a *d-d* gap which is correct for Ti and V oxides where U is relatively small, i.e. the MMCT transition at low energy. This Mott-Hubbard model is also the basis for theories on superexchange [105] and rules for the sign of magnetic interaction [106].

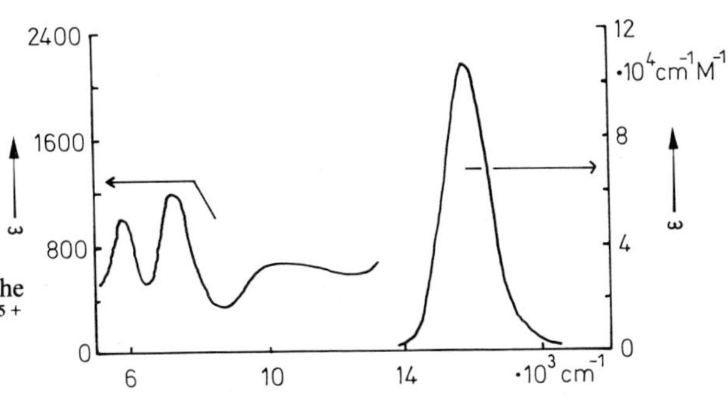

Fig. 17. Absorption spectrum of the ion $[(NH_3)_5Ru(bqd)Ru(NH_3)_5]^{5+}$ (modified from Ref. [102])

However, for the late-transition-metal compounds the gap is related to the electronegativity of the anion and seems to be of the LMCT type. Therefore it is assumed that another excited state, viz. $d_i^n \rightarrow d_i^{n+1} L^{-1}$ plays a role here. Here L^{-1} denotes a hole on the anion, or better in the anion valence band. This is a LMCT state. Its energy is assumed to be Δ.

Calculations [104] show that for $U > \Delta$ (the heavier transition metal ions) the gap is of the charge-transfer type, whereas for $U < \Delta$ (the lighter transition metal ions) the gap is of the *d-d* type. In our nomenclature this may be translated as MMCT $\gtrless$ LMCT. In the charge-transfer semiconductors the holes are light (anion valence band) and the electrons are heavy (*d* bands). Examples are $CuCl_2$, $CuBr_2$, CuO, $NiCl_2$, $NiBr_2$ and NiI_2.

The sulfides of Ni and Cu are metallic in spite of a high value of U. This is due to the fact that the band gap is proportional to the electronegativity of the anion: it closes for the sulfides with a low electronegativity for sulfur.

This development has led to results part of which have been for years interpreted differently:

— the copper-containing high-T_c superconductors do not contain Cu^{3+} but O^-, i.e. holes on the oxygen [107].
— this has also been ascertained for other compositions which formally contain copper in the trivalent state [108]
— in the Cu^{2+}–O^{2-} complex in Ba_2ZnF_6 charge-transfer emission has been observed [109]
— in $Li_xNi_{1-x}O$ the holes compensating the Li(I) impurity are located primarily in the oxygen $2p$ states rather than in the conventionally assumed Ni $3d$ states [110].

Let us now return to MMCT effects in semiconductors. In this class of compounds MMCT may be followed by charge separation, i.e. the excited MMCT state may be stabilized. This is the case if the M species involved act as traps. A beautiful example is the color change of $SrTiO_3$: Fe, Mo upon irradiation [111]. In the dark, iron and molybdenum are present as Fe(III) and Mo(VI). The material is colorless. After irradiation with 400 nm radiation Fe(IV) and Mo(V) are created. These ions have optical absorption in the visible. The Mo(VI) species plays the role of a deep electron trap. The thermal decay time of the color at room temperature is several minutes. Note that the MMCT transition Fe(III) + Mo(VI) $\rightarrow$ Fe(IV) + Mo(V) belongs to the type which was treated above. In the semiconductor the iron and molybdenum species are far apart and the conduction band takes the role of electron transporter. A similar phenomenon has been reported for ZnS: Eu, Cr [112]. There is a photoinduced charge separation Eu(II) + Cr(II) $\rightarrow$ Eu(III) + Cr(I) via the conduction band (see Fig. 18).

Another example comes from the field of semiconductor photoelectrochemistry. Semiconductors in contact with aqueous solutions can drive chemical reactions when irradiated. This is the basis for the photoelectrochemical etching of semiconductors in the electronic industry and for much research aiming at

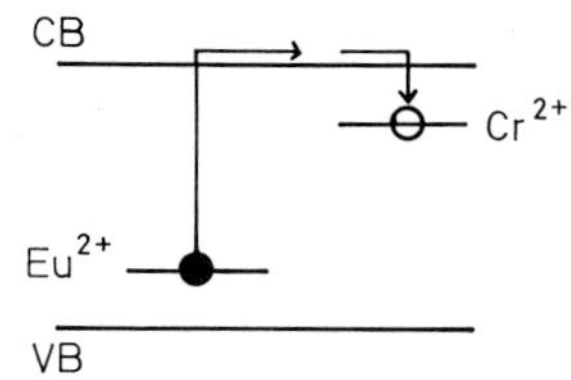

Fig. 18. Processes involved in the charge transfer Eu(II) + Cr(II) → Eu(III) + Cr(I) in ZnS

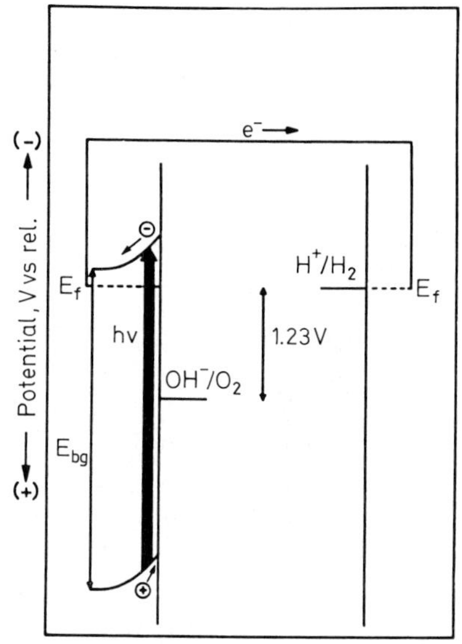

Fig. 19. Energy level diagram for a photoelectrochemical cell decomposing water. The semiconductor is on the left-hand side (E_{bg}: band gap)

solar energy conversion. Figure 19 gives an example for TiO_2 which can split water under ultraviolet excitation. Irradiation creates electrons in the conduction band and holes in the valence band. The electrons move into the bulk, and are collected at the cathode where they reduce H^+ to H_2. The holes move to the surface of the TiO_2 anode where they oxidize OH^- to O_2. The complicated physical processes involved have been reviewed elsewhere [113].

It is possible to sensitize this process for visible light. Two principally different approaches are possible, viz. doping the semiconductor, or covering the surface with a suitable dye. They are illustrated in Figs. 20 and 21, respectively. Possible dyes are Ru(II)-bpy complexes, tetraphenylporphyrine, Cu(I)-triphenyl-phosphine dmp, and others.

If the semiconductor is doped, the excitation transition may be a MMCT transition. An example is Cr^{3+} in TiO_2 or $SrTiO_3$ where irradiation promotes an electron from the Cr^{3+} ion to the conduction band which is essentially titanium 3*d*. This type of transition was discussed in the first part of this section.

If the semiconductor is sensitized with a dye, the process which injects an electron from the dye into the semiconductor has MMCT character. Consider, for example, the dye mentioned above, $Cu(PPh_3)_2dmp^+$, on SnO_2. Irradiation

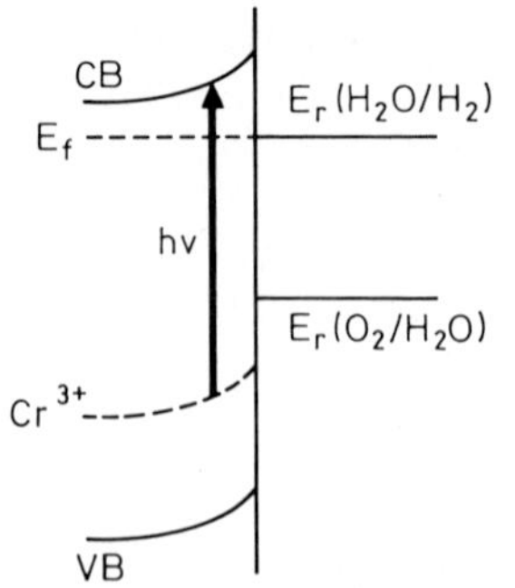

Fig. 20. Energy level diagram for a semiconductor photoanode (e.g. $SrTiO_3$) doped with Cr(III). *VB*: valence band; *CB*: conduction band

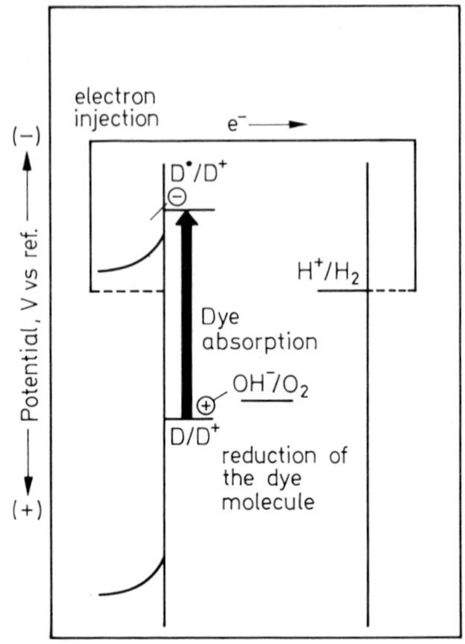

Fig. 21. Energy level diagram for a semiconductor photoanode sensitized by a dye (*D*) layer. The *black arrow* indicates the optical transition on the dye. The semiconductor is on the left-hand side

of the dye excites this Cu(I) complex into the MLCT state. This state injects an electron into the conduction band [114], so that we reach a state with Cu(II)–Sn(III) character, i.e. a state with a certain MMCT character. In conclusion MMCT transitions play a role in a very broad field of semiconducting materials and/or compounds.

7 The Role of CT States in Other Phenomena

In this last paragraph we would like to demonstrate how MMCT excited states, just as other CT states, can influence several types of physical (here often optical) properties. It is interesting to note that these effects have been recognized and understood by "molecular" scientists, whereas up till now they have often been overlooked by workers in the solid-state field.

Therefore, the author who originates from the latter field starts by mentioning the three limiting cases which occur in an ion pair consisting of two species having a weak interaction, following the approach of Balzani, Sabbatini and Scandola [1]. Because of this interaction CT transitions between the species are possible, influencing the properties of the ion pair. The reader will note that there is a strong analogy between this classification and those of the mixed valence compounds (Sect. 5). However, they are not identical. The three limiting cases are the following:

a. The CT state lies at high energy. The absorption spectrum of the pair is the summation of the constituting species. However, the CT state may nevertheless induce nonradiative transitions on the constituents.
b. The CT state lies in the energy range of the spectroscopic levels of the constituents of the pair. In this case the absorption spectrum of the pair will show a stronger absorption than the summation of the constituents. Excitation in the CT state may yield photochemical reactions or radiationless return to the ground state.
c. The CT state is the lowest excited state of the ion pair. A new broad absorption band appears in the spectrum of the pair.

Let us now illustrate this with a couple of examples. Well-known are Co(III) complexes. In the system $[Co(NH_3)_6]^{3+}.I^-$ in aqueous solution excitation with 300 nm (case b) yields $[Co(NH_3)_6]^{2+}.I$ which decomposes into $Co(II)_{aq}$, NH_3 and I_2 [115]. A similar reaction occurs for the ion pair $[Co(III)(NH_3)_6]^{3+}.[Ru(II)(CN)_6]^{4-}$ [116]. It is clear that excitation is now in a MMCT state.

Case c is found in, for example, the ion pair $[Ru(III)(NH_3)_5Cl]^{2+}.[Ru(II)(CN)_6]^{4-}$. It shows a MMCT (intervalence CT) transition at 510 nm [117]. Upon light absorption the ion pair is converted via the MMCT state ($[Ru(II)(NH_3)_5Cl]^{+}.[Ru(III)(CN)_6]^{3-}$) to the binuclear complex $[(NH_3)_5Ru(III)NCRu(II)(CN)_5]^-$.

Luminescence quenching in an ion pair is nicely examplified by the $[Eu \subset 2.2.1]^{3+}.[M(CN)_6]^{4-}$ ion pairs in solution [118]. The $[Eu \subset 2.2.1]^{3+}$ cryptate shows Eu^{3+} luminescence in solution and in the solid state [119], but it is quenched by $[M(II)(CN)_6]^{4-}$ (M = Fe, Ru, Os). At low quencher concentration the quenching process is dynamic in nature and due to a bimolecular electron transfer, which leads to a MMCT excited state: $^*[Eu(III) \subset 2.2.1]^{3+} + [M(II)(CN)_6]^{4-} \rightarrow [Eu(II) \subset 2.2.1]^{2+} + [M(III)(CN)_6]^{3-}$. At higher concentrations ion pairs are formed and a new low-energy MMCT absorption band appears in the visible (case c). The position of this MMCT band shifts to higher energies in the sequence Fe < Os < Ru as is to be expected.

In a similar way the luminescence of the $[Ru(bpy)_3]^{2+}$ ion is quenched by $[CoSiW_{11}O_{39}H_2O]^{6-}$ [120].

In our first examples a photochemical reaction occurs, in the latter ones nonradiative return to the original ground state. These two different processes have been drawn schematically in Fig. 22. The ground state is the ion pair $M^+ . L^-$, the excited CT state is M . L, all other states have been omitted.

The photochemical process will in general not occur in non-molecular solids due to the restriction in the nuclear coordinates. However, the nonradiative return to the ground state does, using MMCT states as an intermediary [35]. This can be nicely illustrated on a molecular solid, viz. the lanthanide-decatungstates which contain complex ions $[Ln(III)W_{10}O_{36}]^{9-}$ [40, 121].

In the case of $EuW_{10}O_{36}^{9-}$ the Eu(III) ions show luminescence with high quantum efficiency. In view of this, $TbW_{10}O_{36}^{9-}$ and $GdW_{10}O_{36}^{9-}$ are also expected to luminescence efficiently. This, however, is not the case. The gadolinium complex shows only tungstate emission and no Gd(III) emission. The excited Gd(III) ion transfers its excitation energy to the tungstate, because the Gd(III) excited levels are at very high energy. Problematic is the absence of any luminescence in the case of $TbW_{10}O_{36}^{9-}$. Obviously the complex as a whole undergoes a nonradiative return to the ground state which is not covered by the Förster-Dexter theory [122], which, by the way, is the usual approach used by solid state scientists. However, the quenching of the luminescence finds an easy explanation in the presence of a low-lying MMCT state, viz. Tb(IV)–W(V).

In the configurational coordinate diagram the quenching process occurs via an excited state which has in this diagram a parabola with a large offset relative to the ground-state parabola [35]. This process is of importance for every complex species consisting of a metal ion which is easily oxidized and another metal ion which is easily reduced. Figure 23 gives a schematic presentation of the quenching process.

There is no essential difference between quenching via a MMCT state or a LMCT state. The latter occurs, for example, in Eu(III) if the LMCT state is either at low energy or if this state shows a large offset in the configurational coordinate diagram [23, 35]. The latter occurs in glasses [123], certain cryptates [124] and lanthanum compounds [125].

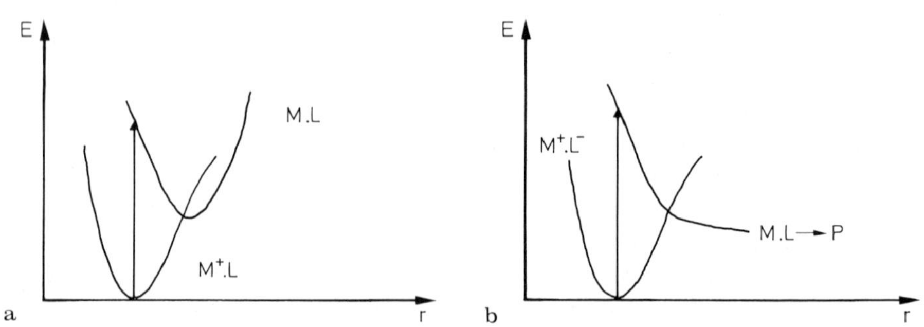

Fig. 22 a and b. Schematic representation of the potential energy curves for a $M^+ . L^-$ ion pair. (**a**) the excited pair M . L returns nonradiatively to the ground state $M^+ . L^-$. (**b**) in competition with the process in (a) a photochemical reaction (P: products) is possible (modified from Ref. [1])

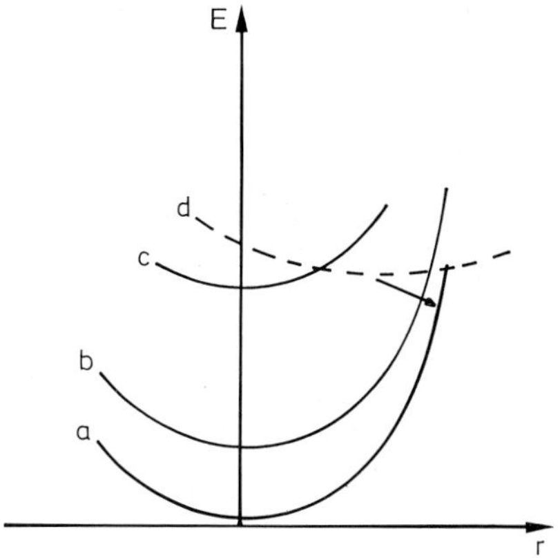

Fig. 23. Schematic representation of quenching by electron transfer. Curves *a*, *b* and *c* are levels of the pair $M^+.L^-$. Level *d* is a charge-transfer state (M.L); it may be a MMCT state (see text). The *arrow* indicates the nonradiative transition from the charge-transfer state *d* to the ground state, so that c → b and c → a emission is quenched

The most striking case in non-molecular solids is YVO_4:Tb(III). Whereas many lanthanide ions emit efficiently in YVO_4 (Eu(III) being the best known), Tb(III) kills all luminescence, even at liquid helium temperature [38, 39]. The excited state, responsible for this quenching, is a MMCT of the type Tb(IV)–V(IV). It is interesting to note that $YTaO_4$:Tb is a luminescent material with a very high quantum efficiency, whereas $YNbO_4$:Tb is in between the tantalate and the vanadate. This shows that the rate of the nonradiative process decreases if the MMCT state shifts to higher energy. The 5th ionisation potential of the host lattice metal increases in the sequence Ta, Nb, V. In line with this, the LMCT state moves to lower energy in the sequence tantalate > niobate ≫ vanadate.

The same arguments as for Tb(III) hold for Ce(III) and Pr(III). The species $PrW_{10}O_{36}^{9-}$ shows only weak luminescence due to this quenching process [40]. In YVO_4 these ions behave as killers of luminescence.

Sometimes solids offer special opportunities to overcome this quenching process. The combinations Tb(III)–VO_4^{3-} and Ce(III)–VO_4^{3-} show efficient lanthanide emission upon vanadate excitation in the host lattice $CaSO_4$ [126] (see also above). Here Dexter's theory holds. In this lattice the lanthanide ions occupy Ca(II) sites and the V(V) ions S(VI) sites. In this way they carry an effectively positive and an effectively negative charge relative to the lattice, so that they attract each other. In $CaSO_4$ the pair concentration is much higher than that of the single species. However, the effective charges shift the MMCT state to high energies because MMCT would even increase the effective charges. In this way the nonradiative processes are suppressed.

Even couples of lanthanide ions show this quenching process. The Ce(III) and Eu(III) ions, for example, quench each other's luminescence [127]. Here a MMCT state with Ce(IV)–Eu(II) character is responsible. In solid $[Ce \subset 2.2.1]^{3+}$ cryptate there occurs energy migration over the cryptate species. Also here $[Eu \subset 2.2.1]^{3+}$ acts as a quencher [128]. The quenching action is restricted to short distances (about 12 Å [129]).

In GdF_3:Ce(III), Eu(III) there is efficient Eu(III) emission upon excitation into the Ce(III) ion [127]. The reason for this is that the cerium and europium concentrations are low, so that the Ce–Eu distance is large. The energy transfer is not a direct process but takes place via the Gd(III) sublattice: Ce(III) → Gd(III) $\xrightarrow{nx}$ Gd(III) → Eu(III). Each arrow indicates a transfer process, and the value of n is ~ 10^4 [35].

We touch here on the distance dependence of the MMCT rate. This problem is widely under study at the moment [130]. This rate depends on the distance r according to $\exp(-\beta r)$. The value of β seems to be in between 1.1 and 1.4 $Å^{-1}$. Elegant work has been performed by Hush, Paddon-Row and Verhoeven [131] who studied molecules in which the reactants are separated by rigid saturated hydrocarbon bridges of various lengths.

This value of β explains also our Ce(III)–Eu(III) results. The radiative rate of Ce(III) is about 10^7 s^{-1}. Since the emission is practically quenched by a Eu(III) neighbor, the MMCT rate will be about 10^9 s^{-1} for r = 4 Å, which is a realistic nearest neighbor distance. For $\beta = 1.2$ $Å^{-1}$ and r = 12 Å this rate has dropped to 10^5 s^{-1}, i.e. very small compared to the radiative rate.

The presence of such a MMCT state does not necessarily lead to quenching of all possible luminescence. An interesting case is $Ba_6Ln_2W_3O_{18}$:Nd(III) (Ln = Gd, Lu, Y) [132]. In our opinion the absorption band at 27 000 cm^{-1} is due to MMCT in the Nd(III)–W(VI) complex [35]. From this state the system returns to lower states. Since efficient Nd(III) emission was observed at ~ 10 000 cm^{-1} from the $^4F_{5/2}$ level of Nd(III), the MMCT state feeds this level with high efficiency.

Also in case of Pb(II)–WO_4^{2-} the MMCT state does not lead to quenching (see above). In this case the quenching temperature of the emission increases relative to that of the constituents, since the parabola offset of this MMCT state is relatively small [50].

As a matter of fact low-lying MMCT states can also influence radiative transition probabilities. The long decay time of the VO_4^{3-} luminescence is considerably shortened by the presence of Bi^{3+} [27] due to a Bi(IV)–V(IV) MMCT state (see also above). Such effects are very well-known for LMCT states in case of transition-metal ions and lanthanide ions [6]. They will not be discussed here any further.

8 Conclusion

In this chapter we have shown that optical metal-to-metal charge-transfer transitions are of large importance in many fields and that they occur very generally. Not only their direct, but also their indirect influence is of great importance. A more unified approach in the different areas of research, and a stronger interaction between the different approaches is desirable.

Acknowledgement. The author has taken advantage from discussions on MMCT states with many colleagues. Especially useful were those with Prof. Christian K. Jørgensen (Geneva), Prof. Andreas Ludi (Bern), and Prof. Nanda Sabbatini (Bologna).

9 References

1. Balzani V, Sabbatini N, Scandola, F (1986) Chem Rev 86: 319
2. Curtis JC, Meyer TJ (1982) Inorg Chem 21: 1562
3. Vogler A, Osman AH, Kunkely H (1985) Coord Chem Rev 64: 159
4. Balzani V (ed) (1987) Supramolecular photochemistry, Reidel, Dordrecht
5. McGlynn SP, Azumi T, Kumar D (1981) Chem Rev 81: 475
6. Reisfeld R, Jørgensen CK (1977) Lasers and Excited States of Rare Earths, Springer, Berlin Heidelberg, New York, p. 46
7. Blasse G, p 355 in Ref. [5].
8. Blasse G, Sabbatini, N (1987) J Solid State Chem 70: 93
9. Ziegler T, Rauk A, Baerends, EJ (1976) Chem Phys 16: 209
10. Hush NS (1967) Progr Inorg Chem 8: 391
11. Meyer TJ (1978) Acc Chem Res 11: 94
12. See e.g. Jonker GH, van Houten S (1961) Halbleiterprobleme, Vieweg, Braunschweig, 6: 118
13. Imbusch GF, Yen WM (1987) In: Yen WM, Levenson MD (eds) Lasers, spectroscopy and new ideas Springer, Berlin Heidelberg, New York, p 248
14. See e.g. Lever ABP (1984) Inorganic electronic spectroscopy, 2nd edn, Elsevier, Amsterdam
15. Blasse G (1967) J Inorg Nucl Chem 29: 1817
16. Nassau K (1978) Am Mineral 63: 219
17. Blasse G, de Korte PHM, Mackor A (1981) J Inorg Nucl Chem 43: 1499
18. Powell RC, Caslavsky JL, Al Shaieb Z, Bowen JM (1985) J Appl Phys 58: 2331
19. See also Blasse G (1981) Comm Inorg Chem 1: 245
20. Townsend MG (1968) Solid State Comm 6: 81
21. See e.g. Phillips CSG, Williams RJP (1965) Inorganic chemistry, Oxford University Press, Oxford
22. Mumm HC, Müller-Buschbaum HK (1988) J Less-Common Metals 142: 85
23. Blasse G, Adv Inorg Chem 35: in press
24. See e.g. Kröger FA (1973) The chemistry of imperfect crystals, North-Holland, Amsterdam
25. Brixner LH (1987) Mat Chem Phys 16: 253
26. Palilla FC, Levine AK (1964) Appl Phys Letters 5: 118; (1966) Appl Opt 5: 1467
27. Blasse G (1980) Structure and Bonding 42: 1
28. Barendswaard W, van Tol J, Weber RT, van der Waals JH (1989) Mol Phys 67: 651
29. Barendswaard W, van der Waals JH (1986) Mol Phys 59: 337
30. Coremans CJM, van der Waals JH, Konijnenburg J, Huizer AH, Varma CAGO (1986) Chem Phys Letters 125: 514
31. Reynolds MC, J Chem Soc 1965, 2991
32. Reinen D (1966) Theor Chim Acta 5: 312
33. Blasse G, Dirksen GJ (1981) Chem Phys Letters 77: 9
34. Rossman GR, Shannon RD, Waring RK (1981) J Solid State Chem 39: 277
35. Blasse G (1988) Progress Solid State Chem 18: 79
36. Paul A (1976) Phys Chem Glasses 17: 7
37. Blasse G, Dirksen GJ, (1981) J Solid State Chem 37: 390
38. de Losh RG, Tien TY, Gibbins EF, Zacmanides PJ, Stadler HL (1970) J Chem Phys 53: 681
39. Blasse G, Bril A (1967) Philips Res Repts 22: 481
40. Blasse G, Dirksen GJ, Zonnevijlle F (1981) J Inorg Nucl Chem 43: 2847
41. Blasse G (1979) In: Gschneidner KA Jr, Le Roy Eyring (eds) Handbook on the physics and chemistry of rare earths, vol 4, North-Holland, Amsterdam, p 268
42. Langenbach-Kuttert B, Sturm J, Gruehn R (1987) Z Anorg Allg Chem 548: 33; Schaffrath U, Gruehn R (1988) Z Anorg Allg Chem 565: 67

43. Greedan JE, McCarthy GJ (1972) Mat Res Bull 7: 531
44. Marinder BO, Wahlström E (1984) Chem Scripta 23: 157
45. Sperka G, Fritzer HP (1988) Solid State Comm 65: 1275
46. Blasse G, van der Heuvel GPM, J. Luminescence (1974) 9: 74
47. Blasse G, Bril A (1968) Philips Res Repts 23: 344; (1969) 24: 275
48. Severin JW, Lammers MJJ, Blasse G, Brixner LH, Torardi CC (1987) J. Solid State Chem 66: 318 (1987)
49. Blasse G, Bril A (1968) J Chem Phys 48: 217
50. Groenink JA, Blasse G (1980) J Solid State Chem 32: 9
51. Blasse G, Boon L (1984) Ber Bunsenges Phys Chem 88: 929
52. Gerault Y, Lecerf A (1979) CR Ac Sci Paris, Série C, 289: 41
53. Jeitschko W, Sleight AW, (1972) Acta Cryst B28: 3174
54. Darriet J (1973) Bull Soc Fr Min Crist 96: 97
55. Kunkely H, Paukner A, Vogler A (1989) Polyhedron 8: 2937
56. Jørgensen CK (1957) Acta Chem Scand 11: 73
57. Reisfeld R, Eyal M, Jørgensen CK (1987) Chimia 41: 117
58. Salje E, Hoppmann G (1981) Phil Mag B43: 105
59. See e.g. Sleight AW (1988) Science 242: 1519
60. See e.g. Brown DB (ed) (1980) Mixed-Valence Compounds, Reidel, Dordrecht
61. Curtis JC, Meyer TJ (1982) Inorg Chem 21: 1562
62. Vogler A, Kunkely H (1988) Inorg Chim Acta 150: 3
63. Clark RJH, p 271 in Ref. [60]
64. Glauser R, Hauser U, Herren F, Ludi A, Roder P, Schmidt E, Siegenthaler H, Wenk F (1973) J Am Chem Soc 95: 8457
65. Table 7.4 in Ref. [14]
66. Ludi A (1988) Chemie in unserer Zeit 22: 123
67. Ludi A, Güdel HU (1973) Structure and Bonding 14: 1
68. Burns RG, Nolet DA, Parkin KM, McCammon CA, Schwartz KB, p 295 in Ref. [60]
69. Braterman PS (1966) J Chem Soc (A) 1471
70. For a review see Day P, p 11 in Ref. [60].
71. Atkinson L, Day P (1969) J Chem Soc (A) 2423
72. Oomen EWJL, Smit WMA, Blasse G (1986) J Phys C: Solid St Phys 19: 3263
73. Vogler A, Paukner A (1989) Inorg Chim Acta 163: 207; Blasse G, Vogler A (1990) Inorg Chim Acta 170: 149
74. Eyring L, chapter 27, Ref. [41], vol 3
75. Hoefdraad HE (1975) J Inorg Nucl Chem 37: 1917
76. Hofmann KA, Hoeschele K (1915) Ber Dtsch Chem Ges 48: 20
77. Bevan DJM, Summerville E, chapter 28, Ref. [41]
78. Krol DM, Ros JPM, Roos A (1980) J Chem Phys 73: 1521
79. Burrows HD, Formosinho SJ, de Graça Miguel M, Pinto Coelho F (1976) J Chem Soc (Faraday I) 72: 163
80. Meyer G (1988) Chem Rev 88: 93
81. Schleid T, Meyer G (1987) Inorg Chim Acta 140: 113
82. Bärnighausen H, cited in Ref. [80]
83. Debnath R, Das SK (1989) Chem Phys Letters 155: 52
84. Wanmaker WL, Bakker C (1959) J Electrochem Soc 106: 1027; Wanmaker WL, Spier HL (1962) J Electrochem Soc 109: 109
85. Pedrini C, Jacquier B (1980) J Phys C: Sol State Phys 13: 4791
86. Texter J, Strome DH, Herman RG, Klier K (1977) J Phys Chem 81: 333
87. Calzaferri G, Forss L (1987) Helv Chim Acta 70: 465
88. Vogler A, Kunkely H (1986) J Am Chem Soc 108: 7211
89. Knotter DM, Blasse G, van Vliet JPM, van Koten G (to be published)
90. Blasse G, Breddels PA, McMillan DR (1984) Chem Phys Letters 109: 24
91. Doumerc JP, Parent C, Zhang Jin Chao, Le Flem G, Ammar A (1988) CR Ac Sci Paris, Série II, 306: 1431
92. Kittel C, Solid state physics, 5th edn, Wiley, New York
93. Mumm HC, Müller-Buschbaum HK (1988) J Less-Common Metals 142: 85
94. Marinder BO, Wahlström E (1984) Chem Scripta 23: 157
95. Gushurst AKI, McMillan DR, Dietrich-Buchecker CO, Sauvage JP (1989) Inorg Chem 28: 4070

96. Seraphin BO (ed) (1979) Topics in Appl Phys 31: 237
97. Clark RJH (1984) Quart Revs 219
98. Robin MB, Day P (1967) Adv Inorg and Radiochemistry 10: 247
99. Hush NS (1967) Progr Inorg Chem 8: 357 and 391
100. Thornton G, Jacobson AJ (1978) Acta Cryst B34: 351
101. Fürholz U, Bürgi HB, Wagner FE, Stebler A, Ammeter JH, Krausz E, Clark RJH, Stead MJ, Ludi A (1984) J Am Chem Soc 106: 121
102. Joss S, Bürgi HB, Ludi A (1985) Inorg Chem 24: 949
103. Joss S, Hasselbach KM, Bürgi HB, Wordel R, Wagner FE, Ludi A (1989) Inorg Chem 28: 1815
104. Zaanen J, Sawatzky GA, Allen JW (1985) Phys Rev Letters 55: 418; Zaanen J, Sawatzky GA (1987) Can J Physics 65: 1262
105. Anderson PW (1963) Solid State Physics 14: 95
106. Goodenough JB (1988) J Phys Chem Solids 6: 287; Kanamori J (1959) J Phys Chem Solids 10: 87
107. Fujimori A, Takayama-Muromach E, Uchida Y, Okai B (1987) Phys Rev B35: 8814
108. Rao GR, Rajumon MK, Sarma DD, Rao CNR (1989) J Chem Soc Chem Comm 1536
109. Matthies H, Recker K, Wallrafen F, Dirksen GJ, Blasse G (1988) Chem Phys Letters 149: 167
110. Kuiper P, Kruizinga G, Ghijsen J, Sawatzky GA, Verweij H (1989) Phys Rev Letters 62: 221
111. Faughnan BW, Kiss ZJ (1968) Phys Rev Letters 21: 1331
112. Godlewski M, Hommel D (1986) Phys Stat Sol (a) 95: 261
113. See e.g. Schiavello M (ed) (1985) Fundamentals and developments of photocatalytic and photoelectrochemical processes, Reidel, Dordrecht; Pelizzetti E (ed) (1986) Homogeneous and heterogeneous photocatalysis, Reidel, Dordrecht
114. Breddels PA, Berdowski PAM, Blasse G (1981) Recl Trav Chim Pays-Bas 100: 439
115. Endicott JF, Ferrandi GJ, Barber JR (1975) J Phys Chem 79: 630
116. Vogler A, Kisslinger J (1982) Angew Chem Int Ed 21: 77
117. Vogler A, Kisslinger J (1982) J Am Chem Soc 104: 2311
118. Sabbatini N, Bonazzi A, Ciano M, Balzani V (1984) J Am Chem Soc 106: 4055; Sabbatini N, Balzani V (1985) J Less-Common Metals 112: 381
119. Blasse G, Buijs M, Sabbatini N (1986) Chem Phys Letters 124: 538; Sabbatini N, Dellonte S, Ciano M, Bonazzi A, Balzani V (1984) Chem Phys Letters 107: 212
120. Ballardini R, Gandolfi MT, Balzani V (1985) Chem Phys Letters 119: 459
121. Blasse G, Dirksen GJ, Zonnevijlle F (1981) Chem Phys Letters 83: 449
122. Dexter DL (1953) J Chem Phys 21: 836
123. Verwey JWM, Dirksen GJ, Blasse G (1988) J Non-cryst Solids 107: 49
124. Hazenkamp MF, Blasse G, Sabbatini N, Ungaro R (1990) Inorg Chim Acta 172: 93
125. Verwey JWM, van der Voort D, Dirksen GJ, Blasse G (to be published)
126. Draai WT, Blasse G (1974) Phys Stat Sol (a) 21: 569; (1974) Chem Phys Letters 25: 167
127. Blasse G (1983) Phys Stat Sol (a) 75: K41
128. Blasse G, Dirksen GJ, Sabbatini N, Perathoner S (1987) Inorg Chim Acta 133: 167
129. Hazenkamp MF, Blasse G (1990) Chem Mater 2: 105
130. Marcus RA, p 45 in Ref. [4].
131. Verhoeven JW, Paddon-Row MN, Hush NS, Oevering H, Heppener M (1986) Pure and Appl Chem 58: 1285; Hush NS, Paddon-Row MN, Cotsaris E, Oevering H, Verhoeven JW, Heppener M (1985) Chem Phys Letters 117: 8; Hush NS, p 53 in Ref. [4].
132. Wolf D, Kemmler-Sack S (1986) Phys Stat Sol (a) 93: 271
133. Jørgensen CK (1970) Progress Inorg Chem 12: 101
134. Jørgensen CK (1963) Acta Chem Scand 17: 1034
135. Jørgensen CK, (1961) Mol Phys 4: 235
136. Allen GC, Wood MB, Dyke JM (1973) J Inorg Nucl Chem 35: 2311
137. Oshio H, Toriumi K, Bandow S, Miyagawa K, Kurita S (1990) J Chem Soc Dalton Trans 1990: 1013
138. Kyle KR, Palke WE, Ford PC (1990) Coord Chem Revs 97: 35

Author Index Volumes 1–76

Ahrland, S.: Factors Contributing to (b)-behavior in Acceptors, Vol. 1, pp. 207–220.

Ahrland, S.: Thermodynamics of Complex Formation between Hard and Soft Acceptors and Donors. Vol. 5, pp. 118–149.

Ahrland, S.: Thermodynamics of the Stepwise Formation of Metal-Ion Complexes in Aqueous Solution. Vol. 15, pp. 167–188.

Allen, G. C., Warren, K. D.: The Electronic Spectra of the Hexafluoro Complexes of the First Transition Series. Vol. 9, pp. 49–138.

Allen, G. C., Warren, K. D.: The Electronic Spectra of the Hexafluoro Complexes of the Second and Third Transition Series. Vol. 19, pp. 105–165.

Alonso, J. A., Balbás, L. C.: Simple Density Functional Theory of the Electronegativity and Other Related Properties of Atoms and Ions. Vol. 66, pp. 41–78.

Andersson, L. A., Dawson, J. H.: EXAFS Spectroscopy of Heme-Containing Oxygenases and Peroxidases. Vol. 74, pp. 1–40.

Ardon, M., Bino, A.: A New Aspect of Hydrolysis of Metal Ions: The Hydrogen-Oxide Bridging Ligand ($H_3O_2^-$). Vol. 65, pp. 1–28.

Armstrong, F. A.: Probing Metalloproteins by Voltammetry. Vol. 72, pp. 137–221.

Augustynski, J.: Aspects of Photo-Electrochemical and Surface Behavior of Titanium(IV) Oxide. Vol. 69, pp. 1–61.

Averill, B. A.: Fe–S and Mo–Fe–S Clusters as Models for the Active Site of Nitrogenase. Vol. 53, pp. 57–101.

Babel, D.: Structural Chemistry of Octahedral Fluorocomplexes of the Transition Elements. Vol. 3, pp. 1–87.

Bacci, M.: The Role of Vibronic Coupling in the Interpretation of Spectroscopic and Structural Properties of Biomolecules. Vol. 55, pp. 67–99.

Baker, E. C., Halstead, G. W., Raymond, K. N.: The Structure and Bonding of 4*f* and 5*f* Series Organometallic Compounds. Vol. 25, pp. 21–66.

Balsenc, L. R.: Sulfur Interaction with Surfaces and Interfaces Studied by Auger Electron Spectrometry. Vol. 39, pp. 83–114.

Banci, L., Bencini, A., Benelli, C., Gatteschi, D., Zanchini, C.: Spectral-Structural Correlations in High-Spin Cobalt(II) Complexes. Vol. 52, pp. 37–86.

Banci, L., Bertini, I., Luchinat, C.: The ^{1}H NMR Parameters of Magnetically Coupled Dimers — The Fe_2S_2 Proteins as an Example. Vol. 72, pp. 113–136.

Bartolotti, L. J.: Absolute Electronegativities as Determined from Kohn-Sham Theory. Vol. 66, pp. 27–40.

Baughan, E. C.: Structural Radii, Electron-cloud Radii, Ionic Radii and Solvation. Vol. 15, pp. 53–71.

Bayer, E., Schretzmann, P.: Reversible Oxygenierung von Metallkomplexen. Vol. 2, pp. 181–250.

Bearden, A. J., Dunham, W. R.: Iron Electronic Configuration in Proteins: Studies by Mössbauer Spectroscopy. Vol. 8, pp. 1–52.

Bergmann, D., Hinze, J.: Electronegativity and Charge Distribution. Vol. 66, pp. 145–190.

Berners-Price, S. J., Sadler, P. J.: Phosphines and Metal Phosphine Complexes: Relationship of Chemistry to Anticancer and Other Biological Activity. Vol. 70, pp. 27–102.

Bertini, I., Luchinat, C., Scozzafava, A.: Carbonic Anhydrase: An Insight into the Zinc Binding Site and into the Active Cavity Through Metal Substitution. Vol. 48, pp. 45–91.

Bertrand, P.: Application of Electron Transfer Theories to Biological Systems. Vol. 75, pp. 1–48.

Blasse, G.: The Influence of Charge-Transfer and Rydberg States on the Luminescence Properties of Lanthanides and Actinides. Vol. 26, pp. 43–79.

Blasse, G.: The Luminescence of Closed-Shell Transition Metal-Complexes. New Developments. Vol. 42, pp. 1–41.

Blasse, G.: Optical Electron Transfer Between Metal Ions and its Consequences. Vol. 76, pp. 153–188.

Blauer, G.: Optical Activity of Conjugated Proteins. Vol. 18, pp. 69–129.

Bleijenberg, K. C.: Luminescence Properties of Uranate Centres in Solids. Vol. 42, pp. 97–128.

Bŏca, R., Breza, M., Pelikán, P.: Vibronic Interactions in the Stereochemistry of Metal Complexes. Vol. 71, pp. 57–97.

Boeyens, J. C. A.: Molecular Mechanics and the Structure Hypothesis. Vol. 63, pp. 65–101.
Bonnelle, C.: Band and Localized States in Metallic Thorium, Uranium and Plutonium, and in Some Compounds, Studied by X-ray Spectroscopy. Vol. 31, pp. 23–48.
Bradshaw, A. M., Cederbaum, L. S., Domcke, W.: Ultraviolet Photoelectron Spectroscopy of Gases Adsorbed on Metal Surfaces. Vol. 24, pp. 133–170.
Braterman, P. S.: Spectra and Bonding in Metal Carbonyls. Part A: Bonding. Vol. 10, pp. 57–86.
Braterman, P. S.: Spectra and Bonding in Metal Carbonyls. Part B: Spectra and Their Interpretation. Vol. 26, pp. 1–42.
Bray, R. C., Swann, J. C.: Molybdenum-Containing Enzymes. Vol. 11, pp. 107–144.
Brooks, M. S. S.: The Theory of 5f Bonding in Actinide Solids. Vol. 59/60, pp. 263–293.
van Bronswyk, W.: The Application of Nuclear Quadrupole Resonance Spectroscopy to the Study of Transition Metal Compounds. Vol. 7, pp. 87–113.
Buchanan, B. B.: The Chemistry and Function of Ferredoxin. Vol. 1, pp. 109–148.
Buchler, J. W., Kokisch, W., Smith, P. D.: Cis, Trans, and *Metal* Effects in Transition Metal Porphyrins. Vol. 34, pp. 79–134.
Bulman, R. A.: Chemistry of Plutonium and the Transuranics in the Biospere. Vol. 34, pp. 39–77.
Bulman, R. A.: The Chemistry of Chelating Agents in Medical Sciences. Vol. 67, pp. 91–141.
Burdett, J. K.: The Shapes of Main-Group Molecules; A Simple Semi-Quantitative Molecular Orbital Approach. Vol. 31, pp. 67–105.
Burdett, J. K.: Some Structural Problems Examined Using the Method of Moments. Vol. 65, pp. 29–90.
Campagna, M., Wertheim, G. K., Bucher, E.: Spectroscopy of Homogeneous Mixed Valence Rare Earth Compounds. Vol. 30, pp. 99–140.
Ceulemans, A., Vanquickenborne, L. G.: The Epikernel Principle. Vol. 71, pp. 125–159.
Chasteen, N. D.: The Biochemistry of Vanadium, Vol. 53, pp. 103–136.
Cheh, A. M., Neilands, J. P.: The γ-Aminolevulinate Dehydratases: Molecular and Environmental Properties. Vol. 29, pp. 123–169.
Ciampolini, M.: Spectra of 3d Five-Coordinate Complexes. Vol. 6, pp. 52–93.
Chimiak, A., Neilands, J. B.: Lysine Analogues of Siderophores. Vol. 58, pp. 89–96.
Clack, D. W., Warren, K. D.: Metal-Ligand Bonding in 3d Sandwich Complexes. Vol. 39, pp. 1–41.
Clark, R. J. H., Stewart, B.: The Resonance Raman Effect. Review of the Theory and of Applications in Inorganic Chemistry. Vol. 36, pp. 1–80.
Clarke, M. J., Fackler, P. H.: The Chemistry of Technetium: Toward Improved Diagnostic Agents. Vol. 50, pp. 57–58.
Cohen, I. A.: Metal–Metal Interactions in Metalloporphyrins, Metalloproteins and Metalloenzymes. Vol. 40, pp. 1–37.
Connett, P. H., Wetterhahn, K. E.: Metabolism of the Carcinogen Chromate by Cellular Constituents. Vol. 54, pp. 93–124.
Cook, D. B.: The Approximate Calculation of Molecular Electronic Structures as a Theory of Valence. Vol. 35, pp. 37–86.
Cooper, S. R., Rawle, S. C.: Crown Thioether Chemistry. Vol. 72, pp. 1–72.
Cotton, F. A., Walton, R. A.: Metal–Metal Multiple Bonds in Dinuclear Clusters. Vol. 62, pp. 1–49.
Cox, P. A.: Fractional Parentage Methods for Ionisation of Open Shells of *d* and *f* Electrons. Vol. 24, pp. 59–81.
Crichton, R. R.: Ferritin. Vol. 17, pp. 67–134.
Daul, C., Schläpfer, C. W., von Zelewsky, A.: The Electronic Structure of Cobalt(II) Complexes with Schiff Bases and Related Ligands. Vol. 36, pp. 129–171.
Dehnicke, K., Shihada, A.-F.: Structural and Bonding Aspects in Phosphorus Chemistry-Inorganic Derivates of Oxohalogeno Phosphoric Acids. Vol. 28, pp. 51–82.
Dobiáš, B.: Surfactant Adsorption on Minerals Related to Flotation. Vol. 56, pp. 91–147.
Doi, K., Antanaitis, B. C., Aisen, P.: The Binuclear Iron Centers of Uteroferrin and the Purple Acid Phosphatases. Vol. 70, pp. 1–26.
Doughty, M. J., Diehn, B.: Flavins as Photoreceptor Pigments for Behavioral Responses. Vol. 41, pp. 45–70.
Drago, R. S.: Quantitative Evaluation and Prediction of Donor-Acceptor Interactions. Vol. 15, pp. 73–139.
Duffy, J. A.: Optical Electronegativity and Nephelauxetic Effect in Oxide Systems. Vol. 32, pp. 147–166.

Dunn, M. F.: Mechanisms of Zinc Ion Catalysis in Small Molecules and Enzymes. Vol. 23, pp. 61–122.
Emsley, E.: The Composition, Structure and Hydrogen Bonding of the β-Diketones. Vol. 57, pp. 147–191.
Englman, R.: Vibrations in Interaction with Impurities. Vol. 43, pp. 113–158.
Epstein, I. R., Kustin, K.: Design of Inorganic Chemical Oscillators. Vol. 56, pp. 1–33.
Ermer, O.: Calculations of Molecular Properties Using Force Fields. Applications in Organic Chemistry. Vol. 27, pp. 161–211.
Ernst, R. D.: Structure and Bonding in Metal-Pentadienyl and Related Compounds. Vol. 57, pp. 1–53.
Erskine, R. W., Field, B. O.: Reversible Oxygenation. Vol. 28, pp. 1–50.
Fajans, K.: Degrees of Polarity and Mutual Polarization of Ions in the Molecules of Alkali Fluorides, SrO, and BaO. Vol. 3, pp. 88–105.
Fee, J. A.: Copper Proteins – Systems Containing the "Blue" Copper Center. Vol. 23, pp. 1–60.
Feeney, R. E., Komatsu, S. K.: The Transferrins. Vol. 1, pp. 149–206.
Felsche, J.: The Crystal Chemistry of the Rare-Earth Silicates. Vol. 13, pp. 99–197.
Ferreira, R.: Paradoxical Violations of Koopmans' Theorem, with Special Reference to the 3d Transition Elements and the Lanthanides. Vol. 31, pp. 1–21.
Fidelis, I. K., Mioduski, T.: Double-Double Effect in the Inner Transition Elements. Vol. 47, pp. 27–51.
Fournier, J. M.: Magnetic Properties of Actinide Solids. Vol. 59/60, pp. 127–196.
Fournier, J. M., Manes, L.: Actinide Solids. 5f Dependence of Physical Properties. Vol. 59/60, pp. 1–56.
Fraga, S., Valdemoro, C.: Quantum Chemical Studies on the Submolecular Structure of the Nucleic Acids. Vol. 4, pp. 1–62.
Fraústo da Silva, J. J. R., Williams, R. J. P.: The Uptake of Elements by Biological Systems. Vol. 29, pp. 67–121.
Fricke, B.: Superheavy Elements. Vol. 21, pp. 89–144.
Frenking, G., Cremer, D.: The Chemistry of the Noble Gas Elements Helium, Neon, and Argon – Experimental Facts and Theoretical Predictions, Vol. 73, pp. 17–96.
Fuhrhop, J.-H.: The Oxidation States and Reversible Redox Reactions of Metalloporphyrins. Vol. 18, pp. 1–67.
Furlani, C., Cauletti, C.: He(I) Photoelectron Spectra of *d*-metal Compounds. Vol. 35, pp. 119–169.
Gázquez, J. L., Vela, A., Galván, M.: Fukui Function, Electronegativity and Hardness in the Kohn-Sham Theory. Vol. 66, pp. 79–98.
Gerloch, M., Harding, J. H., Woolley, R. G.: The Context and Application of Ligand Field Theory. Vol. 46, pp. 1–46.
Gillard, R. D., Mitchell, P. R.: The Absolute Configuration of Transition Metal Complexes. Vol. 7, pp. 46–86.
Gleitzer, C., Goodenough, J. B.: Mixed-Valence Iron Oxides. Vol. 61, pp. 1–76.
Gliemann, G., Yersin, H.: Spectroscopic Properties of the Quasi One-Dimensional Tetracyanoplatinate(II) Compounds. Vol. 62, pp. 87–153.
Golovina, A. P., Zorov, N. B., Runov, V. K.: Chemical Luminescence Analysis of Inorganic Substances. Vol. 47, pp. 53–119.
Green, J. C.: Gas Phase Photoelectron Spectra of *d*- and *f*-Block Organometallic Compounds. Vol. 43, pp. 37–112.
Grenier, J. C., Pouchard, M., Hagenmuller, P.: Vacancy Ordering in Oxygen-Deficient Perovskite-Related Ferrites. Vol. 47, pp. 1–25.
Griffith, J. S.: On the General Theory of Magnetic Susceptibilities of Polynuclear Transitionmetal Compounds. Vol. 10, pp. 87–126.
Gubelmann, M. H., Williams, A. F.: The Structure and Reactivity of Dioxygen Complexes of the Transition Metals. Vol. 55, pp. 1–65.
Guilard, R., Lecomte, C., Kadish, K. M.: Synthesis, Electrochemistry, and Structural Properties of Porphyrins with Metal–Carbon Single Bonds and Metal–Metal Bonds. Vol. 64, pp. 205–268.
Gütlich, P.: Spin Crossover in Iron(II)-Complexes. Vol. 44, pp. 83–195.
Gutmann, V., Mayer, U.: Thermochemistry of the Chemical Bond. Vol. 10, pp. 127–151.
Gutmann, V., Mayer, U.: Redox Properties: Changes Effected by Coordination. Vol. 15, pp. 141–166.

Gutmann, V., Mayer, H.: Application of the Functional Approach to Bond Variations Under Pressure. Vol. 31, pp. 49–66.

Hall, D. I., Ling, J. H., Nyholm, R. S.: Metal Complexes of Chelating Olefin-Group V Ligands. Vol. 15, pp. 3–51.

Harnung, S. E., Schäffer, C. E.: Phase-fixed 3-Γ Symbols and Coupling Coefficients for the Point Groups. Vol. 12, pp. 201–255.

Harnung, S. E., Schäffer, C. E.: Real Irreducible Tensorial Sets and their Application to the Ligand-Field Theory. Vol. 12, pp. 257–295.

Hathaway, B. J.: The Evidence for "Out-of-the Plane" Bonding in Axial Complexes of the Copper(II) Ion. Vol. 14, pp. 49–67.

Hathaway, B. J.: A New Look at the Stereochemistry and Electronic Properties of Complexes of the Copper(II) Ion. Vol. 57, pp. 55–118.

Hellner, E. E.: The Frameworks (Bauverbände) of the Cubic Structure Types. Vol. 37, pp. 61–140.

von Herigonte, P.: Electron Correlation in the Seventies. Vol. 12, pp. 1–47.

Hemmerich, P., Michel, H., Schug, C., Massey, V.: Scope and Limitation of Single Electron Transfer in Biology. Vol. 48, pp. 93–124.

Hider, R. C.: Siderophores Mediated Absorption of Iron. Vol. 58, pp. 25–88.

Hill, H. A. O., Röder, A., Williams, R. J. P.: The Chemical Nature and Reactivity of Cytochrome P-450. Vol. 8, pp. 123–151.

Hilpert, K.: Chemistry of Inorganic Vapors. Vol. 73, pp. 97–198.

Hogenkamp, H. P. C., Sando, G. N.: The Enzymatic Reduction of Ribonucleotides. Vol. 20, pp. 23–58.

Hoffman, B. M., Natan, M. J. Nocek, J. M., Wallin, S. A.: Long-Range Electron Transfer Within Metal-Substituted Protein Complexes. Vol. 75, pp. 85–108.

Hoffmann, D. K., Ruedenberg, K., Verkade, J. G.: Molecular Orbital Bonding Concepts in Polyatomic Molecules – A Novel Pictorial Approach. Vol. 33, pp. 57–96.

Hubert, S., Hussonnois, M., Guillaumont, R.: Measurement of Complexing Constants by Radiochemical Methods. Vol. 34, pp. 1–18.

Hudson, R. F.: Displacement Reactions and the Concept of Soft and Hard Acids and Bases. Vol. 1, pp. 221–223.

Hulliger, F.: Crystal Chemistry of Chalcogenides and Pnictides of the Transition Elements. Vol. 4, pp. 83–229.

Ibers, J. A., Pace, L. J., Martinsen, J., Hoffman, B. M.: Stacked Metal Complexes: Structures and Properties. Vol. 50, pp. 1–55.

Iqbal, Z.: Intra- und Inter-Molecular Bonding and Structure of Inorganic Pseudohalides with Triatomic Groupings. Vol. 10, pp. 25–55.

Izatt, R. M., Eatough, D. J., Christensen, J. J.: Thermodynamics of Cation-Macrocyclic Compound Interaction. Vol. 16, pp. 161–189.

Jain, V. K., Bohra, R., Mehrotra, R. C.: Structure and Bonding in Organic Derivatives of Antimony(V). Vol. 52, pp. 147–196.

Jerome-Lerutte, S.: Vibrational Spectra and Structural Properties of Complex Tetracyanides of Platinum, Palladium and Nickel. Vol. 10, pp. 153–166.

Jørgensen, C. K.: Electric Polarizability, Innocent Ligands and Spectroscopic Oxidation States. Vol. 1, pp. 234–248.

Jørgensen, C. K.: Heavy Elements Synthesized in Supernovae and Detected in Peculiar A-type Stars. Vol. 73, pp. 199–226.

Jørgensen, C. K.: Recent Progress in Ligand Field Theory. Vol. 1, pp. 3–31.

Jørgensen, C. K.: Relationship Between Softness, Covalent Bonding, Ionicity and Electric Polarizability. Vol. 3, pp. 106–115.

Jørgensen, C. K.: Valence-Shell Expansion Studied by Ultra-violet Spectroscopy. Vol. 6, pp. 94–115.

Jørgensen, C. K.: The Inner Mechanism of Rare Earths Elucidated by Photo-Electron Spectra. Vol. 13, pp. 199–253.

Jørgensen, C. K.: Partly Filled Shells Constituting Anti-bonding Orbitals with Higher Ionization Energy than Their Bonding Counterparts. Vol. 22, pp. 49–81.

Jørgensen, C. K.: Photo-electron Spectra of Non-metallic Solids and Consequences for Quantum Chemistry. Vol. 24, pp. 1–58.

Jørgensen, C. K.: Narrow Band Thermoluminescence (Candoluminescence) of Rare Earths in Auer Mantles. Vol. 25, pp. 1–20.

Jørgensen, C. K.: Deep-lying Valence Orbitals and Problems of Degeneracy and Intensities in Photo-electron Spectra. Vol. 30, pp. 141–192.
Jørgensen, C. K.: Predictable Quarkonium Chemistry. Vol. 34, pp. 19–38.
Jørgensen, C. K.: The Conditions for Total Symmetry Stabilizing Molecules, Atoms, Nuclei and Hadrons. Vol. 43, pp. 1–36.
Jørgensen, C. K., Frenking, G.: Historical, Spectroscopic and Chemical Comparison of Noble Gases. Vol. 73, pp. 1–16.
Jørgensen, C. K., Kauffmann, G. B.: Crookes and Marignac – A Centennial of an Intuitive and Pragmatic Appraisal of "Chemical Elements" and the Present Astrophysical Status of Nucleosynthesis and "Dark Matter". Vol. 73, pp. 227–254.
Jørgensen, C. K., Reisfeld, R.: Uranyl Photophysics. Vol. 50, pp. 121–171.
O'Keeffe, M.: The Prediction and Interpretation of Bond Lengths in Crystals. Vol. 71, pp. 161–190.
O'Keeffe, M., Hyde, B. G.: An Alternative Approach to Non-Molecular Crystal Structures with Emphasis on the Arrangements of Cations. Vol. 61, pp. 77–144.
Kahn, O.: Magnetism of the Heteropolymetallic Systems. Vol. 68, pp. 89–167.
Kimura, T.: Biochemical Aspects of Iron Sulfur Linkage in None-Heme Iron Protein, with Special Reference to "Adrenodoxin". Vol. 5, pp. 1–40.
Kitagawa, T., Ozaki, Y.: Infrared and Raman Spectra of Metalloporphyrins. Vol. 64, pp. 71–114.
Kiwi, J., Kalyanasundaram, K., Grätzel, M.: Visible Light Induced Cleavage of Water into Hydrogen and Oxygen in Colloidal and Microheterogeneous Systems. Vol. 49, pp. 37–125.
Kjekshus, A., Rakke, T.: Considerations on the Valence Concept. Vol. 19, pp. 45–83.
Kjekshus, A., Rakke, T.: Geometrical Considerations on the Marcasite Type Structure. Vol. 19, pp. 85–104.
König, E.: The Nephelauxelic Effect. Calculation and Accuracy of the Interelectronic Repulsion Parameters I. Cubic High-Spin d^2, d^3, d^7 and d^8 Systems. Vol. 9, pp. 175–212.
König, E.: Nature and Dynamics of the Spin-State Interconversion in Metal Complexes. Vol. 76, pp. 51–152.
Köpf-Maier, P., Köpf, H.: Transition and Main-Group Metal Cyclopentadienyl Complexes: Preclinical Studies on a Series of Antitumor Agents of Different Structural Type. Vol. 70, pp. 103–185.
Koppikar, D. K., Sivapullaiah, P. V., Ramakrishnan, L., Soundararajan, S.: Complexes of the Lanthanides with Neutral Oxygen Donor Ligands. Vol. 34, pp. 135–213.
Krause, R.: Synthesis of Ruthenium(II) Complexes of Aromatic Chelating Heterocycles: Towards the Design of Luminescent Compounds. Vol. 67, pp. 1–52.
Krumholz, P.: Iron(II) Diimine and Related Complexes. Vol. 9, pp. 139–174.
Kuki, A.: Electronic Tunneling Paths in Proteins. Vol. 75, pp. 49–84.
Kustin, K., McLeod, G. C., Gilbert, T. R., Briggs, LeB. R., 4th.: Vanadium and Other Metal Ions in the Physiological Ecology of Marine Organisms. Vol. 53, pp. 137–158.
Labarre, J. F.: Conformational Analysis in Inorganic Chemistry: Semi-Empirical Quantum Calculation vs. Experiment. Vol. 35, pp. 1–35.
Lammers, M., Follmann, H.: The Ribonucleotide Reductases: A Unique Group of Metalloenzymes Essential for Cell Proliferation. Vol. 54, pp. 27–91.
Lehn, J.-M.: Design of Organic Complexing Agents. Strategies Towards Properties. Vol. 16, pp. 1–69.
Linarès, C., Louat, A., Blanchard, M.: Rare-Earth Oxygen Bonding in the $LnMO_4$ Xenotime Structure. Vol. 33, pp. 179–207.
Lindskog, S.: Cobalt(II) in Metalloenzymes. A Reporter of Structure-Function Relations. Vol. 8, pp. 153–196.
Liu, A., Neilands, J. B.: Mutational Analysis of Rhodotorulic Acid Synthesis in *Rhodotorula pilimanae*. Vol. 58, pp. 97–106.
Livorness, J., Smith, T.: The Role of Manganese in Photosynthesis. Vol. 48, pp. 1–44.
Llinás, M.: Metal-Polypeptide Interactions: The Conformational State of Iron Proteins. Vol. 17, pp. 135–220.
Lucken, E. A. C.: Valence-Shell Expansion Studied by Radio-Frequency Spectroscopy. Vol. 6, pp. 1–29.
Ludi, A., Güdel, H. U.: Structural Chemistry of Polynuclear Transition Metal Cyanides. Vol. 14, pp. 1–21.
Lutz, H. D.: Bonding and Structure of Water Molecules in Solid Hydrates. Correlation of Spectroscopic and Structural Data. Vol. 69, pp. 125.
Maggiora, G. M., Ingraham, L. L.: Chlorophyll Triplet States. Vol. 2, pp. 126–159.

Magyar, B.: Salzebullioskopie III. Vol. 14, pp. 111–140.
Makovicky, E., Hyde, B. G.: Non-Commensurate (Misfit) Layer Structures. Vol. 46, pp. 101–170.
Manes, L., Benedict, U.: Structural and Thermodynamic Properties of Actinide Solids and Their Relation to Bonding. Vol. 59/60, pp. 75–125.
Mann, S.: Mineralization in Biological Systems. Vol. 54, pp. 125–174.
Mason, S. F.: The Ligand Polarization Model for the Spectra of Metal Complexes: The Dynamic Coupling Transition Probabilities. Vol. 39, pp. 43–81.
Mathey, F., Fischer, J., Nelson, J. H.: Complexing Modes of the Phosphole Moiety. Vol. 55, pp. 153–201.
Mauk, A. G.: Electron Transfer in Genetically Engineered Proteins. The Cytochrome *c* Paradigm. Vol. 75, pp. 131–158.
Mayer, U., Gutmann, V.: Phenomenological Approach to Cation-Solvent Interactions. Vol. 12, pp. 113–140.
McLendon, G.: Control of Biological Electron Transport via Molecular Recognition and Binding: The "Velcro" Model. Vol. 75, pp. 159–174.
Mildvan, A. S., Grisham, C. M.: The Role of Divalent Cations in the Mechanism of Enzyme Catalyzed Phosphoryl and Nucleotidyl. Vol. 20, pp. 1–21.
Mingos, D. M. P., Hawes, J. C.: Complementary Spherical Electron Density Model. Vol. 63, pp. 1–63.
Mingos, D. M. P., Johnston, R. L.: Theoretical Models of Cluster Bonding. Vol. 68, pp. 29–87.
Mingos, D. M. P., Zhenyang, L.: Non-Bonding Orbitals in Co-ordination Hydrocarbon and Cluster Compounds. Vol. 71, pp. 1–56.
Mingos, D. M. P., Zhenyang, L.: Hybridization Schemes for Co-ordination and Organometallic Compounds. Vol. 72, pp. 73–112.
Moreau-Colin, M. L.: Electronic Spectra and Structural Properties of Complex Tetracyanides of Platinum, Palladium and Nickel. Vol. 10, pp. 167–190.
Morgan, B., Dophin, D.: Synthesis and Structure of Biometric Porphyrins. Vol. 64, pp. 115–204.
Morris, D. F. C.: Ionic Radii and Enthalpies of Hydration of Ions. Vol. 4, pp. 63–82.
Morris, D. F. C.: An Appendix to Structure and Bonding. Vol. 4 (1968). Vol. 6, pp. 157–159.
Mortensen, O. S.: A Noncommuting-Generator Approach to Molecular Symmetry. Vol. 68, pp. 1–28.
Mortier, J. W.: Electronegativity Equalization and its Applications. Vol. 66, pp. 125–143.
Müller, A., Baran, E. J., Carter, R. O.: Vibrational Spectra of Oxo-, Thio-, and Selenometallates of Transition Elements in the Solid State. Vol. 26, pp. 81–139.
Müller, A., Diemann, E., Jørgensen, C. K.: Electronic Spectra of Tetrahedral Oxo, Thio and Seleno Complexes Formed by Elements of the Beginning of the Transition Groups. Vol. 14, pp. 23–47.
Müller, U.: Strukturchemie der Azide. Vol. 14, pp. 141–172.
Müller, W., Spirlet, J.-C.: The Preparation of High Purity Actinide Metals and Compounds. Vol. 59/60, pp. 57–73.
Mullay, J. J.: Estimation of Atomic and Group Electronegativities. Vol. 66, pp. 1–25.
Murrell, J. N.: The Potential Energy Surfaces of Polyatomic Molecules. Vol. 32, pp. 93–146.
Naegele, J. R., Ghijsen, J.: Localization and Hybridization of 5f States in the Metallic and Ionic Bond as Investigated by Photoelectron Spectroscopy. Vol. 59/60, pp. 197–262.
Nag, K., Bose, S. N.: Chemistry of Tetra- and Pentavalent Chromium. Vol. 63, pp. 153–197.
Neilands, J. B.: Naturally Occurring Non-porphyrin Iron Compounds. Vol. 1, pp. 59–108.
Neilands, J. B.: Evolution of Biological Iron Binding Centers. Vol. 11, pp. 145–170.
Neilands, J. B.: Methodology of Siderophores. Vol. 58, pp. 1–24.
Nieboer, E.: The Lanthanide Ions as Structural Probes in Biological and Model Systems. Vol. 22, pp. 1–47.
Novack, A.: Hydrogen Bonding in Solids. Correlation of Spectroscopic and Crystallographic Data. Vol. 18, pp. 177–216.
Nultsch, W., Häder, D.-P.: Light Perception and Sensory Transduction in Photosynthetic Prokaryotes. Vol. 41, pp. 111–139.
Odom, J. D.: Selenium Biochemistry. Chemical and Physical Studies. Vol. 54, pp. 1–26.
Oelkrug, D.: Absorption Spectra and Ligand Field Parameters of Tetragonal 3*d*-Transition Metal Fluorides. Vol. 9, pp. 1–26.
Oosterhuis, W. T.: The Electronic State of Iron in Some Natural Iron Compounds: Determination by Mössbauer and ESR Spectroscopy. Vol. 20, pp. 59–99.

Orchin, M., Bollinger, D. M.: Hydrogen-Deuterium Exchange in Aromatic Compounds. Vol. 23, pp. 167–193.
Peacock, R. D.: The Intensities of Lanthanide $f \leftrightarrow f$ Transitions. Vol. 22, pp. 83–122.
Penneman, R. A., Ryan, R. R., Rosenzweig, A.: Structural Systematics in Actinide Fluoride Complexes. Vol. 13, pp. 1–52.
Powell, R. C., Blasse, G.: Energy Transfer in Concentrated Systems. Vol. 42, pp. 43–96.
Que, Jr., L.: Non-Heme Iron Dioxygenases. Structure and Mechanism. Vol. 40, pp. 39–72.
Ramakrishna, V. V., Patil, S. K.: Synergic Extraction of Actinides. Vol. 56, pp. 35–90.
Raymond, K. N., Smith, W. L.: Actinide-Specific Sequestering Agents and Decontamination Applications. Vol. 43, pp. 159–186.
Reedijk, J., Fichtinger-Schepman, A. M. J., Oosterom, A. T. van, Putte, P. van de: Platinum Amine Coordination Compounds as Anti-Tumour Drugs. Molecular Aspects of the Mechanism of Action. Vol. 67, pp. 53–89.
Reinen, D.: Ligand-Field Spectroscopy and Chemical Bonding in Cr^{3+}-Containing Oxidic Solids. Vol. 6, pp. 30–51.
Reinen, D.: Kationenverteilung zweiwertiger $3d^n$-Ionen in oxidischen Spinell-, Granat- und anderen Strukturen. Vol. 7, pp. 114–154.
Reinen, D., Friebel, C.: Local and Cooperative Jahn-Teller Interactions in Model Structures. Spectroscopic and Structural Evidence. Vol. 37, pp. 1–60.
Reisfeld, R.: Spectra and Energy Transfer of Rare Earths in Inorganic Glasses. Vol. 13, pp. 53–98.
Reisfeld, R.: Radiative and Non-Radiative Transitions of Rare Earth Ions in Glasses. Vol. 22, pp. 123–175.
Reisfeld, R.: Excited States and Energy Transfer from Donor Cations to Rare Earths in the Condensed Phase. Vol. 30, pp. 65–97.
Reisfeld, R., Jørgensen, C. K.: Luminescent Solar Concentrators for Energy Conversion. Vol. 49, pp. 1–36.
Reisfeld, R., Jørgensen, C. K.: Excited States of Chromium(III) in Translucent Glass-Ceramics as Prospective Laser Materials. Vol. 69, pp. 63–96.
Russo, V. E. A., Galland, P.: Sensory Physiology of *Phycomyces Blakesleeanus.* Vol. 41, pp. 71–110.
Rüdiger, W.: Phytochrome, a Light Receptor of Plant Photomorphogenesis. Vol. 40, pp. 101–140.
Ryan, R. R., Kubas, G. J., Moody, D. C., Eller, P. G.: Structure and Bonding of Transition Metal-Sulfur Dioxide Complexes. Vol. 46, pp. 47–100.
Sadler, P. J.: The Biological Chemistry of Gold: A Metallo-Drug and Heavy-Atom Label with Variable Valency. Vol. 29, pp. 171–214.
Schäffer, C. E.: A Perturbation Representation of Weak Covalent Bonding. Vol. 5, pp. 68–95.
Schäffer, C. E.: Two Symmetry Parameterizations of the Angular-Overlap Model of the Ligand-Field. Relation to the Crystal-Field Model. Vol. 14, pp. 69–110.
Scheidt, W. R., Lee, Y. J.: Recent Advances in the Stereochemistry of Metallotetrapyrroles. Vol. 64, pp. 1–70.
Schmid, G.: Developments in Transition Metal Cluster Chemistry. The Way to Large Clusters. Vol. 62, pp. 51–85.
Schmidt, P. C.: Electronic Structure of Intermetallic B 32 Type Zintl Phases. Vol. 65, pp. 91–133.
Schmidtke, H.-H., Degen, J.: A Dynamic Ligand Field Theory for Vibronic Structures Rationalizing Electronic Spectra of Transition Metal Complex Compounds. Vol. 71, pp. 99–124.
Schneider, W.: Kinetics and Mechanism of Metalloporphyrin Formation. Vol. 23, pp. 123–166.
Schubert, K.: The Two-Correlations Model, a Valence Model for Metallic Phases. Vol. 33, pp. 139–177.
Schultz, H., Lehmann, H., Rein, M., Hanack, M.: Phthalocyaninatometal and Related Complexes with Special Electrical and Optical Properties. Vol. 74, pp. 41–146.
Schutte, C. J. H.: The Ab-Initio Calculation of Molecular Vibrational Frequencies and Force Constants. Vol. 9, pp. 213–263.
Schweiger, A.: Electron Nuclear Double Resonance of Transition Metal Complexes with Organic Ligands. Vol. 51, pp. 1–122.
Sen, K. D., Böhm, M. C., Schmidt, P. C.: Electronegativity of Atoms and Molecular Fragments. Vol. 66, pp. 99–123.
Shamir, J.: Polyhalogen Cations. Vol. 37, pp. 141–210.
Shannon, R. D., Vincent, H.: Relationship Between Covalency, Interatomic Distances, and Magnetic Properties in Halides and Chalcogenides. Vol. 19, pp. 1–43.
Shriver, D. F.: The Ambident Nature of Cyanide. Vol. 1, pp. 32–58.

Siegel, F. L.: Calcium-Binding Proteins. Vol. 17, pp. 221–268.
Simon, A.: Structure and Bonding with Alkali Metal Suboxides. Vol. 36, pp. 81–127.
Simon, W., Morf, W. E., Meier, P. Ch.: Specificity of Alkali and Alkaline Earth Cations of Synthetic and Natural Organic Complexing Agents in Membranes. Vol. 16, pp. 113–160.
Simonetta, M., Gavezzotti, A.: Extended Hückel Investigation of Reaction Mechanisms. Vol. 27, pp. 1–43.
Sinha, S. P.: Structure and Bonding in Highly Coordinated Lanthanide Complexes. Vol. 25, pp. 67–147.
Sinha, S. P.: A Systematic Correlation of the Properties of the f-Transition Metal Ions. Vol. 30, pp. 1–64.
Schmidt, W.: Physiological Bluelight Reception. Vol. 41, pp. 1–44.
Smith D. W.: Ligand Field Splittings in Copper(II) Compounds. Vol. 12, pp. 49–112.
Smith D. W., Williams, R. J. P.: The Spectra of Ferric Haems and Haemoproteins, Vol. 7, pp. 1–45.
Smith, D. W.: Applications of the Angular Overlap Model. Vol. 35, pp. 87–118.
Solomon, E. I., Penfield, K. W., Wilcox, D. E.: Active Sites in Copper Proteins. An Electric Structure Overview. Vol. 53, pp. 1–56.
Somorjai, G. A., Van Hove, M. A.: Adsorbed Monolayers on Solid Surfaces. Vol. 38, pp. 1–140.
Speakman, J. C.: Acid Salts of Carboxylic Acids, Crystals with some "Very Short" Hydrogen Bonds. Vol. 12, pp. 141–199.
Spiro, G., Saltman, P.: Polynuclear Complexes of Iron and Their Biological Implications. Vol. 6, pp. 116–156.
Strohmeier, W.: Problem und Modell der homogenen Katalyse. Vol. 5, pp. 96–117.
Sugiura, Y., Nomoto, K.: Phytosiderophores – Structures and Properties of Mugineic Acids and Their Metal Complexes. Vol. 58, pp. 107–135.
Sykes, A. G.: Plastocyanin and the Blue Copper Proteins. Vol. 75, pp. 175–224.
Tam, S.-C., Williams, R. J. P.: Electrostatics and Biological Systems. Vol. 63, pp. 103–151.
Teller, R., Bau, R. G.: Crystallographic Studies of Transition Metal Hydride Complexes. Vol. 44, pp. 1–82.
Therien, M. J., Chang, J., Raphael, A. L., Bowler, B. E., Gray, H. B.: Long-Range Electron Transfer in Metalloproteins. Vol. 75, pp. 109–130.
Thompson, D. W.: Structure and Bonding in Inorganic Derivatives of β-Diketones. Vol. 9, pp. 27–47.
Thomson, A. J., Williams, R. J. P., Reslova, S.: The Chemistry of Complexes Related to *cis*-$Pt(NH_3)_2Cl_2$. An Anti-Tumor Drug. Vol. 11, pp. 1–46.
Tofield, B. C.: The Study of Covalency by Magnetic Neutron Scattering. Vol. 21, pp. 1–87.
Trautwein, A.: Mössbauer-Spectroscopy on Heme Proteins. Vol. 20, pp. 101–167.
Tressaud, A., Dance, J.-M.: Relationships Between Structure and Low-Dimensional Magnetism in Fluorides. Vol. 52, pp. 87–146.
Tributsch, H.: Photoelectrochemical Energy Conversion Involving Transition Metal d-States and Intercalation of Layer Compounds. Vol. 49, pp. 127–175.
Truter, M. R.: Structures of Organic Complexes with Alkali Metal Ions. Vol. 16, pp. 71–111.
Umezawa, H., Takita, T.. The Bleomycins: Antitumor Copper-Binding Antibiotics. Vol. 40, pp. 73–99.
Vahrenkamp, H.: Recent Results in the Chemistry of Transition Metal Clusters with Organic Ligands. Vol. 32, pp. 1–56.
Valach, F., Koreň, B., Sivý, P., Melník, M.: Crystal Structure Non-Rigidity of Central Atoms for Mn(II), Fe(II), Fe(III), Co(II), Co(III), Ni(II), Cu(II) and Zn(II) Complexes. Vol. 55, pp. 101–151.
Wallace, W. E., Sankar, S. G., Rao, V. U. S.: Field Effects in Rare-Earth Intermetallic Compounds. Vol. 33, pp. 1–55.
Warren, K. D.: Ligand Field Theory of Metal Sandwich Complexes. Vol. 27, pp. 45–159.
Warren, K. D.: Ligand Field Theory of f-Orbital Sandwich Complexes. Vol. 33, pp. 97–137.
Warren, K. D.: Calculations of the Jahn-Teller Coupling Constants for d^x Systems in Octahedral Symmetry via the Angular Overlap Model. Vol. 57, pp. 119–145.
Watson, R. E., Perlman, M. L.: X-Ray Photoelectron Spectroscopy. Application to Metals and Alloys. Vol. 24, pp. 83–132.
Weakley, T. J. R.: Some Aspects of the Heteropolymolybdates and Heteropolytungstates. Vol. 18, pp. 131–176.
Wendin, G.: Breakdown of the One-Electron Pictures in Photoelectron Spectra. Vol. 45, pp. 1–130.
Weissbluth, M.: The Physics of Hemoglobin. Vol. 2, pp. 1–125.

Weser, U.: Chemistry and Structure of some Borate Polyol Compounds. Vol. 2, pp. 160–180.
Weser, U.: Reaction of some Transition Metals with Nucleic Acids and Their Constituents. Vol. 5, pp. 41–67.
Weser, U.: Structural Aspects and Biochemical Function of Erythrocuprein. Vol. 17, pp. 1–65.
Weser, U.: Redox Reactions of Sulphur-Containing Amino-Acid Residues in Proteins and Metalloproteins, an XPS-Study. Vol. 61, pp. 145–160.
West, D.X., Padhye, S.B., Sonawane, P.B.: Structural and Physical Correlations in the Biological Properties of Transitions Metal Heterocyclic Thiosemicarbazone and S-alkyldithiocarbazate Complexes. Vol. 76, pp. 1–50.
Willemse, J., Cras, J. A., Steggerda, J. J., Keijzers, C. P.: Dithiocarbamates of Transition Group Elements in "Unusual" Oxidation State. Vol. 28, pp. 83–126.
Williams, R. J. P.: The Chemistry of Lanthanide Ions in Solution and in Biological Systems. Vol. 50, pp. 79–119.
Williams, R. J. P., Hale, J. D.: The Classification of Acceptors and Donors in Inorganic Reactions. Vol. 1, pp. 249–281.
Williams, R. J. P., Hale, J. D.: Professor Sir Ronald Nyholm. Vol. 15, pp. 1 and 2.
Wilson, J. A.: A Generalized Configuration-Dependent Band Model for Lanthanide Compounds and Conditions for Interconfiguration Fluctuations. Vol. 32, pp. 57–91.
Winkler, R.: Kinetics and Mechanism of Alkali Ion Complex Formation in Solution. Vol. 10, pp. 1–24.
Wood, J. M., Brown, D. G.: The Chemistry of Vitamin B_{12}-Enzymes. Vol. 11, pp. 47–105.
Woolley, R. G.: Natural Optical Activity and the Molecular Hypothesis. Vol. 52, pp. 1–35.
Wüthrich, K.: Structural Studies of Hemes and Hemoproteins by Nuclear Magnetic Resonance Spectroscopy. Vol. 8, pp. 53–121.
Xavier, A. V., Moura, J. J. G., Moura, I.: Novel Structures in Iron-Sulfur Proteins. Vol. 43, pp. 187–213.
Zumft, W. G.: The Molecular Basis of Biological Dinitrogen Fixation. Vol. 29, pp. 1–65.